Mid-Ocean Ridges
Dynamics of processes associated with creation of new ocean crust

The mid-ocean ridge system is the longest continuous feature of the Earth's surface. The vast majority of the ocean floor is created at mid-ocean ridges, with volumes of seafloor volcanism vastly exceeding that of any type of volcano on land. The ridge system provides the means for much of the heat loss from the interior of the earth, a large part of which is carried by flowing sea water that moves through the rocks of the seafloor. Chemicals as well as heat are extracted by this process leading to spectacular areas of hydrothermal venting. This process affects (some say controls) ocean chemistry and global climate.

This book collects together multidisciplinary chapters that illuminate some of the most important problems that arise as ocean crust is created at mid-ocean ridges. The chapters range from studies of the mantle and magma generation within it, through tectonics of mid-ocean ridges, to the physical, chemical and biological dynamics of hydrothermal systems.

The book will be of importance to specialists and researchers wishing to become informed of the latest developments in all aspects of the science of mid-ocean ridges. It will prove especially useful for new scientists entering the field.

J. R. CANN is Professor of Earth Sciences in the University of Leeds.
H. ELDERFIELD is University Reader in Geochemistry at the University of Cambridge.
A. LAUGHTON is Senior Visiting Research Fellow, Southampton Oceanography Centre and formerly Director of the Institute of Oceanographic Sciences.

Mid-Ocean Ridges

Dynamics of processes associated with
creation of new ocean crust

Edited by
J. R. Cann, H. Elderfield and A. Laughton

CAMBRIDGE
UNIVERSITY PRESS

PUBLISHED BY THE PRESS SYNDICATE OF THE UNIVERSITY OF CAMBRIDGE
The Pitt Building, Trumpington Street, Cambridge, United Kingdom

CAMBRIDGE UNIVERSITY PRESS
The Edinburgh Building, Cambridge CB2 2RU, UK   www.cup.cam.ac.uk
40 West 20th Street, New York, NY 10011-4211, USA   www.cup.org
10 Stamford Road, Oakleigh, Melbourne 3166, Australia
Ruiz de Alarcón 13, 28014 Madrid, Spain

© The Royal Society 1999

This book is in copyright. Subject to statutory exception
and to the provisions of relevant collective licensing agreements,
no reproduction of any part may take place without
the written permission of Cambridge University Press.

All papers (except that by Mills and Tivey) first published by
the Royal Society 1997 as *Phil. Trans. A* **355**, 213–486

First published by Cambridge University Press 1999

Printed in the United Kingdom at the University Press, Cambridge

A catalogue record of this book is available from the British Library

Library of Congress Cataloguing in Publication data
Mid-ocean ridges: dynamics of processes associated with creation of
new ocean crust / edited by J. R. Cann, H. Elderfield, and A. Laughton.
p. cm.
Papers from a discussion meeting held in London, Mar. 6–7, 1996.
All but one paper were originally published in the
Philosophical transactions of the Royal Society (A, 355, 1997).
ISBN 0 521 58522 8
1. Mid-ocean ridges–Congresses.   2. Earth–Crust–Congresses.
3. Submarine geology–Congresses.   I. Cann, J. R. (Johnson Robin), 1937–   .
II. Elderfield, H. (Henry), 1943–   .   III. Laughton, A. S.
QE511.7.M53   1999
551.1'3–dc21   98-8322 CIP

ISBN 0 521 58522 8 hardback

Contents

Preface	vii
D. K. BLACKMAN AND J.-M. KENDALL Sensitivity of teleseismic body waves to mineral texture and melt in the mantle beneath a mid-ocean ridge	1
M. C. SINHA, D. A. NAVIN, L. M. MACGREGOR, S. CONSTABLE, C. PEIRCE, A. WHITE, G. HEINSON AND M. A. INGLIS Evidence for accumulated melt beneath the slow-spreading Mid-Atlantic Ridge	17
P. D. ASIMOW, M. M. HIRSCHMANN AND E. M. STOLPER An analysis of variations in isentropic melt productivity	39
P. B. KELEMEN, G. HIRTH, N. SHIMIZU, M. SPIEGELMAN AND H. J. B. DICK A review of melt migration processes in the adiabatically upwelling mantle beneath oceanic spreading ridges	67
R. S. WHITE Rift-plume interaction in the North Atlantic	103
M.-H. CORMIER The ultrafast East Pacific Rise: instability of the plate boundary and implications for accretionary processes	125
D. A. BUTTERFIELD, I. R. JONASSON, G. J. MASSOTH, R. A. FEELY, K. K. ROE, R. E. EMBLEY, J. F. HOLDEN, R. E. MCDUFF, M. D. LILLEY AND J. R. DELANEY Seafloor eruptions and evolution of hydrothermal fluid chemistry	153
A. SCHULTZ AND H. ELDERFIELD Controls on the physics and chemistry of seafloor hydrothermal circulation	171
Y. FOUQUET Where are the large hydrothermal sulphide deposits in the oceans?	211
R. A. MILLS AND M. K. TIVEY Sea water entrainment and fluid evolution within the TAG hydrothermal mound: evidence from analyses of anhydrite	225
K. G. SPEER Thermocline penetration by buoyant plumes	249
S. K. JUNIPER AND V. TUNNICLIFFE Crustal accretion and the hot vent ecosystem	265
H. W. JANNASCH Biocatalytic transformations of hydrothermal fluids	281
Index	293

Preface

The mid-ocean ridge system is the longest continuous feature of the Earth's surface and is where all of the ocean floor is created with volumes of seafloor volcanism vastly exceeding that of any type of volcano on land. It provides the means for much of the heat loss from the interior of the Earth, a large part of which is carried by flowing sea water that moves through the rocks of the seafloor. Chemicals as well as heat are extracted by this process leading to spectacular areas of hydrothermal venting on the seabed at the ridge axis with seafloor sulphide deposits and biological communities. The process affects (some say controls) ocean chemistry and global climate.

In recent years, the recognition that a major coordinated effort was needed in order to understand the fundamental dynamical processes operating at mid-ocean ridges led the scientific community to mount a series of major national and international programmes. The major UK component of this effort has been the British Mid-Ocean Ridge Initiative (BRIDGE) which was established as a NERC Community Research Project, and many other countries have active programmes in this area of science. For this reason a Royal Society Discussion Meeting was organized by the editors and this was held in London on 6 and 7 March 1996. Much of the work presented at the Discussion Meeting has been the product of these initiatives. In addition, a poster session held concurrently with the meeting highlighted achievements of the BRIDGE programme.

This volume contains thirteen of the fourteen papers presented at the Discussion Meeting, twelve of which were originally published in the *Philosophical Transactions of The Royal Society* (A, **355**, 1997). The papers presented were grouped into four inter-related sessions.

Session 1: Imaging of magma production beneath mid-ocean ridges using geophysical and geochemical approaches (papers given by Phipps Morgan [not included in this volume], Blackman, Sinha and Stolper).

Session 2: The production of oceanic crust (papers given by Keleman, White and Cormier).

Session 3: Submarine hydrothermal fluids and deposits (papers given by Butterfield, Schultz, Fouquet and Mills [not included in the *Phil. Trans.* issue]).

Session 4: Water column and biological consequences of hydrothermal activity (papers given by Speer, Juniper and Jannasch).

What emerges strongly is the inter-disciplinary and cross-disciplinary nature of much work. Work in any one scientific discipline has benefited greatly by the framework of such an approach and is often essential. It is also clear that many advances relied on technological innovation. Further, it is very clear that the subject is still firmly within an observational phase. Research vessels, drill ships and submersibles are all needed.

It is sad that the publication of this report coincides with the end of the BRIDGE programme, whose scientists have contributed much to the success of the meeting. A start has been made at understanding some of the major processes associated with the creation of new ocean crust and it is hoped that the initiative and technological development derived from BRIDGE will be exploited, perhaps within a broader framework, to address fundamental scientific problems in ocean sciences. As noted in the foreword to an early BRIDGE document, the ocean floor makes up two-thirds of the surface of the Earth, yet we know less about its shape than we know about the far side of the Moon.

We thank the speakers and co-authors for their contributions to this volume and to them, and the participants in the discussions for their contributions to the meeting. We are also grateful to the staffs of the Royal Society and Cambridge University Press for their organization and help.

J. R. CANN
H. ELDERFIELD
A. LAUGHTON

Sensitivity of teleseismic body waves to mineral texture and melt in the mantle beneath a mid-ocean ridge

By Donna K. Blackman[†] and J.-Michael Kendall

Department of Earth Sciences, University of Leeds, Leeds LS2 9JT, UK

Seismic energy propagating through the mantle beneath an oceanic spreading centre develops a signature due both to the subaxial deformation field and to the presence of melt in the upwelling zone. Deformation of peridotite during mantle flow results in strong preferred orientation of olivine and significant seismic anisotropy in the upper 100 km of the mantle. Linked numerical models of flow, texture development and seismic velocity structure predict that regions of high anisotropy will characterize the subaxial region, particularly at slow-spreading mid-ocean ridges. In addition to mineral texture effects, the presence of basaltic melt can cause travel-time anomalies, the nature of which depend on the geometry, orientation and concentration of the melt. In order to illustrate the resolution of subaxial structure that future seismic experiments can hope to achieve, we investigate the teleseismic signature of a series of spreading centre models in which the mantle viscosity and melt geometry are varied. The P-wave travel times are not very sensitive to the geometry and orientation of melt inclusions, whether distributed in tubules or thin ellipsoidal inclusions. Travel time delays of 0.1–0.4 s are predicted for the melt distribution models tested. The P-wave effects of mineral texture dominate in the combined melt-plus-texture models. Thus, buoyancy-enhanced upwelling at a slow spreading ridge is characterized by 0.7–1.0 s early P-wave arrival times in a narrow axial region, while the models of plate-driven-only flow predicts smaller advances (less than 0.5 s) over a broader region. In general S-wave travel times are more sensitive to the melt and show more obvious differences between melt present as tubules as opposed to thin disks, especially if a preferred disk orientation exists. Mineral texture and the preferred alignment of melt inclusions will both produce shear-wave splitting, our models predict as much as 4 s splitting in some cases.

---

1. Introduction

Decompression of the mantle as it upwells beneath an oceanic spreading centre results in melting of peridotite and the production of basaltic magma. Deformation of peridotite minerals results from viscous shearing of the mantle as it flows in response to plate separation at the ridge axis. Both the presence of lower density melt and the preferred orientation of olivine grains that are strongly anisotropic will influence the seismic signature of the subaxial region. Modelling wave propagation through such

[†] Present address: Scripps Institute of Oceanography, La Jolla, CA 92093, USA.

complex anisotropic regions allows us to explore the range of travel time anomalies that would result from previous models of subaxial structure. These predictions illustrate the degree to which surface measurements can provide insight into complexities at depth.

This paper builds on previous work that linked geodynamic modelling, the corresponding development of mineral textures and raytracing through the resultant anisotropic mantle at a mid-ocean ridge (Blackman *et al.* 1997). That paper focused on the seismic effect of flow induced olivine alignment. Our intent here is to complete the self-consistency of the previous models for which the seismic effect of the predicted melt distribution had been ignored. Kendall (1994) made predictions about the travel time signature of some simple subaxial melt models but the assumptions made about melt distribution were only approximate, not directly tied to any specific flow field. Here we improve the assumptions about melt properties, using published values of seismic velocity for basaltic melt and exploring various melt geometries representing a range from evenly distributed 'pores' through films aligned along grain faces.

Current understanding of melt production, segregation and migration through the mantle is incomplete so our approach is to explore several models that bracket physically reasonable melt distributions. This allows us to show that although different melt geometries may produce rather different reductions in local P-wave velocity, the difference in overall travel time anomalies predicted for various cases is rather small. The overall effects for S-waves are slightly larger. Furthermore, significant shear-wave splitting is predicted if the melt lies in vertically aligned thin disk-like inclusions. Detailed results are presented following brief reviews of the geodynamic setting, the observational evidence on melt distribution in the mantle and the theoretical framework for seismic modelling of melt inclusions.

2. Mantle flow and melting at oceanic spreading centres

Spreading of lithospheric plates occurs mainly as a result of forces distant from the axis of a mid-ocean ridge (Forsyth & Uyeda 1975; Houseman 1983) so the assumption that upwelling, at least in the upper 100–200 km of the mantle, is a passive response to plate separation is foremost in our models of subaxial structure. Temperatures in the broad upwelling zone exceed values far off axis by 100–200 °C (Lachenbruch 1973; Sleep 1975; Phipps Morgan & Forsyth 1988). About 70–80 km below the seafloor, the rising peridotite begins to exceed the solidus temperature which increases with depth at a rate of about $3.25\,°\text{C}\,\text{km}^{-1}$ (see, for example, Hess 1992). The exact nature of the melt as it is initially produced and its mode of migration thereafter is uncertain. The fact that 6–7 km of basaltic crust is steadily produced at the ridge axis (see, for example, Spudich & Orcutt 1980) means that at least this amount eventually migrates to the surface on a time scale appropriate for the local spreading rate. If the melt is retained in the interstices of the residual peridotite then buoyancy forces, due to its lower density, can enhance the upwelling rate of the mantle (melt + matrix) (Rabinowicz *et al.* 1984; Scott & Stevenson 1989). Faster plate spreading rate (i.e. greater passive upwelling rate) and higher mantle viscosity reduce the ability of buoyancy forced flow to dominate the plate driven flow (Parmentier & Phipps Morgan 1990). At slow spreading rates (10–20 mm yr^{-1} half-rate) buoyancy driven flow dominates for asthenosphere viscosity of 5×10^{18} Pa s whereas passive flow dominates for a viscosity of 10^{20} Pa s (figure 1).

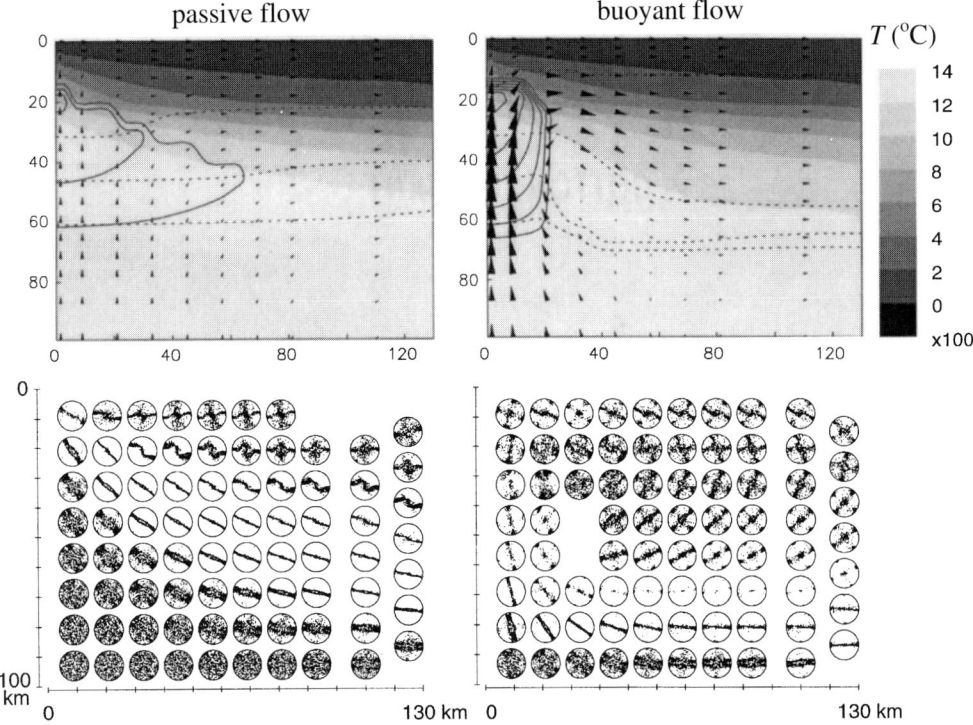

Figure 1. Models of mantle structure beneath a slow-spreading mid-ocean ridge; axis is at left in diagrams. Top panels show flow vectors scaled by rate of flow, 18 mm yr^{-1} half spreading rate is shown by arrowheads in the lithosphere (less than 700 °C indicated by shading). Plate driven flow dominates when asthenosphere viscosity is 10^{20} Pa s (left panels). Buoyancy forces, due to the presence of melt, enhances upwelling rates, which causes focusing towards the axis, when viscosity is 5×10^{18} Pa s (right panels). Solid contours show melt at increments of 1% volume fraction. Up to 6% melt fraction is generated in the buoyant model compared with a maximum of 3% in the passive flow model. Dashed contours show upwardly increasing degree of depletion in the residual mantle due to the removal of melt. Equal-area projection pole figures in the lower panels show the orientation of seismically fast a-axis for olivine crystals within an aggregate at each grid point. Mineral texture is random at 100 km depth but preferred orientations develop as the aggregates are subjected to the strain field of the flow model shown in the top panels (Blackman et al. 1997; Chastel et al. 1993). Pole figure girdles indicate the orientation of the plane in which the a-axes align, the slow b-axis orients perpendicular to this plane. In the passive case, the fast plane subparallels the base of the lithosphere; in the buoyant case strong vertical alignment develops beneath the ridge axis and high shearing due to focussing of the flow results in a layer of subhorizontal a-axis planes that are rafted off axis at depth.

The question of how much melt is retained in the matrix has seismic ramifications not only in terms of the melt distribution itself but also in terms of the development of preferred orientation in olivine since the deformation field is quite different for passive versus buoyancy enhanced flow. If the melt migrates essentially immediately, as suggested by recent geochemical modelling (McKenzie 1984; O'Nions & McKenzie 1993; Spiegelman 1996), melt may not significantly influence the mantle flow pattern. If volume fractions up to 1–2% are retained and asthenosphere viscosity is on the order of 10^{18} Pa s, some buoyancy driven flow will still evolve. There continues to be some debate about subaxial mantle viscosity structure (Karato 1986; Kohlstedt 1992; Hirth & Kohlstedt 1992, 1996) and the degree of melt retention (see, for example, Waff & Holdren 1981; Waff & Faul 1992). Therefore, it is useful to illustrate the seismic signatures that would result from both passive and buoyancy enhanced flow models and to investigate how their differing melt distributions and deformation fields affect the seismic predictions.

Whether melt is retained until shallower depths (10–20 km) or is segregated and migrates immediately, the geometry of the melt inclusions as they form and the mode of melt transport will affect the nature of wave propagation through the melt zone. Interconnected tubules of melt distributed along grain intersections (Waff & Bulau 1982; Toramaru & Fujii 1986) allow melt transport via porous flow and may reduce isotropic P- and S-wave velocities. In contrast, a network of fractures transporting melt (Nicolas 1989; Nielson & Wilshire 1993; Ceuleneer et al. 1996) may produce an anisotropic signature, particularly if the fractures are aligned (next section). Nielson & Wilshire (1993) note that clear evidence for melt transport via large scale percolation has not yet been found but Quick & Gregory (1995) comment that the networks of dykes observed in ophiolites probably represent the shallow levels of the magma supply system rather than its base. Again, we investigate the seismic signature of a range of possible melt geometries: evenly distributed melt, representing porous flow; melt in thin tubules such as might occur at grain edges; and disk-like inclusions that can be aligned, representing melt-filled fracture planes or melt films that mimic interconnected networks when the aspect ratio is large.

3. Modelling the effect of melt on seismic properties

The effects of melt on seismic waves are not only a function of melt fraction, but also a strong function of the melt inclusion geometry and orientation. Experimental studies of melting in peridotites have yet to come to consensus on melt geometry (cf. Schmeling 1985; Forsyth 1992). Melt may lie along grain faces and can be approximated by ellipsoidal inclusions, it may reside in interconnected tubules along grain edges, or it may reside in melt pockets. In order to isolate their seismic responses we treat each in isolation.

Five possible melt orientations and geometries are considered. The first is a simple isotropic reduction of velocities which is obtained from a scaled combination of the host rock velocities and the melt velocities which is controlled by the melt-fraction (hereafter referred to as the *scaled* model).

The next two models consider the melt to be in 'penny-shaped' inclusions. Support for this comes from experimental work with ultramafic partial melts by Waff & Faul (1992) and Faul et al. (1994). They found that more than 75% of the melt occurred in ellipsoidal inclusions with small aspect ratios. O'Connell & Budiansky (1974) and Schmeling (1985) showed that these inclusions can be approximated as oblate spheroids (disks) when estimating their effects on elastic properties. We adopt the approach of Hudson (1980) to estimate effective elastic constants of material filled with such disk-like inclusions. We use the Hudson (1980) expressions which are accurate to second order in $\varepsilon = Na^3/V$ (where N is the number density of the inclusions, a is the mean radius of the inclusions and V is volume). The theories of O'Connell & Budiansky (1977) and Schmeling (1985) are more accurate than Hudson (1980), but they are only valid for randomly oriented inclusions and we are interested in potential anisotropic effects due to the alignment of such inclusions. Furthermore, the crack densities and aspect ratios we consider are within the accuracy limits of the Hudson (1980) theory. We consider the isotropic case of randomly aligned disks (*random disks*) and the anisotropic case of disks aligned with the disk-face

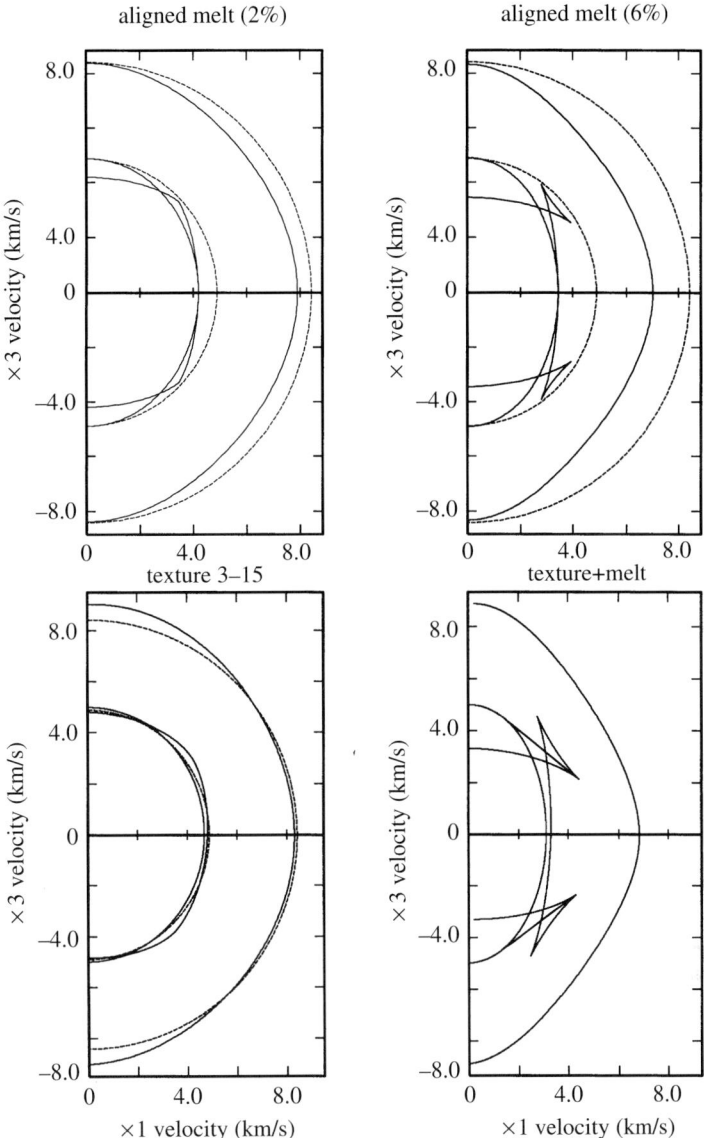

Figure 2. Wave surfaces predicted for anisotropic melt and texture models; point source at 0, 0 for each plot. The dashed lines show the P- (outer) and S- (inner) surfaces for a reference isotropic model of randomly oriented olivine crystals. Upper panels: melt distributed in disk-like inclusions with the normal to the disk-face oriented in the direction of plate spreading. Melt fractions of 0.02 and 0.06 are shown. Lower left: anisotropy due to the lattice preferred orientation of olivine crystals predicted 3 km off axis and 15 km deep for the buoyant flow model (Blackman *et al.* 1997). Lower right: combined effect of oriented olivine and aligned melt inclusions (6% melt). Note the development of S-wave triplications as the melt fraction increases. In the melt models the vertically travelling S-wave polarized normal to the inclusion faces will be slower than that polarized parallel to the faces, which, like the vertically travelling P-wave, is largely unaffected by the aligned melt.

normals oriented in the direction of plate spreading (*aligned disks*). Support for such an orientation comes from the theoretical studies of Phipps Morgan (1987) and Spiegelman (1993).

The final type of inclusion we consider is the tubule which we approximate as a cigar shaped spheroid. Mavko (1980) and others consider the effects of tubules with circular and non-circular cross-sections. Schmeling (1985) showed that the range of possible tubule shapes give similar results to those for ellipsoidal inclusions with aspect ratios between 0.2 and 0.5 (i.e. not very flat). The elastic moduli are not very sensitive to this range of variations in aspect ratio so we feel it representative to only consider tubules with circular cross-sections. We are interested in the effects of randomly oriented tubules (*random tubules*) and tubules with vertically oriented major-axes (*aligned tubules*). The theory of Tandon & Weng (1984) and Sayers (1992) is used to estimate the effective elastic constants for the situation of oriented tubule inclusions. We use a Voigt average of these elastic constants to estimate the elastic moduli for random orientations of the tubules.

We assume that the matrix rock has the elastic properties of randomly oriented olivine, $v_P = 8421$ m s^{-1}, $v_S = 4487$ m s^{-1} and density $= 3311$ kg m^{-3} (velocities are obtained using Voigt–Ruess–Hill averages for the elastic constants of Kumazawa & Anderson (1969)). The assumed melt properties are $v_P = 2500$ m s^{-1}, $v_S = 0$ and density $= 2700$ kg m^{-3} (Murase & McBirney 1973). Based on the measurements of Faul *et al.* (1994) we use an aspect ratio of 0.05 for the disks. The aspect ratio is defined as the height of the oblate spheroid to its radius. Thus disks have an aspect ratio less than one while tubules are greater than one. The assumed aspect ratio for the tubules is 20. The maximum melt fraction in the buoyancy case is 0.06 which corresponds to an inclusion density of 0.38.

Wavefronts for some of the models we consider are shown in figure 2. As the percentage of melt increases, the velocities decrease and if the melt lies in oriented inclusions the anisotropic effects can be quite dramatic. For vertically aligned disk-like inclusions of melt, the S-wavefronts can develop triplications as the melt fraction increases. The S-wave which is polarized perpendicular to the disk faces will be affected the most, just as P-waves propagating normal to the disk face will be slower than those propagating parallel to the disk faces. Figure 2 also shows the wave surfaces for a region of the texture models with fairly high degree of alignment of olivine crystals. The combination of melt inclusion alignment and mineral alignment produces complex S-wave surfaces.

We have not considered the effects of attenuation or the effects of inclusions being hydraulically connected (sometimes known as 'melt-squirt' (Mavko & Nur 1975)). For waves travelling nearly parallel to the inclusion faces, which is the direction of wave propagation we are considering, such melt interconnectivity is expected to have negligible effects (Thomsen 1995).

The theories for estimating the effective elastic properties of partial melt assume the melt has a spatial distribution which is small compared to the seismic wavelength. It is not possible to distinguish between melt in large but sparse inclusions and melt in many small inclusions.

4. Results

(a) *Influence of melt on velocities*

The velocity dependence on geometry, orientation and concentration of the melt is shown in figure 3. Randomly oriented disks (flat ellipsoidal inclusions) have the largest effect on P-wave velocities and both the P- and S-velocities are sensitive to the aspect ratio (figure 3, Schmeling 1985). If the disks are vertically aligned, the vertically travelling P-wave shows little sensitivity to the inclusions regardless of the aspect ratio. Likewize, the vertically travelling S-wave polarized parallel to the disk faces is not sensitive to the melt. Considerable shear-wave splitting is possible, though, as the orthogonally polarized S-wave is very sensitive to the melt.

Velocities for the tubule model are not very sensitive to the aspect ratio of the tubules. Randomly oriented tubules have a larger effect on the P-wave velocity in the vertical direction than that due to aligned tubules. The S-waves are not sensitive to the alignment or aspect ratio of the tubules and agree well with the results of simply scaling the melt and matrix S-velocities with a melt-fraction weight average. The vertically aligned tubules will not produce splitting in vertically travelling S-waves.

The curves in figure 3 are in line with previous laboratory and theoretical studies. Faul et al. (1994) compute 1.8% and 3.3% reductions in P- and S-wave velocity, respectively, for a melt geometry (penny-shaped ellipses) representing thin films along grain boundaries versus 1% and 2.3% reductions for a geometry (cigar shaped ellipsoids) representing melt along grain edges.

It is important to note that although one model of melt geometry may predict an effective velocity of the melt + matrix that is reduced more than twice as much as another model, all models predict changes of the order of a few percent. Dramatic as 'twice the reduction' may sound, it is the accumulation of travel time delay during transit through the melt zone, which extends on the order of a few tens of kilometres and in which v_P is about 8 km s^{-1} (4 km s^{-1} for S-waves), that will be measured at the surface.

(b) Travel time predictions for melt-only models

Asymptotic ray theory is used to track wavefronts through three dimensional models which have up to 21 independent elastic constants (Kendall & Thomson 1989; Guest & Kendall 1993). The travel time delays due to the subaxial melt fractions predicted by the passive and buoyant flow models primarily reflect the amount of melt present and the overall shape of the melt regions. For all melt geometries, the maximum delay is at the axis and the delay for the passive model is less than half that predicted for the buoyant model. This mainly reflects the fact that melt fractions in the passive model are generally 1–2%, reaching 3% only in a narrow axial region at the top of the melt zone. In the buoyant model values of 2–5% are characteristic and 6% melt fractions are reached at 18–25 km depths within 18 km of the axis (figure 1).

The largest travel time delay for near-vertical P-waves is produced by the randomly oriented melt-filled disks. Arrivals at the axis are predicted to be 0.13 s and 0.33 s later than off-axis arrivals for the passive and buoyant cases, respectively (figure 4). In contrast, aligned disks produce the least P-wave travel time delay for near vertical rays. The travel times for random and aligned disks bracket the case where the velocity is a simple scaling of the melt and matrix velocities. The difference between the predicted travel times for the disk and tubule models is less than 0.1 s for the P-waves and would generally not be resolvable in teleseismic signals recorded at the seafloor (Blackman et al. 1995). In summary, near-vertical P-waves propagating through randomly oriented inclusions are more strongly affected by the melt than P-waves traveling vertically through a medium with vertically aligned melt inclusions.

The S-wave travel time delays for the melt geometries that produce an isotropic medium are at least three times as great as the P-wave delays (figure 4, lower panels). The aligned disks produce significant shear wave splitting. The slowest S-wave (polarized perpendicular to the disk faces) arrives 1.7 s (passive) and 4.2 s (buoyant)

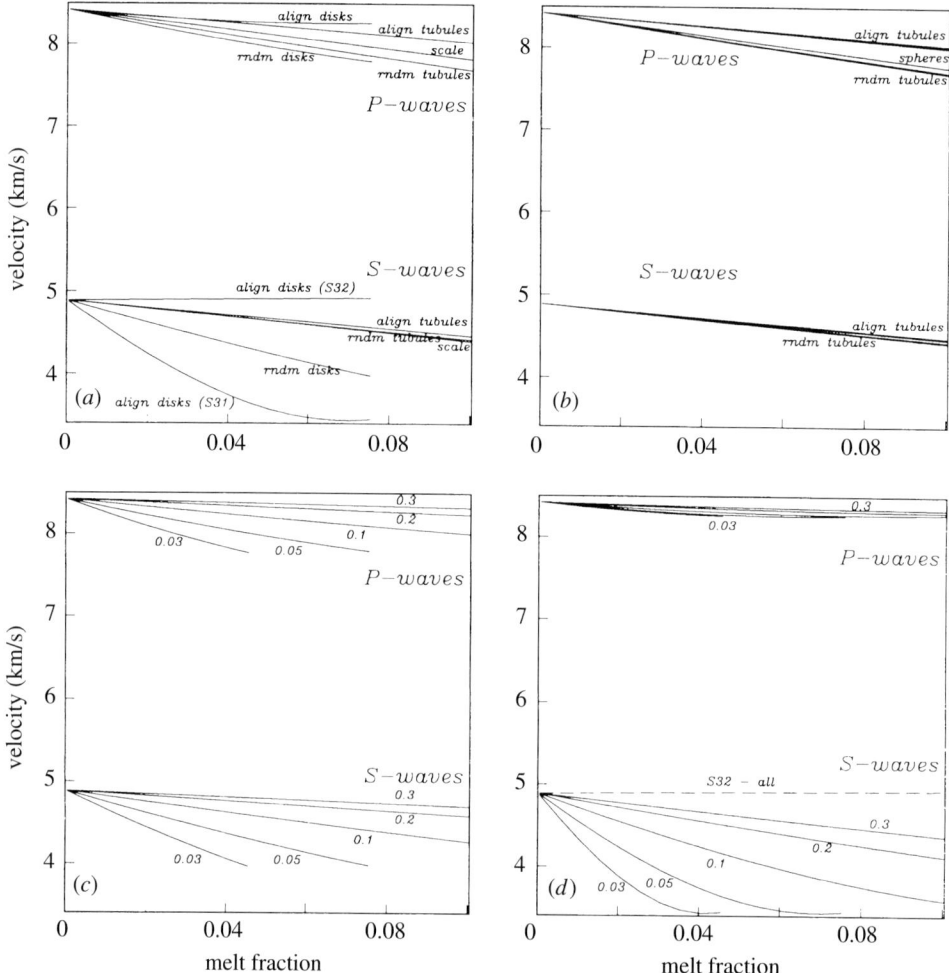

Figure 3. Effects of melt inclusion geometry on seismic velocity as a function of melt fraction. All models have a background matrix of randomly oriented olivine crystals. (a) Comparison of models: disk models, both vertically aligned and randomly oriented, have an aspect ratio of 0.05. When disks are aligned vertically, anisotropic velocities for vertically travelling waves are shown. The S-wave polarized normal to the disk faces is denoted $S31$ while that polarized parallel to the face is denoted $S32$. Velocities are plotted only in regions where the disk theory (Hudson 1980) is valid. The tubules have an aspect ratio of 20. The case where the velocities are determined simply from a melt-fraction weighted average of the melt and matrix velocities is marked *scale*. (b) Randomly aligned and vertically aligned tubules models. Aspect ratios of 100, 20, 10, 5 and 1 (spheres) are considered, but only for the case of P-waves traveling through media with spherical inclusions is there discernable difference between models. (c) Randomly oriented disks modelled with aspect ratios of 0.03, 0.05, 0.1, 0.2 and 0.3. (d) Vertically aligned disks with various labelled aspect ratios. The results for S-waves polarized perpendicular to the cracks are shown in solid lines while the results for S-waves polarized parallel to the disks plot on a single line for all models (dashed and marked $S32$). Note the evidence of the Hudson theory breaking down at small aspect ratio and high melt fractions (e.g. aspect ratio of 0.03 and melt fraction of 0.04).

later at the axis than the fastest S-wave (polarized parallel to the disk faces). This is the maximum possible splitting as it assumes all of the melt lies in aligned disk-like inclusions.

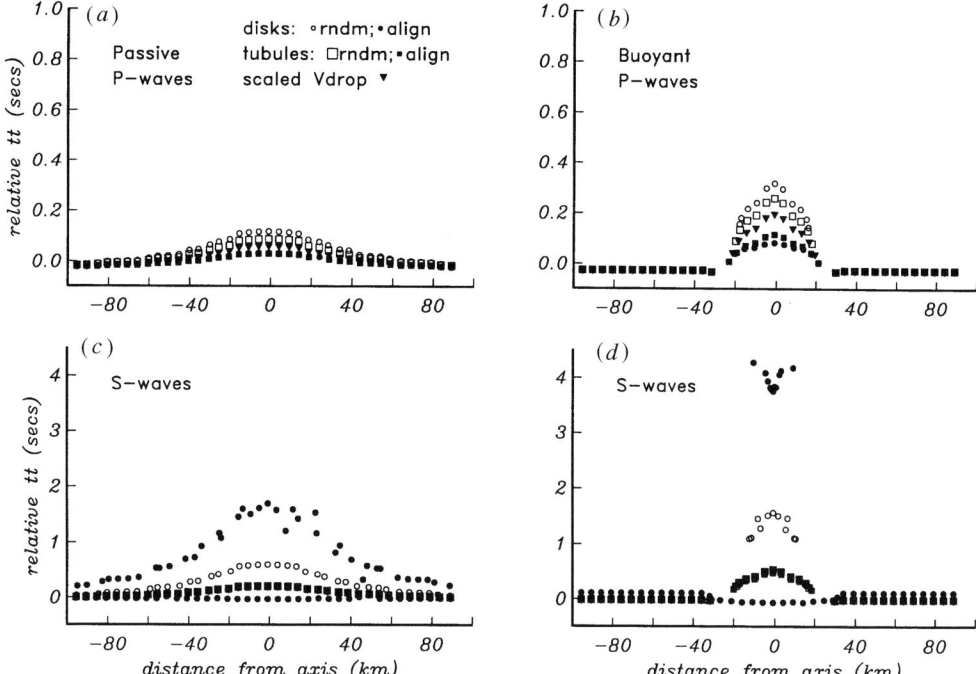

Figure 4. Relative travel times of near-vertical rays that transit subaxial mantle models containing basaltic melt (solid contours in figure 1) within homogeneous, isotropic olivine: passive flow model (left), buoyant flow model (right), P-wave travel times (upper panels), and S-wave travel times (lower). Predictions for different models of the geometry of melt within the matrix (see text) are indicated by different symbols as labelled. Maximum P-wave delay at the axis occurs for randomly oriented melt disks. Maximum S-wave delay occurs at the axis as well and is accompanied by shear-wave splitting if melt-filled disks are vertically aligned. Delay magnitudes are greater in the buoyant model which has higher melt fractions and a narrower melt zone than the passive flow model. The minor travel time advance at the axis for the fast shear wave (S32) in the aligned disk case is due to the effective density of the melt zone being lower than that in the off-axis region while the elasticity for this polarization is largely unaffected by the melt (velocity is proportional to the elasticity divided by the density).

The models are parametrized on a regularly spaced grid. In regions where the melt fraction changes quickly there is a degree of roughness in the spline interpolation of the elastic constants for the models. Some of the scatter in the travel times can be attributed to this, but most is due to the complex ray trajectories which result from wave propagation through regions with sharp velocity gradients. Nevertheless, the overall features of the travel time curves are robust.

Rays travelling from teleseismic events at ranges greater than about 90° will be near vertically incident on the seafloor at a mid-ocean ridge but rays from events at closer range will arrive at shallower angles. Figure 5 shows the predicted P-wave travel time anomalies for the suite of melt geometry models for both the passive and buoyancy enhanced flow cases. The waves propagate in a left to right sense and have an initial trajectory 30° from the vertical at the bottom of the model. As expected, the relative travel times across the profile are asymmetric with respect to the ridge axis but the maximum delay is still at the axis. This is due to the effects of ray bending towards the lower axial velocities. The size of the axial delay for the passive case is slightly greater than that predicted for near-vertical arrivals, but it

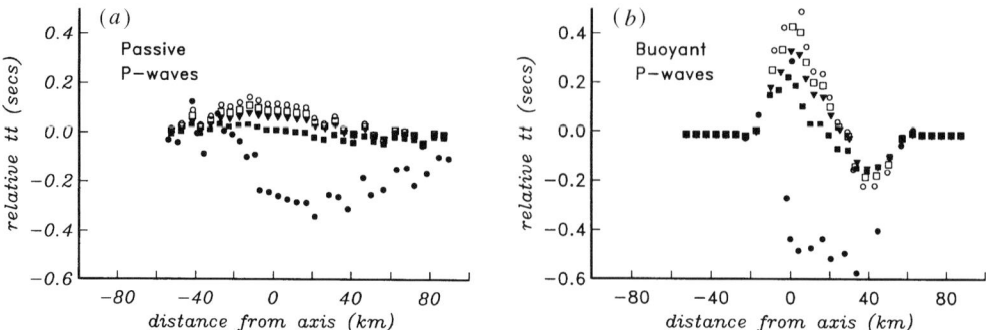

Figure 5. Relative travel times of P-waves incident at 30° from vertical, travelling from left to right, for the passive (a) and buoyant (b) models of subaxial melt distribution. Velocity structure is the same as in figure 4 as are symbols indicating different melt geometry models. See text for discussion of travel time advance for vertically aligned melt-filled disk case (solid circles).

is more than twice as great in the buoyant case. Early travel times are predicted in the axial region to the right of the ridge axis (roughly 0–50 km off axis) for all melt geometry models in the buoyant case. Only the aligned disk model produces such a travel time advance in the passive model. Although it is initially surprising to see early arrivals for rays that transit the melt zone, the signal is a result of rays being refracted into the lower-velocity axial region thereby shortening the raypaths and travel times. The velocity gradients are the most extreme in the aligned disk model due to the significant anisotropy.

The travel time patterns for S-waves propagating with initial take-off angles of 30° are not shown, but are similar to those for the P-waves. The isotropic models show an asymmetry about the axis, but the near-axis delays and advances are more pronounced due to the slower S-wave velocities. The buoyant case, for example, shows up to 1 s delay to the source-side of the axis and 0.3 s advance on the other side of the axis. The anisotropic aligned-disk models show large degrees of shear-wave separation. The fast S-wave for the passive case is roughly 0.3 s early at the axis, whilst the slow S-wave is over 2.2 s late compared to the off axis arrivals. Predictions for the buoyant model show much larger and more complex effects.

(c) Travel time predictions for melt + texture models

We now combine the effects of mineral texture (Blackman et al. 1997) and melt distribution. Near vertical P-waves that propagate through the texture models (anisotropy due to mineral alignment) arrive at the axis earlier than off axis (lower curves in the top panels of figure 6). The axial advance in the passive case is about 0.5 s relative to 50 km off axis. The axial advance in the buoyant case is more dramatic, reaching 1.0 s relative to the same distance off axis. For both models the off axis delay reflects textural anisotropy in which the slow direction for P-waves is near vertical as illustrated in figure 1 (see Blackman et al. (1997) for more complete discussion). The greater P-wave advance of the buoyant axial arrivals, relative to the passive case, also reflects the vertical alignment of the plane of fast olivine axes due to focusing and enhancement of the upwelling. Blackman et al. (1993, 1995) present sparse data from a teleseismic experiment on the southern Mid-Atlantic Ridge that suggests that travel time delays may indeed increase with distance from the ridge axis for stations located outside the median valley.

Combinations of the melt models discussed in the previous section and the textural

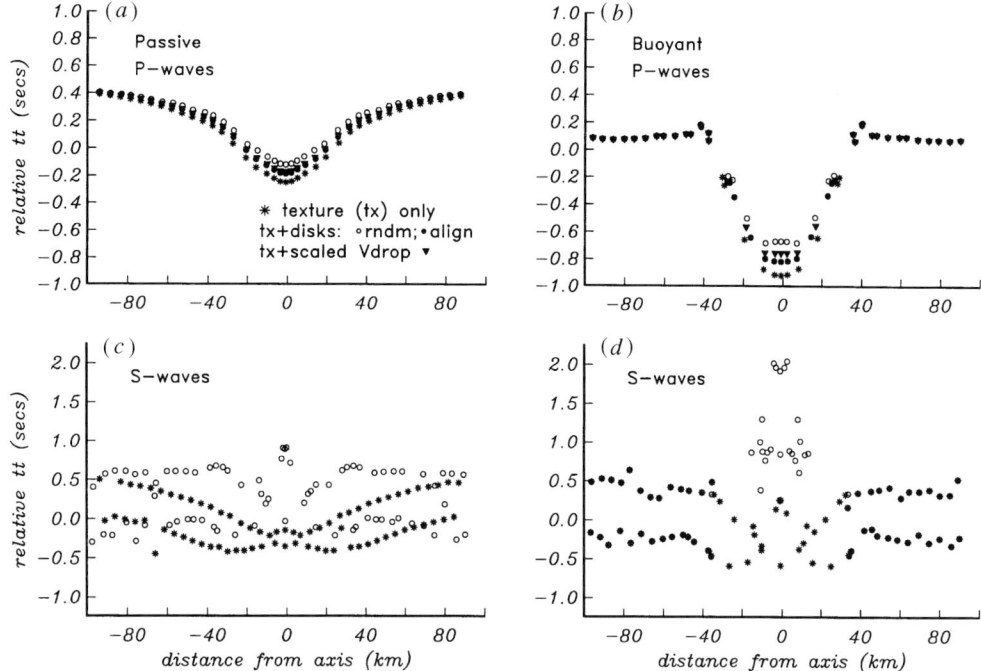

Figure 6. Relative travel times of near-vertical rays through combined melt + textural anisotropy models of subaxial structure for passive (left panels) and buoyant (right panels) flow model. P-waves arrive on axis earlier than off axis due to preferred orientation of olivine crystals (figure 1) which slows vertical propagation off axis and, in the buoyant case, speeds vertical propagation below the axis. Addition of melt to the texture models reduces the size of the axial travel time advance a small amount for P-waves. An axial S-wave delay overprints the texture-only (∗) signal as illustrated for the randomly oriented melt disk geometry (open circles). The shear-waves polarized perpendicular to the ridge axis are the late arrivals off axis (—∗—), but they are earliest to arrive at the ridge axis.

anisotropy models illustrate the relative importance of each mechanism on travel times. For P-waves in the melt alone models, randomly oriented disks of melt in the buoyant model can produce axial delays up to 0.4 s. The P-wave results for the combined textural and melt models (upper panels of figure 6) show that the texture effects dominate. The relative axial delay due to the melt is muted by the off-axis delays due to the horizontal alignment of the a-axis of olivine. For the models tested, the reduction in the texture-controlled early axial travel times caused by presence of melt is less than 0.3 s for the buoyant case and less than 0.1 s for the passive case.

Rays paths for S-waves in the combined melt + texture models can be complicated but the overall pattern found for the texture travel times is evident for the combined models. The polarity of the fastest arriving shear wave off axis is opposite that of the earliest S-wave at the axis (figure 6, lower panels). The amount of predicted shear wave splitting is greater in the buoyant case than in the passive case where the axial splitting would probably not be resolvable in data. Addition of melt distributed in randomly oriented melt-filled disks to the texture model modifies the axial S-wave travel time anomaly in a fashion that is predictable from the melt-only model. An axial delay of up to 2.5 s is predicted for the buoyant model, whereas for the passive model the relative delay is quite a bit smaller due to the lower melt fractions. The amount of shear wave splitting at the axis increases somewhat over the texture-only model even though the randomly oriented disk model does not give anisotropic elastic constants. This is simply because there is an overall reduction of velocity in the melt zone therefore S-wavefronts have more time to accrue separation.

5. Discussion

The travel time delay that we predict to be associated with a given melt model depends on our initial assumption that the melt produced during upwelling is retained in the interstices of the matrix during most of the mantle ascent beneath a spreading centre. This is useful in that it allows us to estimate the maximum effect that melt in a series of distribution models might have on teleseismic travel times. If melt migrates rapidly upon production or does not accumulate to more than 1% volume fraction then the melt-related axial delay can only be smaller than what we predict.

Currently, teleseismic data with enough resolution to detect a delay associated with subaxial melt is limited. Forsyth (1997) reports that ocean bottom hydrophones deployed on the Mid-Atlantic Ridge 35° N recorded PKP arrivals that show a relative travel time delay that can be attributed to the presence of a narrow melt zone about 20–30 km deep beneath the spreading centre. Stations located within the median valley had travel time delays, after correction for variable topography and range differences between stations, up to about 0.3 s relative to stations located 16 km off axis. The size of this signal is close to our predictions for the buoyancy enhanced flow model with up to 6% melt fraction distributed in randomly oriented disks or tubules. The narrowness of the travel time anomaly observed by Forsyth (1997) would suggest that melt is more focused than our flow modelling predicts. Blackman et al. (1995) note that there may be a 0.2 s P-wave delay for a central American earthquake that is received by an ocean bottom seismometer 10 km west of the East Pacific rise axis. It is possible that the delay is due to melt in the mantle beneath the ridge axis but the experiment geometry does not allow local heterogeneity at the station to be ruled out.

Melt distributed in aligned disk-like inclusions has the largest effect on near vertical shear waves that transit the subaxial region but this reflects the fact that the plane of the disks are oriented vertically. A different alignment would produce different results. It is not certain that melt films will preferentially occupy grain faces that are vertical in the upwelling zone but Phipps Morgan (1987) and Spiegelman (1993) show that melt paths should be near vertical in this region if pressure gradients and oriented permeability in the matrix control the migration. If melt occupies an interconnected network it is reasonable to assume that the films along faces that are not oriented vertically will migrate towards the vertical faces if the pressure drop that is driving upward flow of melt along the vertical faces is sufficient. At a larger scale, this type of behaviour is predicted (Sleep 1984) and observed (Ceuleneer & Rabinowicz 1992; Takahashi 1992) where vertical dykes or melt-filled fractures drain regions of lower melt fraction.

The travel time contributions of lithospheric velocity structure and crustal structure have been ignored in our calculations in order to emphasize the effect of the melt and the textural anisotropy. Thermal heterogeneity in the mantle is included indirectly in terms of the location and presence of the melt zone. An additional thermal signal due to the small decrease in velocity between the cool lithosphere and the underlying asthenosphere would reduce the predicted delays 100 km off axis by less than 0.1 s and 0.4 s, for P- and S-waves, respectively (Kendall 1994). Oceanic crustal structure is generally rather uniform compared to the continents (Spudich & Orcutt 1980) but variations in its thickness and density structure do occur. Differences in crustal thickness of 3 km (see, for example, Michael et al. 1994; Tolstoy et al. 1993) across the profile for which we predict travel time anomalies would introduce less than 0.25 s variations. The axial region is generally highly fractured and hydrothermal circulation may occur to depths of a few kilometres. The observed axis-parallel orientation of most median valley fissures would result in significant crustal anisotropy at the shallowest levels. MacDonald et al. (1994) report more than 50% anisotropy in the upper few hundred metres of the crust at the Juan de Fuca ridge. To get a rough estimate of the shear-wave splitting effects due to crustal anisotropy we assume the velocities and anisotropy reported in MacDonald et al. (1994) throughout a 1 km thick region of aligned fractures at the ridge axis. Shear wave splitting of about 0.4 s would result. This is less than 20% of the splitting magnitude that we predict to occur for mantle melt fractions of about 2% distributed in vertically oriented disks.

The modelling shows that clear P- and S-arrivals with a range of slownesses are required to discriminate between passive and buoyancy driven flow using ocean-bottom teleseismic observations. Probably the most robust feature of the models is the style of anomalous travel time behaviour across the axis. The buoyant case is characterized by larger travel time effects in a narrower axial region, while the passive case shows smaller effects over a broader axial region. Inferring the geometry of melt inclusions will be difficult, especially with P-wave data. Our predictions indicate that S-waves will show more obvious differences between melt stored in tubules rather than disks especially if a preferred disk orientation exists. Untangling these effects from those due to mineral alignment will require high quality and well distributed data.

References

Blackman, D. K., Orcutt, J. A. Forsyth, D. W. & Kendall, J.-M. 1993 Seismic anisotropy in the mantle beneath an oceanic spreading center. *Nature* **366**, 675–677.

Blackman, D. K., Orcutt, J. A. & Forsyth, D. W. 1995 Recording teleseismic earthquakes using ocean bottom seismographs at mid-ocean ridges. *Bull. Seism. Soc. Am.* **185**, 1648–1664.

Blackman, D. K., Kendall, J.-M., Dawson, P. R., Wenk, H.-R., Boyce, D. & Phipps Morgan, J. 1997 Teleseismic imaging of subaxial flow at mid-ocean ridges: travel time effects of anisotropic mineral texture in the mantle. *Geophys. J. Int.* (In the press.)

Ceuleneer, G. & Rabinowicz, M. 1992 Mantle flow and melt migration beneath oceanic ridges: models derived from observations in ophiolites. In *Mantle flow and melt generation at mid-ocean ridges* (ed. J. Phipps Morgan et al.), pp. 123–154. (Geophysical Monograph 71.) Washington, DC: AGU.

Ceuleneer, G., Monnereau, M. & Amri, I. 1996 Thermal structure of a fossil mantle diapir inferred from the distribution of mafic cumulates. *Nature* **379**, 149–153.

Chastel, T. B., Dawson, P. R., Wenk, H.-R. & Bennett, K. 1993 Anisotropic convection with implications for the upper mantle. *J. Geophys. Res.* **98**, 17 575–17 771.

Faul, U. H., Toomey, D. R. & Waff, H. S. 1994 Intergranular basaltic melt is distributed in thin, elongated inclusions. *Geophys. Res. Lett.* **21**, 29–32.

Forsyth, D. W. & Uyeda, S. 1975 On the relative importance of the driving forces of plate motion. *Geophys. Jl R. Astr. Soc.* **43**, 163–200.

Forsyth, D. W. 1992 Geophysical constraints on mantle flow and melt generation beneath mid-ocean ridges. In *Mantle flow and melt generation at mid-ocean ridges* (ed. J. Phipps Morgan et al.), pp. 103–122. (Geophysical Monograph 71.) Washington, DC: AGU.

Forsyth, D. W. 1997 Partial melting beneath a Mid-Atlantic Ridge segment detected by teleseismic PKP delays. *Geophys. Res. Lett.* (In the press.)

Guest, W. S. & Kendall, J.-M. 1993 Modelling waveforms in anisotropic inhomogeneous media using ray and Maslov theory: applications to exploration seismology. *Can. J. Expl. Geophys.* **29**, 78–92.

Hess, P. C. 1992 Phase equilibria constraints on the origin of ocean floor basalts. In *Mantle flow and melt generation at mid-ocean ridges* (ed. J. Phipps Morgan et al.), pp. 67–102. (Geophysical Monograph 71.) Washington, DC: AGU.

Hirth, G. & Kohlstedt, D. L. 1992 Experimental constraints on the dynamics of the partially molten upper mantle: deformation in the diffusion creep regime. *J. Geophys. Res.* **100**, 1981–2001.

Hirth, G. & Kohlstedt, D. L. 1996 Water in the oceanic upper mantle: implications for rheology, melt extraction and the evolution of the lithosphere. *Earth Planet. Sci. Lett.* **144**, 93–108.

Houseman, G. 1983 The deep structure of ocean ridges in a convecting mantle. *Earth Planet. Sci. Lett.* **64**, 283–294.

Hudson, J. A. 1980 Overall properties of a cracked solid. *Math. Proc. Camb. Phil. Soc.* **88**, 371–384.

Karato, S. 1986 Does partial melting reduce the creep strength of the upper mantle? *Nature* **319**, 309–310.

Kendall, J.-M. 1994 Teleseismic arrivals at a mid-ocean ridge: effects of mantle melt and anisotropy. *Geophys. Res. Lett.* **21**, 301–304.

Kendall, J.-M. & Thomson, C. J. 1989 A comment on the form of the geometrical spreading equations, with some examples of seismic ray tracing in inhomogeneous, anisotropic media. *Geophys. J. Int.* **99**, 401–413.

Kohlstedt, D. 1992 Structure, rheology and permeability of partially molten rocks at low melt fractions. In *Mantle flow and melt generation at mid-ocean ridges* (ed. J. Phipps Morgan et al.), pp. 103–122. (Geophysical Monograph 71.) Washington, DC: AGU.

Kumazawa, M. & Anderson, O. L. 1969 Elastic moduli, pressure derivatives and temperature derivatives of a single-crystal olivine and a single crystal forsterite. *J. Geophys. Res.* **74**, 5961–5972.

Lachenbruch, A. 1973 A simple mechanical model for oceanic spreading centers. *J. Geophys. Res.* **78**, 3395–3417.

MacDonald, M. A., Webb, S. C., Hildebrand, J. A., Cornuelle, B. D. & Fox, C. F. 1994 Seismic structure and anisotropy of the Juan de Fuca ridge at 45° N. *J. Geophys. Res.* **99**, 4857–4873.

Mavko, G. M. & Nur, A. 1975 Melt squirt in the asthenosphere. *J. Geophys. Res.* **80**, 1444–1448.

Mavko, G. M. 1980 Velocity and attenuation in partially molten rocks. *J. Geophys. Res.* **85**, 5173–5289.

McKenzie, D. 1984 Generation and compaction of partially molten rock. *J. Petrol.* **25**, 713–765.

Michael, P. J. et al. 1994 Mantle control of a dynamically evolving spreading center: Mid-Atlantic ridge 31–34° S. *Earth Planet Science Lett.* **121**, 451–468.

Murase, T. & McBirney, A. R. 1973 Properties of some common igneous rocks and their melts at thigh temperatures. *Geol. Soc. Am. Bull.* **84**, 3563–3592.

Nicolas, A. 1989 *Structure of ophiolites and dynamics of oceanic lithosphere*, p. 367. Deventer: Kluwer.

Nielson, J. E. & Wilshire, H. G. 1993 Magma transport and metasomatism in the mantle: a critical review of current geochemical models. *Am. Mineralogist* **78**, 1117–1134.

O'Connell, R. J. & Budiansky, B. 1977 Viscoelastic properties of fluid-saturated cracked solids, *J. Geophys. Res.* **82**, 5719–5735.

O'Connell, R. J. & Budiansky, B. 1974 Seismic velocities in dry and saturated cracked solids, *J. Geophys. Res.* **79**, 5412–5426.

O'Nions, R. K. & McKenzie, D. 1993 Estimates of mantle thorium/uranium ratios from TH, U and Pb isotope abundances in basaltic melts. *Phil. Trans. R. Soc. Lond.* A **342**, 65–77.

Parmentier, E. M. & Phipps Morgan, J. 1990 The spreading rate dependence of three-dimensional spreading center structure. *Nature* **348**, 325–328.

Phipps Morgan, J. 1987 Melt migration beneath mid-ocean spreading centers. *Geophys Res. Lett.* **14**, 1238–1241.

Phipps Morgan, J. & Forsyth, D. W. 1988 Three-dimensional flow and temperature perturbations due to a transform offset: effects on oceanic crustal and upper mantle structure. *J. Geophys. Res.* **93**, 2955–2966.

Quick, J. E. & Gregory, R. T. 1995 Significance of melt-wall rock reaction: a comparative anatomy of three ophiolites. *J. Geology* **103**, 187–198.

Rabinowicz, M., Nicolas, A. & Vigneresse, J. 1984 A rolling mill effect in the asthenosphere beneath oceanic spreading centers. *Earth Planet. Sci. Lett.* **67**, 97–108.

Sayers, C. 1992 Elastic anisotropy of short-fibre reinforced composites. *Int. J. Solids Structures.* **29**, 2933–2944.

Schmeling, H. 1985 Numerical models on the influence of partial melt on elastic, anelastic and electrical properties of rocks. I. Elasticity and anelasticity. *Phys. Earth Planet. Int.* **41**, 34–57.

Scott, D. & Stevenson, D. 1989 A self-consistent model of melting magma migration, and buoyancy-driven circulation beneath mid-ocean ridges. *J. Geophys. Res.* **94**, 2973–2988.

Sleep, N. H. 1975 Sensitivity of heat flow and gravity to the mechanism of sea-floor spreading. *J. Geophys. Res.* **74**, 542–549.

Sleep, N. H. 1984 Tapping of magmas from ubiquitous mantle heterogeneities: an alternative to mantle plumes? *J. Geophys. Res.* **89**, 10 029–10 041.

Spiegelman, M. 1996 Tracers on the move: the sensitivity of trace element geochemistry to melt transport. *Eos* **76**, 594.

Spiegelman, M. 1993 Physics of melt extraction: theory, implications and applications. *Phil. Trans. R. Soc. Lond.* A **342**, 23–41.

Spudich, P. & Orcutt, J. A. 1980 A new look at the seismic velocity structure of the oceanic crust. *Rev. Geophys. Space Physics* **18**, 627–645.

Takahashi, N. 1992 Evidence for melt segregation towards fractures in the Horoman mantle peridotite complex. *Nature* **359**, 52–55.

Tandon, G. P. & Weng, G. J. 1984 The effect of aspect ratio of inclusions on the elastic properties of unidirectionally aligned composites. *Polymer Composites* **5**, 327–333.

Thomsen, L. 1995 Elastic anisotropy due to aligned cracks in porous rock. *Geophys. Prosp.* **43**, 805–829.

Tolstoy, M., Harding, A. J. & Orcutt, J. A. 1993 Evolution of crustal structure at 34S on the mid-Atlantic ridge. *Eos* **74**, 646.

Toramaru, A. & Fujii, N. 1986 Connectivity of melt phase in a partially molten peridotite. *J. Geophys. Res.* **91**, 9239–9252.

Waff, H. S. & Bulau, J. R. 1982 Experimental studies of near-equilibrium textures in partially molten silicates at high pressure. *Adv. Earth Planet. Sci.* **12**, 229–236.

Waff, H. S. & Faul, U. H. 1992 Effects of crystalline anisotropy on fluid distribution in ultramafic partial melts. *J. Geophys. Res.* **97**, 9003–9014.

Waff, H. S. & Holdren, G. R. 1981 The nature of grain boundaries in dunite and lherzolite xenoliths: implications for magma transport in refractory upper mantle material. *J. Geophys. Res.* **86**, 3677–3683.

Evidence for accumulated melt beneath the slow-spreading Mid-Atlantic Ridge

BY M. C. SINHA[1], D. A. NAVIN[2], L. M. MACGREGOR[1], S. CONSTABLE[3], C. PEIRCE[2], A. WHITE[4], G. HEINSON[4] AND M. A. INGLIS[2]

[1] Bullard Laboratories, Department of Earth Sciences, University of Cambridge, Cambridge CB3 0EZ, UK
[2] Department of Geological Sciences, University of Durham, Durham DH1 3LE, UK
[3] Institute of Geophysics and Planetary Physics, Scripps Institution of Oceanography, University of California San Diego, La Jolla, CA 92093, USA
[4] School of Earth Sciences, Flinders University of South Australia, Adelaide 5001, Australia

The analysis of data from a multi-component geophysical experiment conducted on a segment of the slow-spreading (20 mm yr^{-1}) Mid-Atlantic Ridge shows compelling evidence for a significant crustal magma body beneath the ridge axis. The role played by a crustal magma chamber beneath the axis in determining both the chemical and physical architecture of the newly formed crust is fundamental to our understanding of the accretion of oceanic lithosphere at spreading ridges, and over the last decade subsurface geophysical techniques have successfully imaged such magma chambers beneath a number of intermediate and fast spreading (60–140 mm yr^{-1} full rate) ridges. However, many similar geophysical studies of slow-spreading ridges have, to date, found little or no evidence for such a magma chamber beneath them.

The experiment described here was carefully targeted on a magmatically active, axial volcanic ridge (AVR) segment of the Reykjanes Ridge, centred on 57° 43′ N. It consisted of four major components: wide-angle seismic profiles using ocean bottom seismometers; seismic reflection profiles; controlled source electromagnetic sounding; and magneto-telluric sounding. Interpretation and modelling of the first three of these datasets shows that an anomalous body lies at a depth of between 2 and 3 km below the seafloor beneath the axis of the AVR. This body is characterized by anomalously low seismic P-wave velocity and electrical resistivity, and is associated with a seismic reflector. The geometry and extent of this melt body shows a number of similarities with the axial magma chambers observed beneath ridges spreading at much higher spreading rates. Magneto-telluric soundings confirm the existence of very low electrical resistivities in the crust beneath the AVR and also indicate a deeper zone of low resistivity within the upper mantle beneath the ridge.

---

1. Introduction

Current explanations for the mode of formation of many of the lithological and structural features observed in young oceanic crust, and in ophiolites, require the presence of an axial magma chamber at shallow depths of no more than a few kilometres below the seabed. The importance of the crustal magma chamber concept is that it provides the setting in which the gradual crystallization of an initially primitive melt into the range of observed petrologies can occur (Sinton & Detrick 1992); it provides a reservoir from which melt can be injected into the newly forming crust as dykes or erupted onto the seafloor as lavas; and it provides a location from which deeper crustal rocks can form by a process of slow crystallization.

Geophysical observations to date have provided valuable constraints on the geometry and properties of crustal magma chambers beneath the faster spreading ridges in the Pacific (Detrick et al. 1987, 1993a; Collier & Sinha 1992; Harding et al. 1989; Vera et al. 1990), and have been extremely influential in the development of models for the magmatic accretion of crystalline oceanic crust at spreading rates of greater than about 60 mm yr^{-1} (Sinton & Detrick 1992; Henstock et al. 1993). However, there has been only one report of direct geophysical evidence for a crustal melt body beneath a slow spreading ridge (Calvert 1995). This observation relies on faint reflections seen on heavily processed multichannel seismic reflection profiles, the data from which had been previously processed and interpreted as evidence against any melt body at that location (Detrick et al. 1990). Numerous other geophysical experiments at a wide variety of locations on slow spreading ridges (see, for example, Fowler 1976; Fowler & Keen 1979; Purdy & Detrick 1986; Bunch & Kennett 1980; Smallwood et al. 1995) have failed to find evidence for a significant crustal melt body. Arguments based on thermal modelling have shown that the heat flux into the crust associated with magma delivery from the mantle is insufficient to maintain a steady-state crustal melt body, even of small dimensions, at slow spreading rates (Sleep 1975; Kusznir & Bott 1976). This apparent contradiction between the need for crustal melt bodies to explain structural and petrological observations, and the geophysical evidence that in most places they are not present, is best resolved by the proposition that beneath slow spreading ridges, melt-filled magma chambers are transient features. If this is the case, then the process of magmatic accretion at slower spreading rates must be episodic or cyclic, and observations of crustal melt bodies will be dependent on identifying a spreading segment that is at the appropriate stage of the cycle.

In the absence of any reliable images of a crustal melt body *in situ* beneath a slow spreading ridge to date, it remains impossible to resolve important questions concerning the differences and similarities between ridge processes at different spreading rates. Of particular relevance is the question of whether accretion involves a melt body of broadly similar geometry at all spreading rates, with the major differences being confined to the temporal variability of the process (short-lived, transient melt bodies at slow rates, compared to long-lived or steady-state bodies at faster rates); or whether the geometry of the melt body, and hence the whole mechanism of crustal accretion and architecture at all scales of the crystalline crust, depends strongly on spreading rate.

In this paper we describe the results of the first integrated geophysical experiment to provide convincing evidence for a magma chamber beneath a slow spreading ridge. The experiment was centred on an AVR segment of the Reykjanes Ridge—the section of the Mid-Atlantic Ridge between the tip of the Reykjanes Peninsula on the south coast of Iceland, at 63° 30' N, and the Charlie–Gibbs fracture zone at 52° N (inset figure 1). Our subseafloor geophysical experiment consisted of four integrated components: wide-angle seismic profiles shot to determine crustal seismic velocity structure and layer thickness; seismic reflection profiles shot to measure off-axis sediment thickness (and hopefully image crustal seismic reflectivity); controlled

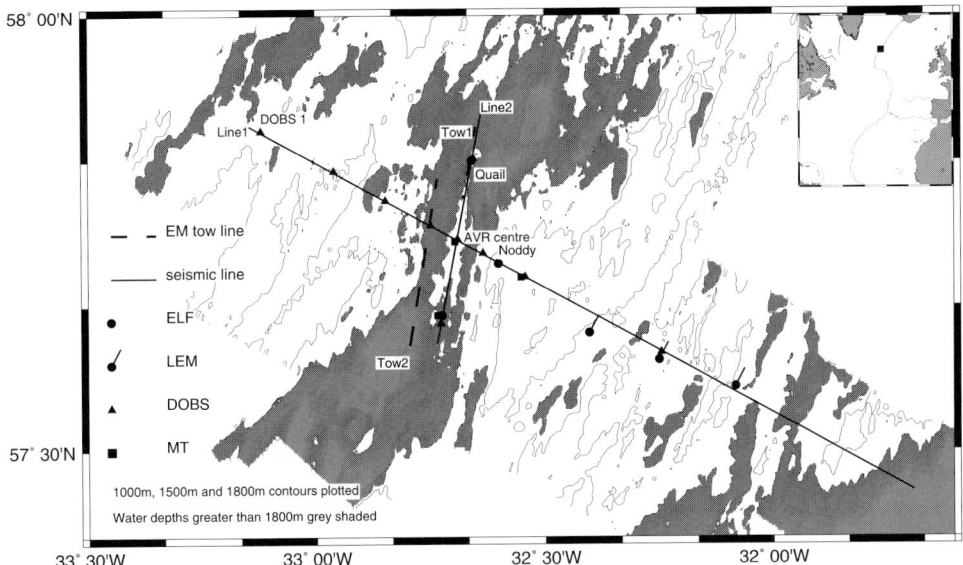

Figure 1. Experimental geometry and location. Water depths of greater than 1800 m are indicated by shading. The 1800 m bathymetric contour outlines the median valley of the ridge, and the axial volcanic ridge within the valley. Seismic (solid) and electromagnetic tow (dashed) lines are shown together with instrument locations (see key). Inset: map of the North Atlantic region showing the location of the study area.

source electromagnetic soundings to determine crustal electrical resistivity structure; and magneto-telluric soundings to investigate the deeper resistivity structure of the mantle.

2. Tectonic setting

At its northern end, the Reykjanes Ridge is profoundly affected by the proximity of the Iceland hot spot, which results in anomalously shallow water depths, the absence of a median valley and a crustal composition dominated by plume geochemistry. The hot spot influence decreases with increasing distance from Iceland, with both the geochemical and bathymetric anomalies declining steadily (Schilling et al. 1983; Keeton et al. 1997) and a discernible rift valley appearing at about 58° 40′ N. The Reykjanes Ridge is segmented into a series of elongated AVRs with dimensions of up to a few tens of kilometres in length, arranged *en echelon* so that in general they overlap each other (Keeton et al. 1997; Searle & Laughton 1981; Searle et al. 1994). Individual AVRs are aligned approximately orthogonal to the spreading direction, but the overall trend of the ridge is oblique to this so that all offsets between AVR segments are right lateral. There are also no transform offsets between 57° N and the Reykjanes Peninsula—a distance of almost 950 km. Side-scan sonar imagery of the seafloor at the axis of the Reykjanes Ridge (Murton & Parson 1993) reveals a complex interplay between constructive volcanism (leading to the construction of AVRs), tectonic extension (leading to the break up of AVRs by normal faults aligned orthogonally to the spreading direction), and at distances of more than about 700 km from Iceland, the development of major inward-facing normal faults that run subparallel to the overall trend of the ridge, and form the bounding walls of the developing median valley. Comparisons between AVRs reveal that constructive volcanism dominates at some, while at others there is little evidence of recent magmatism and tectonic features dominate. This observation leads to the inference that AVRs follow a life cycle in which occasional influxes of magma from the mantle are separated by long periods of magmatic dormancy (Parson *et al.* 1993).

The AVR chosen for the present study extends along an approximately N–S trend for some 35 km, between 57° 33′ N and 57° 53′ N (figure 1). Our selection was based on the assumption that the rate of melt injection into the crust (and hence the likelihood of finding a crustal melt reservoir) will be at a maximum at the point in time when the AVR is still undergoing vigorous construction. In contrast, a more mature AVR that has reached its maximum topographic development is probably already at the end of the magmatically robust phase of its life cycle. We selected, using all the available side-scan sonar and bathymetry data, the AVR which showed the greatest evidence of fresh volcanic construction (hummocky topography, bright back scattering and fresh-looking lava flows extending for distances of several kilometres), and the least evidence of the recent normal faulting and fissuring that characterizes most AVRs, and that appears to indicate post-magmatic, tectonic extension. By comparison with a number of other larger AVRs, which we interpret as being at, or close to, the end of the magmatically robust stage of their lives, our selected AVR is a relatively unimpressive feature topographically.

The site of our experiment is approximately 840 km from the Reykjanes Peninsula, and over 1100 km from the centre of the Iceland plume. At this latitude, a clear median valley is present, which extends for more than 100 km northwards towards Iceland before an axial topographic high appears. Also, as we will show in the next section, away from the axis the crustal thickness, determined seismically, is between 6 and 8 km—typical of normal Atlantic oceanic crust. We can infer from this observation that although, at the site of our experiment, the Iceland plume is close enough to cause anomalously shallow water depths and a ridge topography that is atypical of other parts of the Mid-Atlantic Ridge, it is sufficiently distant that it causes no increase in the overall rate of melt production. In §4, we will argue, based on the normal melt production rate, that what distinguishes this locality from other segments of slow-spreading ridges is not its position relative to the Iceland hot spot, but the fact that it is in the most magmatically active stage of its life cycle.

3. Experimental results

(a) The wide-angle seismic experiment

For the wide-angle seismic experiment we deployed an array of 11 digital ocean bottom seismometers (DOBSs) along two profiles (figure 1). Line 1 was 100 km in length and ran orthogonally to the overall trend of the Reykjanes Ridge, crossing the axis near the centre of the AVR. Line 2 was 35 km in length and ran along the AVR axis. Explosive shots at 1 or 2 km intervals and airgun shots at 100 m intervals were fired along both lines. Example explosive and airgun record sections from DOBS 1, which was located at the WNW end of line 1, are shown in figure 2a. The observed data from all instruments on both lines show clear crustal diving ray arrivals (Pg), while data from line 1 also show arrivals that have been refracted in the uppermost mantle (Pn).

The travel times and amplitudes of the seismic data along each profile have been analysed using synthetic seismogram modelling (figures 2b, c), using the ray-theory based Maslov algorithm (Chapman & Drummond 1982). The resulting P-wave velocity model along line 1 is shown in figure 2d. Both airgun and explosive record sections from all DOBSs on this profile were used to constrain the modelling. At distances of greater than 10 km from the axis, the seismic velocity structure is typical of that of normal oceanic crust (White et al. 1992). Beneath a thin and discontinuous layer of sediments, up to 200 m thick, the seismic velocity varies between 2.8 and 3.5 km s^{-1} at the top of the crystalline crust. A steep velocity gradient is present in the top 2.0–2.5 km of the crust, including a change in gradient at ca. 500 m depth (located approximately at the 4.5 km s^{-1} velocity contour) (Navin et al. 1996). The velocity reaches 6.5 km s^{-1} at a depth of 2.0–2.5 km below the seafloor. Below this level, which we interpret as the boundary between seismic layers 2 and 3, P-wave velocity increases more slowly with depth, reaching about 7 km s^{-1} just above the crust–mantle boundary, which occurs at between 6 and 8 km below the seabed. The uppermost mantle has a velocity of 7.9 km s^{-1}. This part of the structure is constrained primarily by travel times of lower crustal and upper mantle phases, but modelled amplitudes are also consistent with the data. Uncertainties in the model indicated by the minimum perturbations that significantly degrade the fit of synthetic to observed seismograms are typically ±0.1 km s^{-1} in velocity, and ±0.5 km in depth.

The axial region has a distinctively different structure however. It is marked topographically by a shallow rift valley some 30 km wide and up to 800 m deep, within which lies the AVR. The walls of the valley are formed by a succession of large inward-facing normal fault scarps, which are clearly visible in swath bathymetry data collected during the experiment (Sinha et al. 1994; Keeton et al. 1997). In order to fit the travel times of crustal diving rays passing through the axial region, P-wave velocities must be depressed relative to the off-axis structure over a zone approximately 20 km wide. The seafloor velocity at the AVR is about 2.4 km s^{-1}, and crustal velocities are between 1.0 and 1.5 km s^{-1} lower than the off-axis velocities at equivalent depths throughout most of layers 2 and 3.

Perhaps the most intriguing feature of the seismic data from line 1 is the amplitude behaviour of crustal diving rays that turn at mid-crustal depths beneath the AVR axis. The amplitudes of such arrivals are anomalously low, creating prominent shadow zones on all eight record sections from the across-axis line (line 1). Unlike other variations in amplitude observed along the record sections, these anomalies cannot be explained either in terms of the velocity structure described above, or by the focusing or defocusing effects of seafloor topography. An indication of the nature of the feature that causes these shadow zones is provided by the DOBS record sections from line 2, the along-axis profile. These sections show a prominent cut-off in crustal diving ray arrivals occurring at source-receiver offsets of between 8 and 12 km. Such a feature is characteristic of a low seismic velocity zone at depth. Travel time and amplitude modelling of the data from both lines (figures 2b, c) has shown that a thin layer of extremely low-velocity material is required at between 2 and 3 km depth below the axis of the AVR to explain the observed P-wave shadow zones. Only models with such a feature generate synthetic seismograms which match the observed travel times and amplitude variations with offset along both lines. In figure 2d, this low-velocity body is represented by a thin, lens-shaped feature, 3 km in across-axis extent, 200 m thick and with a P-wave velocity of 3 km s^{-1}. Figure 2b shows synthetic seismograms calculated with the Maslov code for this model while, for comparison, figure 2c shows

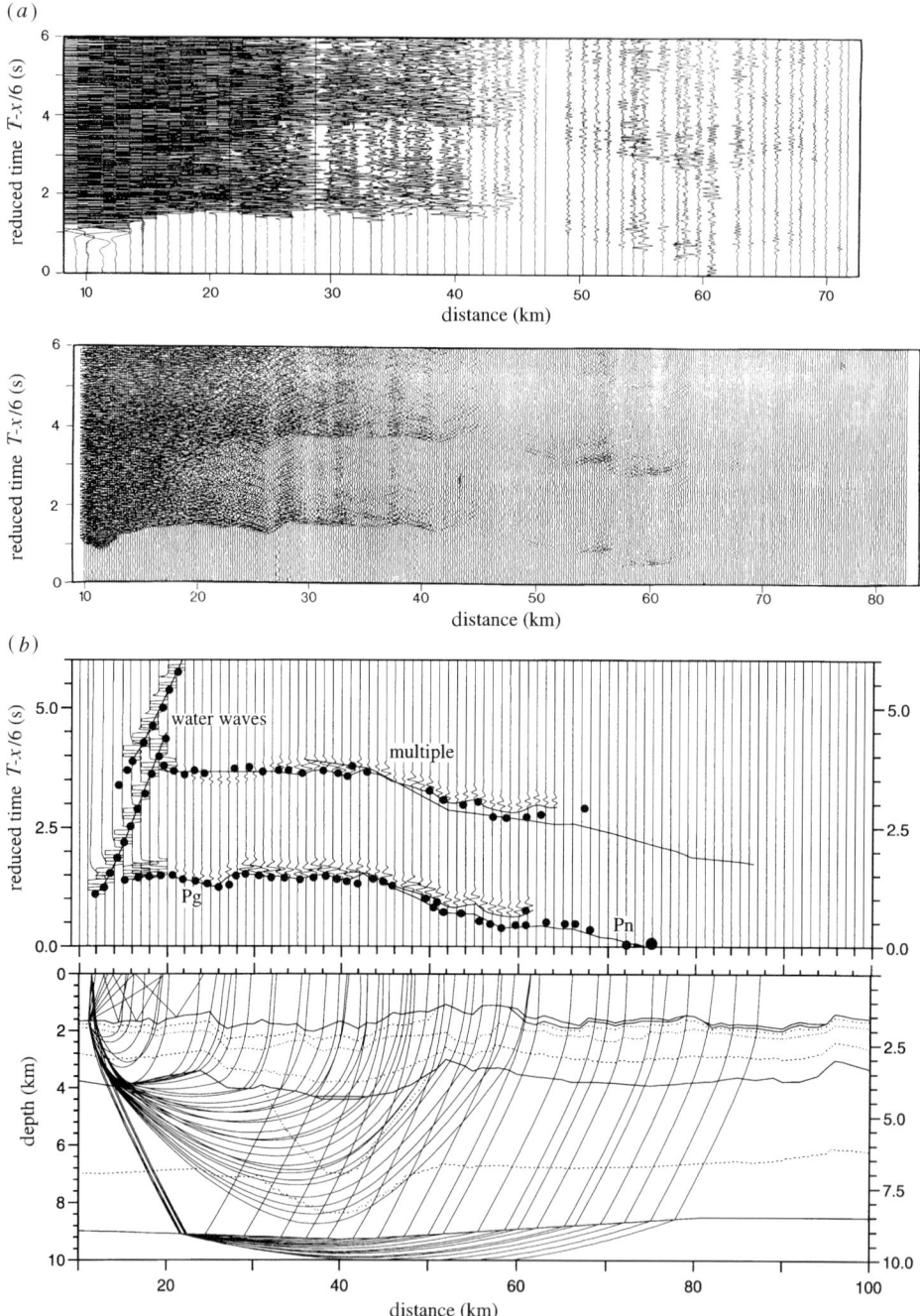

Figure 2. (a) Example explosive (upper) and airgun (lower) record sections from DOBS 1 of the wide-angle seismic experiment, showing water-borne, crustal and upper mantle arrivals. A distinct shadow zone for P-waves is visible at ranges between 43 and 49 km, and is associated with the spreading axis. (b) Synthetic record section computed using the Maslov ray-theory algorithm, showing the fit of observed to synthetic data for DOBS 1. Observed travel time picks are shown by dots whose size gives an indication of the picking error. The lower panel shows a representative subset of rays used to calculate the synthetics. Dotted lines show the 4.5, 5.5, 6.0, 6.5 and 6.8 km s^{-1} isovelocity contours. (continued opposite)

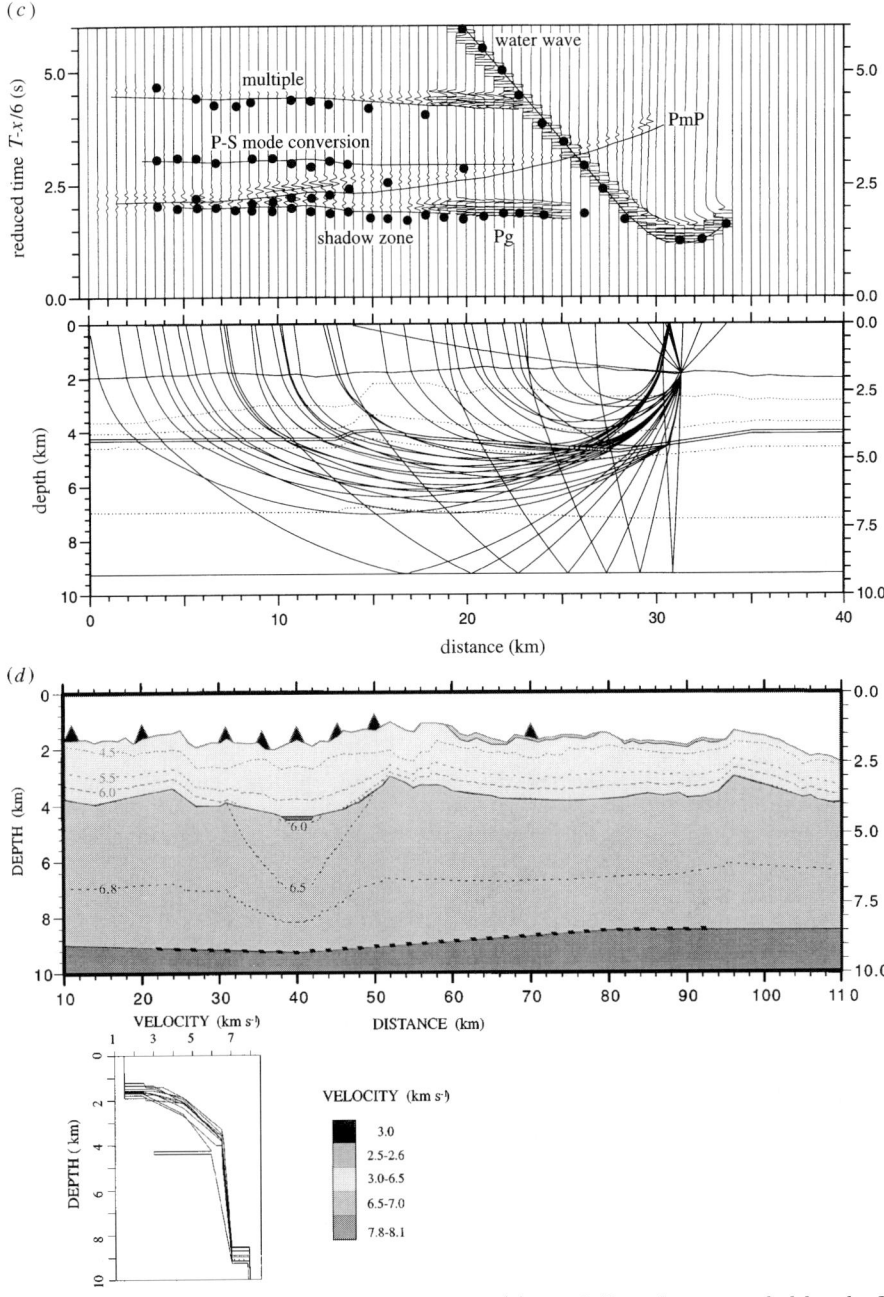

Figure 2 (cont.). (c) Synthetic record section, as in (b), modelling data recorded by the DOBS located at the northern end of the along-axis line (line 2). Note the modelling of the P-wave shadow zone associated with the subaxis low-velocity body. PmP is the wide-angle reflection from the Moho discontinuity. (d) Across-axis (line 1) seismic velocity–depth model deduced from the modelling. Contours are in km s^{-1} and DOBS positions are shown by triangles. Interfaces shown by solid lines indicate (from top to bottom): seafloor; sediment–basement boundary (where present); layer 2–layer 3 boundary; and crust–mantle boundary. The entire length of the model is constrained by many crossing ray paths. The length of the Moho constrained by the modelling is shown by the diagonal dashed line. Bottom left: velocity–depth profiles taken at intervals across the model showing crustal velocity gradients and depth to the low-velocity zone.

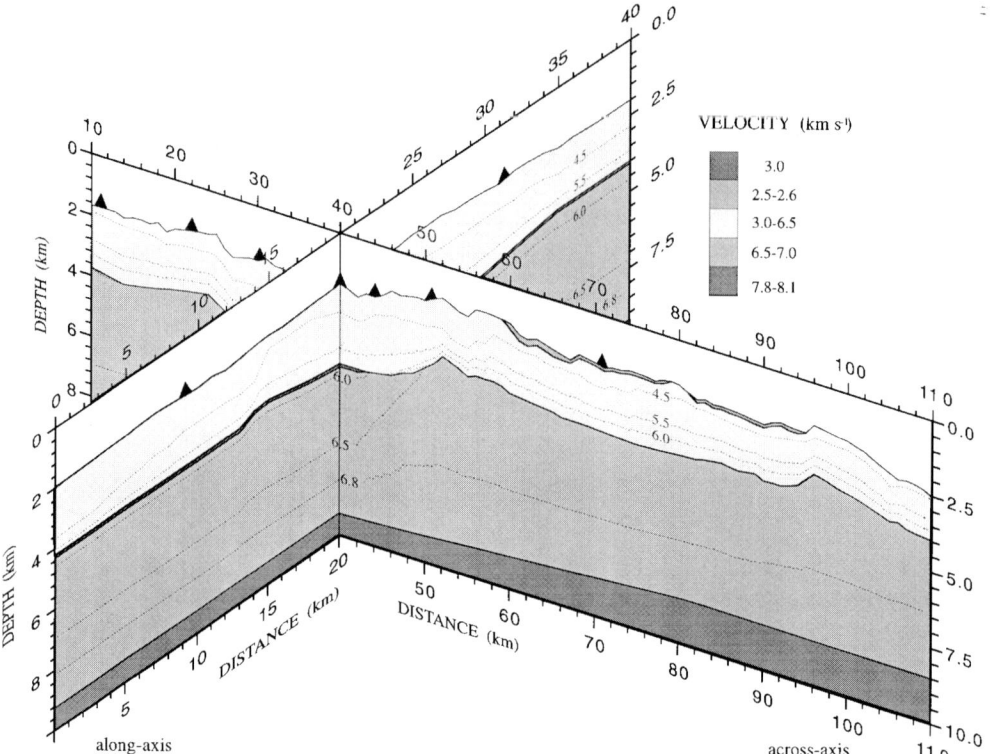

Figure 3. Final along- (line 1) and across-axis (line 2) velocity–depth models, showing the good correlation at the intersection point. DOBS locations are shown by triangles. Velocity–depth contours are in km s^{-1}. Note the limited lateral, and continuous along-axis, extent of the low-velocity body.

synthetics calculated for the along-axis model. The velocity models independently derived by the modelling procedure for the across and along-axis profiles, lines 1 and 2, are in excellent agreement at the intersection of the two lines (figure 3).

(b) The seismic reflection experiment

The coincident seismic reflection data were collected simultaneously with the airgun component of the wide-angle dataset, using an 800 m long eight-channel hydrophone streamer. The principal reason for collecting these reflection data was to measure off-axis sediment thickness, in order to provide shallow constraints on wide-angle velocity models. The 100 m shot spacing, short streamer length and consequent low stack fold, combined with the fact that the source array parameters were optimized for long-range wide-angle work, mean that the resulting data are far from ideal for observing crustal reflectivity. However, careful processing of the data from the along-axis line (line 2) has revealed a set of seismic reflections at between 3.0 and 4.0 s two-way travel time (TWTT), corresponding in depth—after allowing for variations in water depth, and using crustal velocities derived from the wide-angle data—to 2.0 to 2.5 km below the seafloor. Part of the along-axis reflection profile is shown in figure 4a. The travel time of the crustal reflector is much smaller than that of the first multiple of the seafloor reflection—which is calculated to arrive at much greater travel times of between 4.5 and 5.0 s, where it can indeed be observed in figure 4a. On the other hand, its travel times are in excellent agreement with those predicted to the low-velocity body by calculating the synthetic reflection section expected from the wide-angle velocity model along line 2. Both the TWTTs of the reflector, and the large (up to 1 s) variations in TWTT to it, are matched accurately along the entire length of the line (figure 4b). We therefore interpret the reflector as an intracrustal reflection from the top surface of the low-velocity body.

Amplitude modelling of this reflection event suggests that the low-velocity body may be bound by narrow gradient zones rather than distinct interfaces (figure 4b). However, the comparatively low data quality prevents any further quantitative analysis and investigation of this apparent contradiction with the wide-angle models. The contradiction may simply result from the limitations of the wide-angle modelling method, since it proved impossible to incorporate the gradient zone into the models and still be able to succesfully ray-trace through them.

The seismic reflection profile along line 2 shows some gaps in the magma body reflection. Whether these are due to the limited data quality, or represent real along-axis discontinuities in the melt body, is unclear. Calculation of the expected reflection profile from the wide-angle model using a continuous melt body along-axis reveals that gaps in reflector continuity may be associated with severe changes in seabed topography (figure 4b). However, three across-axis reflection profiles, collected as part of this programme, all cross the AVR axis at points that coincide with gaps in the reflector on the along-axis profile. The reflector is not visible on any of the across-axis profiles. Thus our observations may indicate that the melt body is discontinuous along-axis within the length-scale of the AVR, but this cannot be unambiguously resolved with the currently available data.

The presence of a thin layer of very low-velocity material beneath the axis, and a larger region of more moderately depressed velocities surrounding and underlying it, is strikingly similar to the sill-like 'melt lens' and surrounding low-velocity zone structure that has been reported for intermediate and fast spreading ridges in the Pacific. We therefore interpret the low-velocity body as a crustal melt accumulation beneath the AVR axis. In the next two sections we will show that this interpretation is consistent with the findings obtained from the modelling of the electromagnetic data.

(c) The controlled-source electromagnetic experiment

The controlled-source electromagnetic (CSEM) technique involves the use of a deep-towed transmitter, which emits low-frequency continuous wave signals that are recorded by seafloor electric field recording instruments. The amplitude of the signal at the receiver provides information about the degree of attenuation of the signal during its propagation through the seafloor from source to receiver; and that in turn varies with the electrical resistivity structure of the crust. For this experiment, we deployed an array of short-arm (ELF) and long-wire (LEM) receivers (Webb *et al.* 1985) (figure 1). All of these instruments are horizontal component electric field recorders. The ELFs record two orthogonal horizontal components from a pair of 12 m dipoles, and can be used to measure signals at source-receiver ranges of up to 10–15 km. The LEMs record a single horizontal component from a 300 m dipole, and can be used at source-receiver ranges of several tens of kilometres. In this paper, we will limit our analysis to the shorter range (up to 15 km) data recorded close to the axis by the ELFs. Controlled-source signals were provided by the DASI deep-towed

(a)

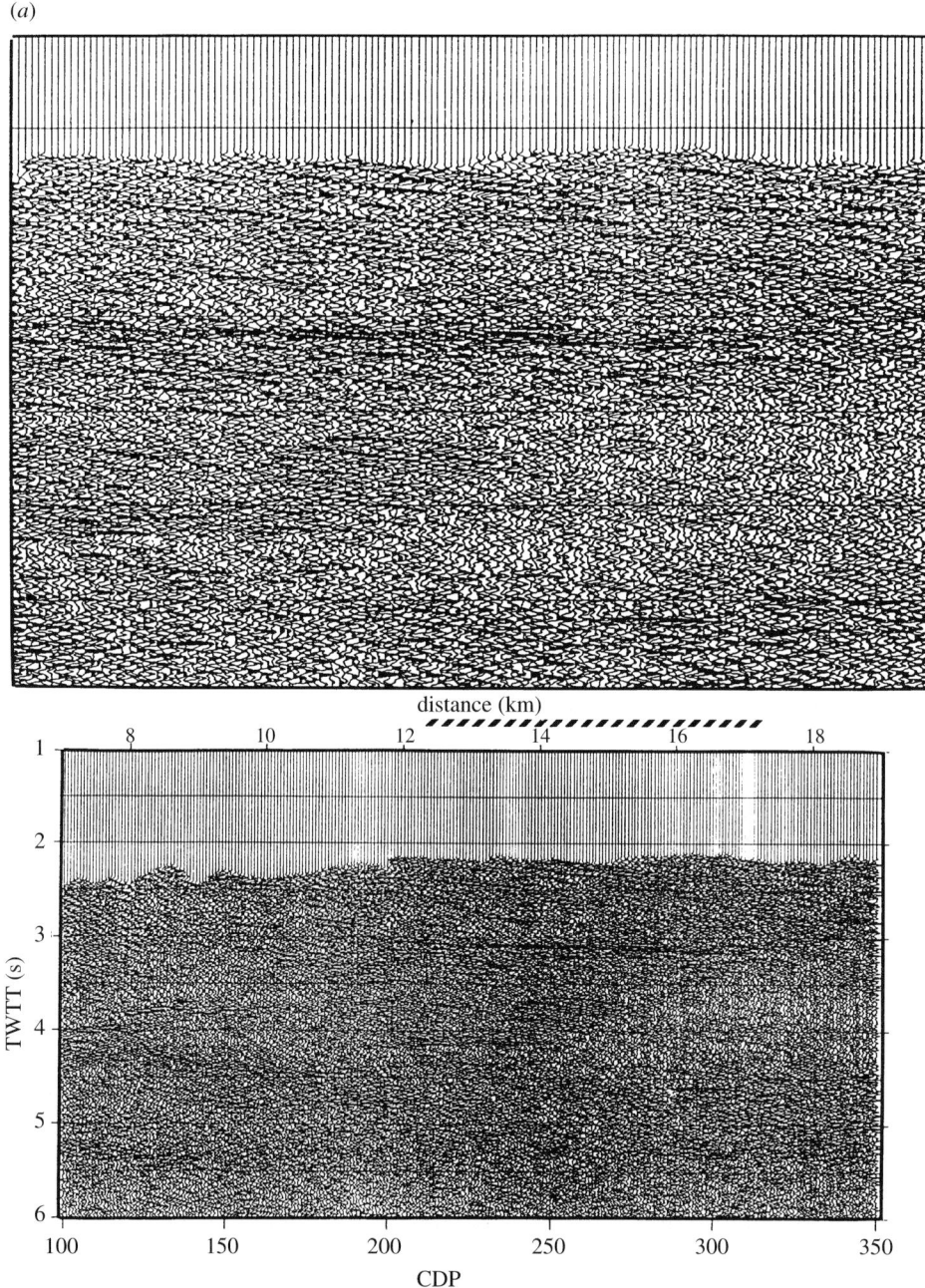

Figure 4. (a) Seismic reflection profile along line 2 (lower diagram). Note the prominent intracrustal reflection event at about 3 s TWTT, which corresponds to the thin sill-like axial low-velocity zone inferred from the wide-angle data, and which we interpret as the top surface of the axial magma chamber. The upper diagram shows an expanded view of the region denoted by the diagonal dashed line. Note how the magma chamber reflection event cross-cuts other prominent crustal reflections. (continued opposite)

transmitter system (Sinha *et al.* 1990) along two tow lines (figure 1). Tow 1 runs along the axis of the northern part of the AVR. Tow 2 runs parallel to, and 4 km

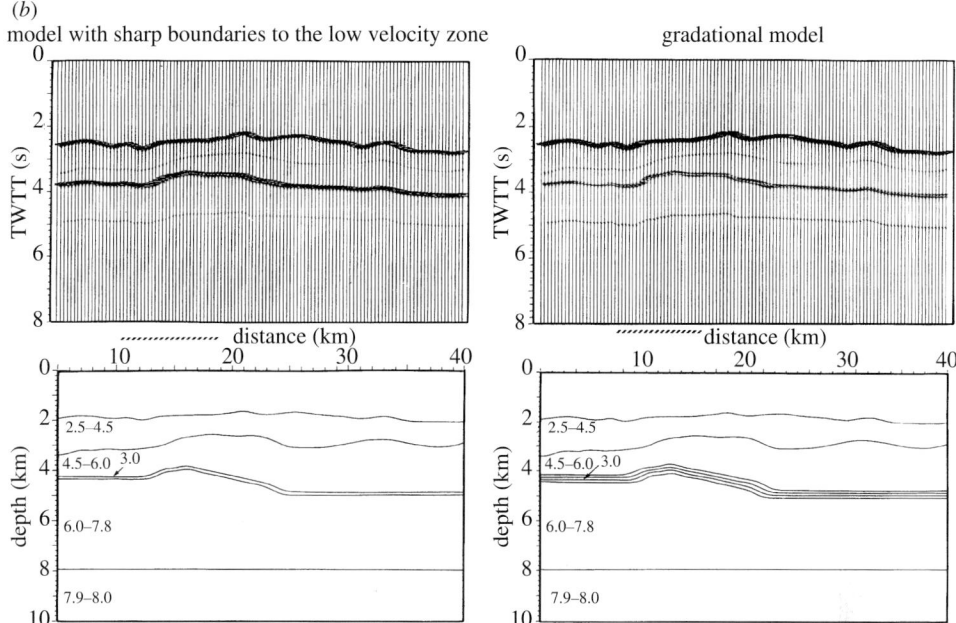

Figure 4 (cont.). (b) Amplitude modelling of the magma body reflector using velocities and layer boundaries obtained from the wide-angle velocity model derived for line 2, for both a distinct upper and lower boundary (left diagram) to the low-velocity body and an upper and lower gradient zone (right diagram). Note how the gradient zone modelled amplitudes match the observed variation with distance along the profile better than those obtained using the sharp boundary model (cf. figure 5a). Layer velocities are in km s^{-1}.

west of, the AVR. Transmission frequencies were 0.35 Hz and 11 Hz on tow 1 and 0.75 Hz on tow 2.

The best control on the shallow resistivity structure of the crust is provided by short range data. In this case, the shortest source-receiver ranges are provided by the DASI tow 1 transmissions into ELF Quail, at the northern end of the AVR. Since these data are from along-strike transmissions, as well as at short ranges (up to 4 km), they are least likely to be perturbed by lateral variations in structure, and so have been initially interpreted using a one-dimensional (resistivity varying only with depth) approach. A resistivity-depth profile derived from these data using a smooth one-dimensional Occam inversion (Constable et al. 1987) is shown in figure 5a, and the fit of the model to the observed data is shown in figure 5b. The inversion is unable to fit the data without some bias in the residuals—the modelled amplitudes are consistently too large for the lower frequency—and this may be due to two or possibly three-dimensional features in the true structure. However, inversions of subsets of the data—e.g. using only data from north or south of the receiver, or only one of the frequencies—yield very similar structures to the one shown, providing some confidence that the overall structure is in fact well resolved by the one-dimensional approximation, at least locally beneath tow 1 and down to a depth of between 1 and 2 km.

The error bars shown in figure 5b correspond to estimates of ±1 standard deviation. Errors in this type of data arise from two sources. First, there are measurement errors

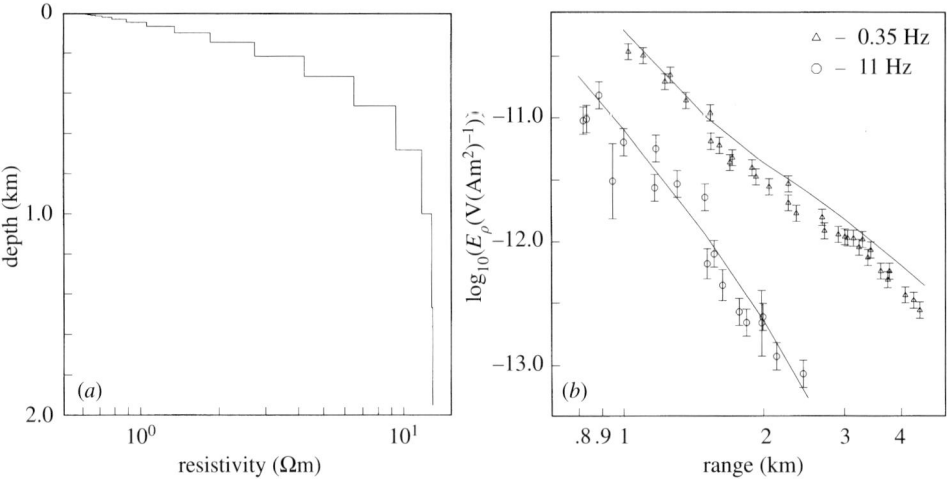

Figure 5. (a) Resistivity–depth profile for the upper part of the crust beneath the AVR axis. This structure was obtained by one-dimensional smooth inversion using the Occam algorithm of the amplitudes recorded by ELF Quail during DASI tow 1. (b) Circles and diamonds with one standard deviation error bars show the amplitudes of the $E\rho$ component of the horizontal electric field recorded by ELF Quail during transmitter tow 1, at two different frequencies. Superimposed (solid lines) are the computed $E\rho$ responses at these frequencies of the one-dimensional model shown in (a).

arising from contamination of the receiver signal by noise, and uncertainties in source parameters (such as height above the seafloor) and source-receiver geometry. Second, there are variations in the signal amplitude due to fine scale heterogeneity in the resistivity structure close to the seafloor beneath the moving transmitter. The latter are equivalent to the 'static shift' errors observed in other types of electromagnetic experiment. Measurement errors of the first type can be readily quantified, and have been shown (Evans et al. 1994) to be smaller than the observed scatter of seafloor data. Since, in practice, it is not possible to separate the two sources of scatter in the observed data, they are both treated simply as stochastic uncertainties in the measured amplitudes. Thus the error bars shown in figure 5b represent a combination of two effects—first a contribution from genuine measurement errors, and second the influence of fine scale heterogeneity close to the transmitter, corresponding to unresolved static shift effects.

The resistivities in figure 5a are very low for crystalline rocks, ranging from 1 Ω m or less at the seabed to no more than 10–20 Ω m at 1 km depth. Both the low overall resistivity values and the steep gradient in resistivity in the top 1 km of the crust are required in order to fit the data. Such low resistivities in a basaltic crust indicate a large proportion of conductive fluid permeating the rock matrix. If we assume that, in the upper 1 km of the crust, the resistivity is primarily controlled by the penetration of sea water into fractured basalts, then these resistivities imply very high porosities (more than 10%) at the seabed, and significant porosity (at least 2%) and a highly interconnected sea water (or hydrothermal fluid) phase persisting to a depth of at least 1 km. The resistivities are lower than those found over the same depth range at the axis of the East Pacific Rise (EPR) near 13° N (Evans et al. 1994), implying even higher porosities here than at the EPR.

While the model shown in figure 5a provides a good fit to the short range tow 1 data, it is not consistent with data observed at longer (5–15 km) ranges. The amplitudes predicted by the one-dimensional model are consistently much too low (by at least an order of magnitude) to fit the observed amplitudes at intermediate source-receiver ranges shown in figures 6b, c. This indicates that away from the axis, and/or at depths of greater than 1–2 km, the structure departs significantly from that shown in figure 6a. Attempts to invert all the data at short (figure 5b) and intermediate (figures 6b, c) ranges simultaneously using the one-dimensional approach fail to converge; while separate inversions of the three subsets of intermediate range data shown in figures 6b, c produce models that are all characterized by high resistivities in the upper crust, and a sharp reduction in resistivity at a depth of between 1 and 5 km.

The response of a CSEM experiment to sub-surface structure depends not only on the structure itself, but also on the experimental geometry. In the case where a horizontal electric dipole is used for both source and receiver, as here, two distinct geometric modes exist. If both dipoles are collinear, then only the radial component ($E\rho$) of the source field is seen. If the vector from source to receiver is orthogonal to both the source and receiver dipoles, then only the azimuthal component ($E\varphi$) of the source field is seen. MacGregor & Sinha (1996) have shown that, in a one-dimensional earth structure, the two modes respond very differently to the presence of a low resistivity layer at depth. In the $E\rho$ (collinear dipole geometry) case, a decrease in resistivity at depth leads to an increase in signal amplitude at moderate ranges, due to an effect that has been described as current channelling (Unsworth 1991) or a wave-guide phenomenon (Chave & Cox 1982; Chave et al. 1990). In the $E\varphi$ (source-receiver vector orthogonal to dipoles) case, the corresponding increase in amplitude is much smaller, or even, in some cases, absent altogether. The result of these two different behaviours is a characteristic 'splitting' of the $E\rho$ and $E\varphi$ amplitudes at the same range, with the $E\rho$ amplitude appearing anomalously large compared to that of $E\varphi$.

The ELF data at 5–15 km range show evidence of exactly this type of splitting. The along-axis transmissions (ELF Quail, DASI tow 2 and ELF Noddy, DASI tow 1) are dominated by the $E\rho$ component, since the source dipole is aligned parallel to the axis on both tows. The across-axis transmissions (ELF Noddy, DASI tow 2) are dominated by the $E\varphi$ component. As described above, the $E\rho$ and $E\varphi$ amplitudes are split, with the along-axis ($E\rho$) amplitudes being much larger than the across-axis ($E\varphi$) amplitudes. The split is greatest for the shortest range Noddy tow 2 data, which correspond to almost pure $E\varphi$, and diminishes as the Noddy tow 2 range increases—corresponding to an increasing $E\rho$ contribution to the signal at this instrument.

Taken together, the data indicate that any model that fits the observations must have three general features. First, the shallow structure beneath the AVR axis must be similar to that shown in figure 5a. Second, the overall resistivity of the crust must increase away from the axis, to explain the high amplitudes observed at 5–15 km range. Third, there must be a significant down turn in resistivity at depth beneath the axis, to account for the splitting of the $E\rho$ and $E\varphi$ component amplitudes. We have investigated the structure further by two-dimensional forward modelling, using the finite-element code of Unsworth et al. (1993). Figure 6a shows a two-dimensional model incorporating the general features described above. In addition, in order to produce a self-consistent solution that takes account of the seismic as well as the electromagnetic data, the down turn in resistivity has been constrained to be in the same location, and to have the same dimensions, as the seismic low-velocity body. Figure 6b shows the fit of this model to the observed data. The modelled

(a)

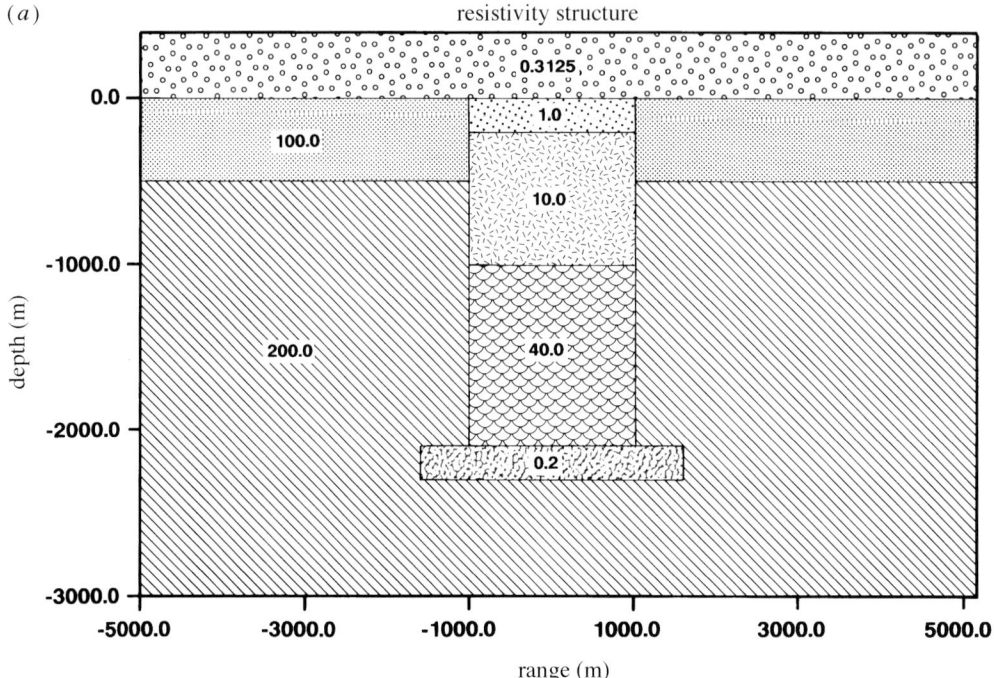

Figure 6. (a) A two-dimensional electrical resistivity model of the crust beneath the axial region, which is consistent with both the CSEM data and the structural constraints imposed by the seismic data. Note that the seismic low-velocity zone in figure 2d is represented here as a body with the same dimensions and location, and with an extremely low resistivity of 0.2 Ω m. (continued opposite)

amplitudes take account of the varying source-receiver geometry in the along-strike as well as across-strike directions. It can be seen that this model provides a generally good fit, reproducing both the overall amplitudes for all source-receiver pairs, and the splitting of $E\rho$ and $E\varphi$ amplitudes. For comparison, figure 6c shows the fit of the same model if the low-resistivity body is not included. Although the modelled response has amplitudes that are of the same order of magnitude as the data, the splitting is not reproduced: indeed, the split in the modelled amplitudes has the wrong polarity.

Based on the modelling presented here, we conclude that the resistivity structure beneath the experiment area must be broadly similar to that shown in figure 6a. Very low resistivities, and a steep increase in resistivity with depth, occur in the upper 1 km of the crust beneath the AVR axis and are due to penetration of sea water or hydrothermal fluids into a highly fractured upper crust. At mid-crustal depths beneath the AVR, there is a sharp downturn in resistivity. Almost certainly this feature coincides with the seismic low-velocity zone, and is caused by a crustal magma body containing a high proportion of interconnected melt. The melt body in the model has a very low resistivity (0.2 Ω m) corresponding to the probable value for a pure basalt melt. A smaller melt body, or a similar sized body with significantly higher resistivity, is unable to reproduce the observed amplitude splitting. A larger body of slightly higher resistivity (for example, a region containing a 25% melt fraction would be expected to have resistivities of the order of 1–5 Ω m, depending on the connectivity of the melt fraction) could also fit the data. At short distances

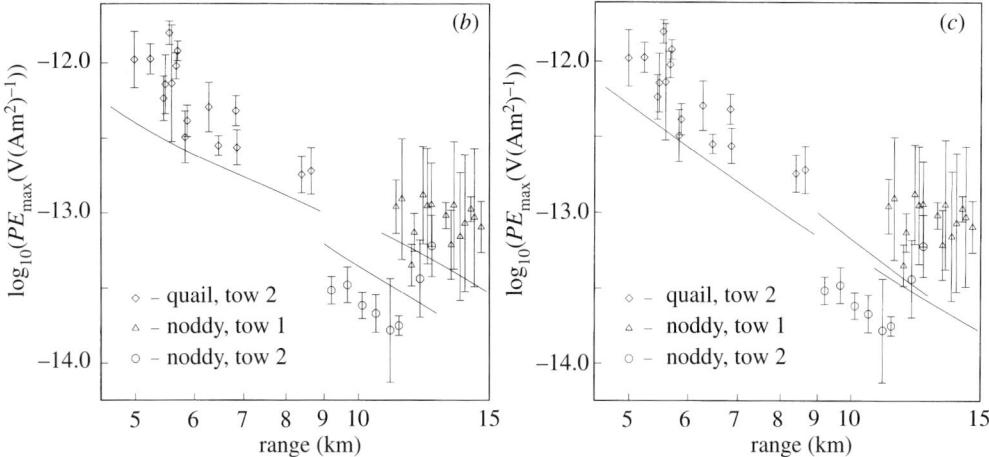

Figure 6 (cont.). (b) Plot of observed amplitudes versus range for the 5–15 km range data. The amplitudes shown (circles, diamonds and triangles with one standard deviation error bars) are those of the major axis of the polarization ellipse from the two-component, short-arm instruments. Note also the much higher amplitudes of $E\rho$ (along-axis) transmissions (Quail tow 2 and Noddy tow 1), compared with $E\varphi$ (across-axis) transmissions (Noddy tow 2), at ranges greater than 5 km. This splitting of amplitudes between the two components is characteristic of a down turn in resistivity at depth. Superimposed on this figure (solid lines) are the predicted responses of the two-dimensional model shown in (a). Note the good fit of the calculated responses to the observed data, including the prominent splitting between $E\varphi$ and $E\rho$ components for ELF Noddy. The changing geometry during the two source tows results in slightly different ratios of contribution to the major axis of the polarization ellipse by the $E\varphi$ and $E\rho$ components as a function of source-receiver range, for all three source-receiver pairs. In this and other figures, the modelled responses take account of this varying geometry, which causes the small, short wavelength variations in modelled amplitudes. (c) As (b), but showing the predicted response for a model that is the same as that in (a) but does not include the low-resistivity melt lens. This model fails to reproduce the observed splitting of the $E\varphi$ and $E\rho$ components.

(less than 5 km) from the axis, resistivities in the upper 1–2 km of the crust must increase significantly compared to those at the axis. Longer range data from the LEM instruments indicate that further off-axis, crustal resistivities must continue to increase rapidly. Further modelling of the CSEM data is still required to define these resistivity variations in greater detail, and to incorporate the long range data from the LEM instruments.

(d) The magneto-telluric experiment

The magneto-telluric (MT) component of the experiment involved the collection of low-frequency horizontal electric field data at the centre of the AVR and at six other sites, and three-component magnetic field data at the AVR centre, its southern limit and a site 15 km east (figure 1). The source-field for the observations was provided by naturally occurring variations of Earth's magnetic field, which induce electric currents in the sea water, crust and mantle. MT impedance estimates were calculated at the latter three sites in the period range 100–10 000 s using the robust response estimation technique of Egbert & Booker (1986) using local magnetic sites for remote referencing. A combination of relatively shallow water and an energetic period of geomagnetic activity enabled the collection of good quality data even at the shorter periods in this range.

Rotation of the MT tensor to find maximum and minimum values was effected using a tensor decomposition method (Lilley 1995). This showed a very stable orientation for the maximum value, which is coincident with the overall strike of the Reykjanes Ridge and corresponds to the TE mode. Analysis of the orthogonal, across-strike (TM) mode showed that it was not susceptible to one-dimensional inversion. Numerical modelling of the ridge suggests that this is largely due to the seafloor topography in the vicinity of the AVR, which results in electric charge accumulations on the fault scarps bounding the median valley. In addition there may be some effect due to adjacent bounding coastlines (Heinson & Constable 1992). The TE mode response is much less affected because of its alignment with the predominant geological strike of the ridge. Two-dimensional forward modelling using finite-elements (Wannamaker et al. 1986) suggests that the TE mode should be close to the one-dimensional response of the underlying ridge resistivity structure (Evans & Everett 1992). For the AVR centre site a one-dimensional Occam inversion (Constable et al. 1987) was used to obtain a smooth resistivity model beneath the AVR. Figure 7 shows the TE mode response data with error estimates, the optimum one-dimensional resistivity profile and the model response superimposed on the data. The most striking features of the data are the drop in phase at short periods (high frequencies) and the inflection in both apparent resistivity and phase at periods of 1000–3000 s. As is seen in the resulting profile, these correspond to a low-resistivity region close to the seafloor and a low-resistivity layer at depths between 50 and 100 km.

Of the features shown in the inversion in figure 7b, the most robust is the low-resistivity region close to the seafloor. Modelled crustal resistivities of the order of 1 to 10 Ω m are in agreement with the CSEM estimates for the top few kilometres beneath the AVR shown in figure 6a. The MT method lacks fine scale resolution, as skin depths are comparable to crustal thickness even at the shortest periods, but the bulk electrical properties are significant. Such a low-resistivity region is not found in similar one-dimensional inversions of TE mode data from MT sites south and east of the AVR, suggesting that it may reflect either the presence of the magma chamber beneath the AVR, or the high porosities and low resistivities in the upper 1 km of the crust beneath the AVR, or more probably the combination of both. We believe this to be the first clear evidence from seafloor MT work for the presence of conductive fluids within the crust beneath a mid-ocean ridge at any spreading rate.

A low-resistivity layer (5–10 Ω m) appears in the one-dimensional inversion over a depth range of 50–100 km. This feature is less well constrained by the TE mode data, and it is possible that it is due to two-dimensional or three-dimensional effects common to all sites. This is currently being investigated by numerical modelling. However, it is present in the inversions of data from all three MT sites (not surprisingly, since they are only 15 km apart), and persists as the data constraint is relaxed (i.e. the misfit is allowed to increase) in the smooth inversion. If it is real, it may imply a region of asthenospheric melt beneath the ridge, separated from the crust by a less conductive lithosphere which has a resistivity of more than 100 Ω m. We believe that the deeper conductive zone represents the source region of the mantle, in which adiabatic decompression of upwelling asthenosphere results in partial melting of the mantle. Assuming a highly connected melt phase within this asthenospheric layer, the inferred resistivities are consistent with a melt fraction of the order of 5% (Tyburczy & Waff 1983). These results confirm the utility of MT experiments on ridges for detecting and studying mantle melting.

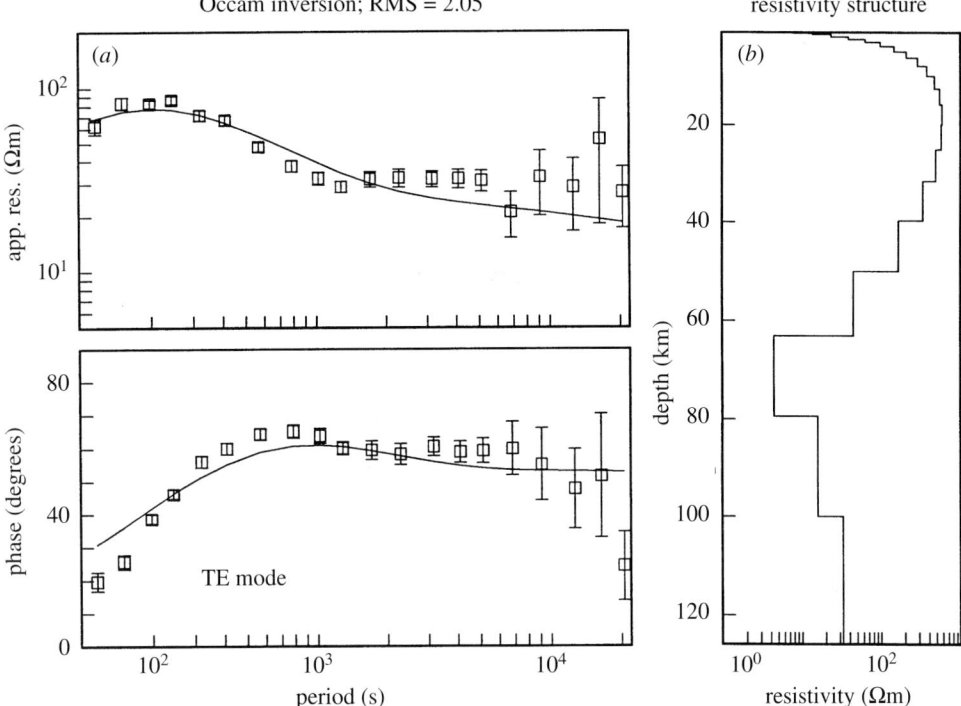

Figure 7. (a) Boxes with one standard deviation error bars show the TE mode apparent resistivity and phase responses for the MT instrument located over the centre of the AVR. Superimposed (solid lines) are the computed responses of the one-dimensional resistivity–depth model shown in (b). (b) Resistivity versus depth curve beneath the axis of the AVR, derived by one-dimensional smooth inversion of the MT data shown in (a).

4. Conclusions

In combination, the geophysical results presented here provide convincing evidence for a crustal melt body beneath the AVR, represented by low seismic velocities, an intra-crustal seismic reflector, and low electrical resistivities. The MT results also provide some evidence of partial melting at deeper levels in the mantle. The most important conclusions that can be drawn from our modelling to date relate to the crustal melt body. This appears to have many broad similarities (depth below the sea surface, vertical and across-axis dimensions, and internal structure consisting of a thin, sill-like body embedded within a larger, axial low-velocity region) to those seen beneath faster spreading ridges in the Pacific. Our data are, though, unable to resolve its along-axis continuity. We conclude that the basic geometry of melt accumulation and emplacement here, beneath the slow-spreading Reykjanes Ridge, retains similar features to those at much higher spreading rates, even though the patterns of temporal and along-axis variations probably differ.

It may be argued that the Reykjanes Ridge is atypical of slow spreading ridges, owing to the influence of the Iceland plume. If this is the case, the similarities with faster spreading ridges may be unique to this location, and may not apply elsewhere. In response to this, we would point out that while hot spot proximity may have a strong influence on some features of the ridge, such as rift valley development (Chen & Morgan 1990), the most important factor relating to the mode of emplacement of the crystalline crust is likely to be the overall flux of melt into the crust. The seismic evidence (figure 3) for normal (6–8 km) thickness crust shows very clearly that the rate of melt production here is no higher than typical values found elsewhere (away from ridge offsets) on the mid-ocean ridge system. The magmatic flux here can, therefore, be no greater than it is elsewhere on the Mid-Atlantic Ridge, where similar thickness crust is produced at the same or higher spreading rates. Other seismic experiments on the Reykjanes Ridge (Bunch & Kennett 1980; Smallwood et al. 1995) have failed to find any strong evidence for a crustal melt body, even though they are located closer to the Iceland hot spot, where mantle temperatures are high enough to cause some increase in overall melt production rates leading to thicker than normal crust (McKenzie & Bickle 1988). We would therefore argue that what distinguishes this locality from other segments of slow spreading ridges is not its position relative to the Iceland hot spot, but the fact that it is in the most magmatically active stage of its life cycle. Following this argument, we would expect that most AVRs on the Reykjanes Ridge—even very close to Iceland—would not be underlain by an axial melt body, since such features are short-lived and hence rare. Conversely, we would expect that other slow spreading ridge segments that, like our AVR, are at the most magmatically active stage of their life cycle, might well be underlain by comparable melt bodies, irrespective of their proximity to, or distance from, a hot spot. Hence the results of this study may be representative of the generality of slow spreading ridge segments, but only at a particular, magmatically robust stage in their magmatic-tectonic cycle.

Having identified a crustal magma body beneath a slow spreading ridge segment, an important objective for future work will be to investigate its along-axis variability—which is only poorly resolved by the experiment described here—and to relate that to the seafloor expression of crustal construction as evidenced by the morphological features of the AVR itself, and the offset discontinuities between it and adjacent spreading segments. Numerous studies of systematic along-axis variations in crustal structure and thickness at the Mid-Atlantic Ridge (see, for example, Sinha & Louden 1983; Detrick et al. 1993b) have shown that the segmentation is a fundamental property of accretion, and is probably related to three-dimensional (or four-dimensional) patterns of upwelling and melt production in the underlying mantle. The outcome of this experiment encourages us to believe that, by means of careful targeting of experiments and the co-ordinated use of multiple geophysical techniques, we may in the future be able to study segmentation by direct observation of the along-axis variations in crustal magmatic structure during the magmatically active stage of a segment's life cycle.

At 2.5 km below the seafloor, the crustal melt body is at the upper end of the range of depths observed on the fast spreading East Pacific Rise (100 to 140 mm yr^{-1}), but is shallower than that observed beneath the intermediate spreading Valu Fa Ridge (Lau Basin, SW Pacific—60 mm yr^{-1}. Recent observations of apparently systematic variations in axial magma chamber depth along the East Pacific Rise have led to the suggestion that magma chamber depth may depend strongly on spreading rate (Phipps Morgan & Chen 1993). Our data refute that suggestion—the melt body is no deeper here than along parts of the East Pacific Rise which are spreading five times faster, and is shallower than at the Valu Fa Ridge, which is spreading three times faster. Spreading rate cannot therefore be the only, or major, determinant of magma chamber depth.

Finally, the low seismic velocities and extremely low resistivities observed in the upper part of the crust imply both a high porosity and a highly connected fluid phase.

In combination with the presence of a shallow crustal magma chamber, these are the conditions which would be expected to drive vigorous hydrothermal circulation systems. A systematic search of the Reykjanes Ridge axis (German *et al.* 1994) between Iceland and 58° N—terminating just north of our study area—failed to find any evidence of high-temperature venting except for the Steinaholl field, just 45 km from the Icelandic coast. It is possible, therefore, that none of the AVR segments (numbering more than 40) between Steinaholl and the 57° 43′ N AVR, have experienced current or very recent magmatic activity. If this is representative of the proportion of segments that are magmatically active at a given time, it would suggest that the magmatically robust phase occupies only a small proportion (a few percent or less) of the total magmatic-tectonic life span of a slow-spreading segment.

We thank the officers and crew of RRS *Charles Darwin*, the sea-going technical staff of NERC Research Vessel Services, and other members of the scientific party of Cruise CD81, without whom the data presented in this paper could not have been collected. We also thank Roger Searle for extremely useful discussions during the selection of our target AVR, and Martyn Unsworth for the use of his two-dimensional EM code. This work was supported by the UK Natural Environment Research Council in the form of research grants, ship time and two Ph.D. studentships, by the US National Science Foundation and by the Australian Research Council.

References

Bunch, A. W. H. & Kennett, B. L. N. 1980 The crustal structure of the Reykjanes Ridge at 59° 30′ N. *Geophys. Jl R. Astr. Soc.* **61**, 141–146.

Calvert, A. J. 1995 Seismic evidence for a magma chamber beneath the slow-spreading Mid-Atlantic Ridge. *Nature* **377**, 410–414.

Chapman, C. H. & Drummond, R. 1982 Body wave seismograms in inhomogeneous media using Maslov asymptotic ray theory. *Bull. Seism. Soc. Am.* **72**, S277–S317.

Chave, A. D. & Cox, C. S. 1982 Controlled electromagnetic sources for measuring electrical conductivity beneath the oceans. 1. Forward problem and model study. *J. Geophys. Res.* **87**, 5327–5388.

Chave, A. D., Flosadottir, A. H. & Cox, C. S. 1990. Some comments on the seabed propagation of ULF/ELF electromagnetic fields. *Radio Science* **25**, 825–836.

Chen, Y. & Morgan, W. J. 1990 Rift valley/no rift valley transition at mid-ocean ridges. *J. Geophys. Res.* **95**, 17 571–17 581.

Collier, J. S. & Sinha, M. C. 1992 Seismic mapping of a magma chamber beneath the Valu Fa Ridge, Lau Basin. *J. Geophys. Res.* **97**, 14 031–14 053.

Constable, S. C., Parker, R. L. & Constable, C. G. 1987 Occam's inversion: a practical algorithm for generating smooth models from electromagnetic sounding data. *Geophysics* **52**, 289–300.

Detrick, R. S., Buhl, P., Vera, E., Mutter, J., Orcutt, J., Madsen, J. & Brocher, T. 1987 Multichannel seismic imaging of a crustal magma chamber along the East Pacific Rise. *Nature* **326**, 35–42.

Detrick, R. S., Mutter, J. C., Buhl, P. & Kim, I. 1990 No evidence from multichannel reflection data for a crustal magma chamber in the MARK area on the Mid-Atlantic Ridge. *Nature* **347**, 61–64.

Detrick, R. S., Harding, A. J., Kent, G. M., Orcutt, J. A., Mutter, J. C. & Buhl, P. 1993*a* Seismic structure of the southern East Pacific Rise. *Science* **259**, 499–503.

Detrick, R. S., White, R. S. & Purdy, G. M. 1993*b* Crustal structure of North Atlantic fracture zones. Seismic structure of the southern East Pacific Rise. *Rev. Geophys.* **31**, 439–458.

Evans, R. L. & Everett, M. E. 1992 Magneto-tellurics and mid-ocean ridge melt transport: a two-dimensional perspective. In *Mantle flow and melt generation at Mid-Ocean Ridges* (ed. J. Phipps Morgan, D. K. Blackman & J. M. Sinton), pp. 353–361. (Geophysical Monograph 71.) Washington, DC: AGU.

Evans, R. L., Sinha, M. C., Constable, S. C. & Unsworth, M. J. 1994 On the electrical nature of the axial melt zone at 13° N on the East Pacific Rise. *J. Geophys. Res.* **99**, 577–588.

Egbert, G. D. & Booker, J. R. 1986 Robust estimation of geomagnetic transfer functions. *Jl R. Astr. Soc.* **87**, 173 194.

Fowler, C. M. R. 1976 Crustal structure of the Mid-Atlantic Ridge crest at 37° N. *Geophys. Jl R. Astr. Soc.* **47**, 459–491.

Fowler, C. M. R. & Keen, C. E. 1979 Oceanic crustal structure—Mid-Atlantic Ridge crest at 45° N. *Geophys. Jl R. Astr. Soc.* **56**, 219–226.

German, C. R. *et al.* (11 authors) 1994 Hydrothermal activity on the Reykjanes Ridge: the Steinaholl vent-field at 63° 06' N. *Earth Planet. Sci. Lett.* **121**, 647–654.

Harding, A. J., Orcutt, J. A., Kappus, M. E., Vera, E. E., Mutter, J. C., Buhl, P., Detrick, R. S. & Brocher, T. M. 1989 The structure of young oceanic crust at 13° N on the East Pacific Rise from expanding spread profiles. *J. Geophys. Res.* **94**, 12 163–12 196.

Heinson, G. S. & Constable, S. C. 1992 The electrical conductivity of the oceanic upper mantle, *Geophys. J. Int.* **110**, 159–179.

Henstock, T. J., Woods, A. W. & White, R. S. 1993 The accretion of oceanic crust by episodic sill intrusion. *J. Geophys. Res.* **98**, 4143–4161.

Keeton J. A., Searle, R. C., Parsons, B., White, R. S., Murton, B. J., Parson, L. M., Peirce C. & Sinha, M. 1997 Bathymetry of the Reaykjanes Ridge. *Mar. Geophys. Res.* (In the press.)

Kusznir, N. J. & Bott, M. H. P. 1976 A thermal study of the formation of oceanic crust. *Geophys. Jl R. Astr. Soc.* **47**, 83–95.

Lilley, F. E. M. 1995 Strike direction: obtained from basic models for 3D magnetotelluric data. In *Three-dimensional electromagnetics* (ed. M. Oristaglio & B. Spies), pp. 359–369. Ridgefield, CT: Schlumberger–Doll Research.

MacGregor, L. M. & Sinha, M. C. 1996 Marine controlled-source EM: effect of source-receiver geometry on the response of 1-D structures. Submitted to *Geophys. J. Int.* (Submitted).

McKenzie, D. P. & Bickle, M. J. 1988 The volume and composition of melt generated by extension of the lithosphere. *J. Petrol.* **29**, 625–675.

Murton, B. J. & Parson, L. M. 1993 Segmentation, volcanism and deformation of oblique spreading centres: a quantitative study of the Reykjanes Ridge. *Tectonophysics* **222**, 237–257.

Navin, D. A., Peirce, C. & Sinha, M. C. 1996 Seismic evidence for a crustal magma chamber beneath the slow-spreading Reykjanes Ridge. *JAG Newsletter* **7**, 20.

Parson, L. M. *et al.* (17 authors) 1993. En echelon axial volcanic ridges at the Reykjanes Ridge: a life cycle of volcanism and tectonics. *Earth Planet. Sci. Lett.* **117**, 73–87.

Phipps Morgan, J. & Chen, Y. J. 1993 The genesis of oceanic crust: magma injection, hydrothermal circulation, and crustal flow. *J. Geopys. Res.* **98**, 6283–6297.

Purdy, G. M. & Detrick, R. S. 1986 Crustal structure of the Mid-Atlantic Ridge at 23° N from seismic refraction studies. *J. Geophys. Res.* **91**, 3739–3762.

Schilling, J. G., Zajac, M., Evans, R., Johnston, T., White, W., Devine, J. D. & Kingsley, R. 1983 Petrologic and geochemical variations along the Mid-Atlantic Ridge from 29° N to 73° N. *Am. J. Sci.* **283**, 510–586.

Sleep, N. H. 1975 Formation of oceanic crust: some thermal constraints. *J. Geophys. Res.* **80**, 4037–4042.

Searle, R. C. & Laughton, A. S. 1981 Fine scale sonar study of tectonics and volcanism on the Reykjanes Ridge. *Oceanol. Acta* **4**, 5–13.

Searle, R. C., Field, P. R. & Owens, R. B. 1994 Segmentation and a non-transform ridge offset on the Reykjanes Ridge near 58° N. *J. Geophys. Res.* **99**, 24 159–24 172.

Sinha, M. C. & Louden, K. E. 1983 The oceanographer fracture zone. I. Crustal structure from seismic refraction studies. *Geophys. Jl R. Astr. Soc.* **75**, 713–736.

Sinha, M. C., Patel, P. D., Unsworth, M. J., Owen, T. R. E. & MacCormack, M. R. G. 1990 An active source electromagnetic sounding system for marine use. *Mar. Geophys. Res.* **12**, 59–68.

Sinha, M., Peirce, C., Constable, S. & White, A. 1994 An integrated geophysical investigation of the axial volcanic region of the Reykjanes Ridge at 57° 45' N—RRS. *Charles Darwin 81 Cruise Report*, p. 39. University of Cambridge.

Sinton, J. M. & Detrick, R. S. 1992 Mid-ocean ridge magma chambers. *J. Geophys. Res.* **97**, 197–216.

Smallwood, J. R., White, R. S. & Minshull, T. A. 1995 Sea-floor spreading in the presence of the Iceland plume: the structure of the Reykjanes Ridge at 61° 40′ N. *J. Geol. Soc.* **152**, 1023–1029.

Tyburczy, J. A. & Waff, H. S. 1983 Electrical conductivity of molten basalt and andesite to 25 kilobars pressure: geophysical significance and implications for charge transport and melt structure. *J. Geophys. Res.* **88**, 2413–2430.

Unsworth, M. J. 1991 Electromagnetic exploration of the oceanic crust with controlled sources. Ph.D. thesis, University of Cambridge.

Unsworth, M. J., Travis, B. J. & Chave, A. D. 1993 Electromagnetic induction by a finite electric dipole source over a two-dimensional Earth. *Geophysics* **58**, 198–214.

Vera, E. E., Mutter, J. C., Buhl, P., Orcutt, J. A., Harding, A. J., Kappus, M. E., Detrick, R. S. & Brocher, T. M. 1990 Structure of 0- to 0.2-m.y. old oceanic crust at 9° N on the East Pacific Rise from expanded spread profiles. *J. Geophys. Res.* **95**, 15 529–15 556.

Wannamaker, P. E., Stodt, J. A. & Rijo, L. 1986 A stable finite element solution for two-dimensional magnetotelluric modelling, *Geophys. Jl R. Astr. Soc.* **88**, 277–296.

Webb, S. C., Constable, S. C., Cox, C. S. & Deaton, T. K. 1985 A sea-floor electric field instrument. *J. Geomagn. Geoelect.* **37**, 1115–1130.

White, R. S., McKenzie, D. & O'Nions, R. K. 1992 Oceanic crustal thickness from seismic measurements and rare earth element inversions. *J. Geophys. Res.* **97**, 19 683–19 715.

An analysis of variations in isentropic melt productivity

By P. D. Asimow[1], M. M. Hirschmann[1,2] and E. M. Stolper[1]

[1] Division of Geological and Planetary Sciences, California Institute of Technology, Pasadena, CA 91125, USA
[2] Department of Geology, University of North Carolina, Chapel Hill, NC 27599, USA

The amount of melt generated per unit pressure drop during adiabatic upwelling, the isentropic melt productivity, cannot be determined directly from experiments and is commonly assumed to be constant or to decrease as melting progresses. From analysis of one- and two-component systems and from calculations based on a thermodynamic model of peridotite partial melting, we show that productivity for reversible adiabatic (i.e. isentropic) depressurization melting is never constant; rather, productivity tends to increase as melting proceeds. Even in a one-component system with a univariant solid–liquid boundary, the $1/T$ dependence of $(\partial S/\partial T)_P$ and the downward curvature of the solidus (due to greater compressibility of liquids relative to minerals) lead to increased productivity with increasing melt fraction during batch fusion (and even for fractional fusion in some cases). Similarly, for multicomponent systems, downward curvature of contours of equal melt fraction between the solidus and the liquidus contributes to an increase in productivity as melting proceeds. In multicomponent systems, there is also a lever-rule relationship between productivity and the compositions of coexisting liquid and residue such that productivity is inversely related to the compositional distance between coexisting bulk solid and liquid. For most geologically relevant cases, this quantity decreases during progressive melting, again contributing to an increase in productivity with increasing melting. These results all suggest that the increases in productivity with increasing melt fraction (punctuated by drops in productivity upon exhaustion of each phase from the residue) predicted by thermodynamic modelling of melting of typical mantle peridotites using MELTS are neither artifacts nor unique properties of the model, but rather general consequences of adiabatic melting of upwelling mantle.

1. Introduction

The amount of melting experienced by upwelling mantle is one of the most important parameters required for understanding the dynamics of basalt production and the observed compositional variability of basalts at mid-ocean ridges and sites of hot spot magmatism. The key parameter is the 'productivity' of the melting process (i.e. the amount of melt production per decrement of pressure (Hirschmann et al. 1994)), which exerts important controls on the dynamics and style of melt extraction, particularly if it varies with depth (Spiegelman 1993; Asimow et al. 1995a). The productivity also relates the geometry of the melting region to the average depth of melt generation and to the total amount of melt produced, and therefore to the thickness of the oceanic crust (Langmuir et al. 1992). The motivation for this paper is that the primary information available from the study of igneous rocks, their compositions and volumes, cannot be interpreted in terms of source dynamics and geometry without an understanding of the factors influencing productivity in upwelling mantle.

Mantle upwelling at mid-ocean ridges and plumes in the absence of melt or fluid flow is usually approximated as an adiabatic process. To the extent that upwelling is slow, relative to mass and thermal transfer in the ascending peridotite, the process can also be envisioned as reversible. Under these conditions, processes occurring in upwelling mantle can be approximated as isentropic (Verhoogen 1965; McKenzie 1984). Once melt or fluid migration is allowed to occur or the effects of viscous deformation of the solids are considered, the process is no longer locally adiabatic or isentropic, and indeed, no simple thermodynamic constraints can be applied to the general case. Nevertheless, certain idealized end-member processes, such as batch fusion or fractional fusion (in which each increment of melt production during upwelling can be approximated as adiabatic and reversible), can be evaluated relatively simply from a thermodynamic perspective. However, despite the apparent simplicity of the problem when posed thermodynamically—e.g. for the reversible adiabatic case, pressure (P), entropy (S) and chemical composition are the independent variables in upwelling mantle and the equilibrium state is one of minimum enthalpy (H)—the general features of productivity during adiabatic depressurization of mantle peridotite (and even of simpler, model systems) are little understood. The difficulty is partly that adiabatic processes are not readily simulated by experiment; i.e. whereas it is relatively simple to do an experiment at fixed or known P, temperature (T), and chemical composition, there is no straightforward way to do an experiment at fixed or known S or H at high pressure. Another difficulty, however, is that there is, to our knowledge, no general treatment of adiabatic melting and its consequences even in simple systems, so there is no framework or background for understanding the behaviour of complex natural systems undergoing this process. Although space limitations prevent us from presenting a complete treatment, the goal of this paper is to expose some of the key parameters entering into the process of *isentropic* melting of the mantle and, by illustrating some of the expected behaviours and what causes them, to help calibrate people's intuition about this important process.

2. Background and previous work

The simplest approach to estimating melt production in upwelling peridotite is to assume that melt fraction increases linearly as pressure decreases; i.e. isentropic productivity is assumed to be constant (Turcotte & Ahern 1978; Klein & Langmuir 1987; Niu & Batiza 1991; Kinzler & Grove 1992). In other cases, plausible assumptions have been made that lead to decreasing productivity with progressive decompression; for example, McKenzie & Bickle (1988) inferred, based on available peridotite melting data, that melt production would be enhanced near the solidus of natural peridotite (similar to the behaviour of a simple system at a eutectic or peritectic; see McKenzie & Bickle 1990), and Langmuir et al. (1992) argued that melting is more productive at high pressure because the solidus and liquidus are closer together. Although both of these effects could be important under some conditions, we show below that decreasing productivity during upwelling is exactly the opposite of what is expected in most natural cases; i.e. productivity is generally smaller at the

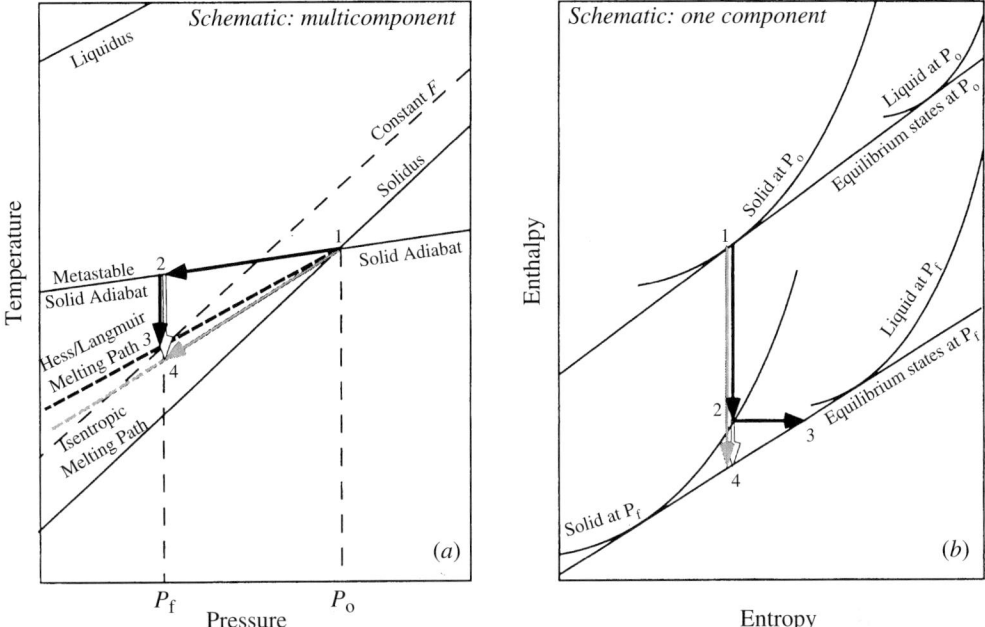

Figure 1. Schematic comparison of enthalpy- and entropy-conservation during adiabatic melting. (a) In P–T space, a parcel of adiabatically upwelling mantle intersects its solidus at P_0, state 1. At lower pressure P_f, the metastable extension of the solid adiabat is state 2. The stable partially molten state 3 is obtained by an adiabatic, isobaric (i.e. isenthalpic) process whereby the enthalpy recovered by cooling from state 2 to state 3 equals that required to melt up to some degree of melting F. State 4 is reached by reversible, adiabatic (i.e. isentropic) melting from state 1. (b) In H–S space, the difference between the two adiabatic processes (the isentropic path from 1 → 4, shown as a gray arrow, versus the path from 1 → 2 → 3, shown as black arrows) is shown for a hypothetical one-component system. The partially molten, stable state 4 is reached by adiabatic and reversible upwelling from the stable solid state 1 (on the solidus at P_0); the direct path from 1 → 4 is accomplished in a series of infinitesimal, reversible, adiabatic decompression steps; this state is clearly the minimum possible H for this S at P_f. State 2 is the metastable solid state at P_f reached by reversible, adiabatic (i.e. constant S) decompression from state 1 at P_0. State 3 is shown to be the stable, partially molten state on the tie-line between solid and liquid at P_f that has the same enthalpy as state 2 (reached by an irreversible, adiabatic, isobaric path at P_f that maximizes S). Clearly, state 3 has higher H and higher S than state 4. Furthermore, application of the lever rule along the tie-line shows that state 3 has higher F than state 4. Returning to (a), note that in a multicomponent system state 3 generally also has higher T than state 4, although in a one-component system both states lie on the solidus and are indistinguishable in P–T space.

initiation of decompression melting and increases with progressive decompression. Note that in this paper we restrict our attention to changes in productivity with progressive melting along particular adiabats; we leave comparisons among adiabats of different potential temperature for future work.

A widely used approach to estimating adiabatic melting paths is based on the assumption that at a given pressure the enthalpy of the metastable solid adiabat (a state that can be readily calculated from an initially stable subsolidus assemblage) and that of the stable partially molten adiabat are equal (Ramberg 1972; Cawthorn 1975; Hess 1992; Langmuir et al. 1992; Longhi 1992; Hart 1993). In practice, these authors balanced the enthalpy required to melt the metastable solid against the enthalpy recovered by cooling to the stable partially molten assemblage. Although these authors presumably intended to calculate the amount of melt produced on reversible, adiabatic upwelling of peridotite, they actually calculated the amount of melt produced on a somewhat different adiabatic path. Consider figure 1, where this melting path (informally labelled the 'Hess–Langmuir melting path') is compared with isentropic upwelling in P–T space (figure 1a) and in the less familiar but more informative H–S plane (figure 1b). We have drawn figure 1a for a multicomponent system, so that the melting paths do not coincide with the solidus, but for simplicity we have drawn figure 1b for a one-component system. In H–S space, the locus of states of a phase at constant pressure fall on a curve whose slope is temperature (i.e. $(\partial H/\partial S)_P = T$). The coexistence of two phases at equilibrium requires equal T and equal P, so it is represented by a tie-line tangent to isobaric curves for the two phases. The isentropic path from stable state 1 on the solidus at P_0 to the stable, partially molten state 4 at P_f is vertical in H–S space. The Hess–Langmuir melting path, however, corresponds to a path on figure 1b from the stable state 1 at P_0 to the metastable solid state 2 at P_f (reached by a reversible adiabatic decompression), followed by a second path to the stable partially molten state 3 at P_f (which has the same enthalpy as state 2). Figure 1b shows that state 3 is at a higher melt fraction (F), higher S and higher H than state 4. For a multicomponent system, state 3 can also be at a higher T than state 4 on the isentropic path. Hence the process that these authors actually approximated contains an *irreversible*, isobaric, adiabatic melting step (i.e. the adiabatic path from state 2 to state 3 at constant P and H leading to maximization of S), which leads to more melting than that produced by reversible adiabatic upwelling. Although the quantitative differences between these two paths are small, particularly at low degrees of melting, this example illustrates the importance of precise definition of the thermodynamics governing the melting process. Note that this treatment is not 'wrong' in that it does follow an adiabatic path, just not the reversible one, and it is possible that such a path could be of petrologic or geophysical interest; e.g. at a solid–solid phase transformation if the transition is kinetically inhibited (Solomatov & Stevenson 1994).

An alternative to this approach would be to estimate melt production during reversible (i.e. isentropic) upwelling by balancing S rather than H in equivalent calculations comparing the metastable solid assemblage and the stable partially molten assemblage (i.e. by breaking the isentropic path from state 1 to state 4 in figure 1b into the sum of paths $1 \to 2$ and $2 \to 4$). Actual conversion of the metastable solid state 2 to final state 4 would require in this case an irreversible non-adiabatic process at constant P and S leading to a minimization of H. The direct path from $1 \to 4$, on the other hand, is accomplished in a series of infinitesimal reversible adiabatic steps. Although both entropy-conserving and enthalpy-conserving calculations of this sort are conceptually simple, rigorous application of this approach to modelling adiabatic productivity in multicomponent systems would in practice be difficult because of the difficulty of incorporating into the calculation the dependence of the thermodynamic parameters (heat capacities, entropy of fusion, etc.) on changes in melt, solid, and system composition, on residual mineralogy, and on temperature and pressure.

There have been several well-defined thermodynamic treatments of isentropic batch melting of decompressing peridotite (McKenzie 1984; Miller et al. 1991; Albarède 1992; Iwamori et al. 1995). These treatments use as inputs parametrizations of experimental data on isobaric productivity (i.e. $(\partial F/\partial T)_P$, where F is the melt fraction), the positions of the solidus and liquidus, and the entropy of fusion (assumed to be constant). These parametrizations are generally poorly constrained (particularly the isobaric productivity near the solidus) and the isentropic productivity functions that have been presented are consequently highly variable. For example, McKenzie (1984) favoured models that yield roughly constant or strongly decreasing productivity during upwelling, while Iwamori et al. (1995) and Miller et al. (1991) presented models with complex productivity functions that largely reflect their fits to the solidus and liquidus and to the isobaric productivity function. Note that these treatments cannot easily incorporate the effects on melting of changes in the compositions or abundances of residual phases, of pressure dependent solid–solid phase changes (Asimow et al. 1995a), or of changing bulk composition, and thus, like the simple enthalpy- or entropy-balances described in the preceding paragraphs, they do not provide insight into the influence of these features of peridotite phase equilibria on productivity, which are likely to be substantial. Moreover, these treatments are not well-suited to evaluating the productivity of fractional fusion.

We have adopted in our work (Hirschmann et al. 1994; Baker et al. 1995; Asimow et al. 1995a, b) an approach based on a self-consistent thermodynamic model of multicomponent liquid-crystal equilibria. Using a modification of the MELTS code (Ghiorso & Sack 1995), we minimize directly the enthalpy for a specific bulk composition at a given P and S. Because this treatment incorporates internally consistent thermochemical models for the liquid and solid phases in mantle peridotites, it implicitly takes into account the phase and compositional changes that occur on melting of peridotite without having to incorporate them into parametrizations for the solidus and liquidus, the isobaric productivity etc., which in natural peridotite are unlikely to be fit by simple or general functional forms. In addition, because the model is not linked to any particular bulk composition on which experiments have been conducted, it is equally applicable to batch and fractional melting and can be applied to a range of fertile through depleted peridotite compositions. Though the accuracy of MELTS predictions is at present imperfect, it has been shown to capture even some relatively subtle features of available melting experiments on peridotite (Hirschmann et al. 1994; Baker et al. 1995), and it thus is a promising vehicle for modelling peridotite phase equilibria and melting energetics. Contrary to the results of all previous treatments, MELTS predicts that isentropic productivity strongly increases with progressive melting; e.g. initial melting of a fertile peridotite is predicted to be extremely unproductive, with near-solidus isentropic productivity values near 0.25% kbar^{-1}, rising to values of ca. 3% kbar^{-1} near the exhaustion of clinopyroxene (figure 7; see also Hirschmann et al. 1994; Asimow et al. 1995a).

3. Isentropic melting in simple systems

Given the wide range in productivity functions that have been proposed for melting during adiabatic decompression—ranging from constant (Turcotte & Ahern 1978; Klein & Langmuir 1987; Scott & Stevenson 1989; Niu & Batiza 1991; Sparks & Parmentier 1991; Kinzler & Grove 1992), to decreasing as melting proceeds (McKenzie 1984; McKenzie & Bickle 1988; McKenzie & Bickle 1990; Langmuir et al. 1992; Longhi 1992), to increasing as melting proceeds (Hirschmann et al. 1994; Asimow et al. 1995a, b), to complex and irregular (McKenzie 1984; McKenzie & O'Nions 1991; Miller et al. 1991; Iwamori et al. 1995)—it is fair to say that this phenomenon is poorly understood. In order to develop a more complete understanding of the relationship between melting energetics, phase equilibria, and productivity, in this section we examine the behaviour of melting during isentropic upwelling of simple model systems. The melting behaviour of these systems is easy to understand, yet surprisingly rich in insights that can be generalized to multicomponent systems. These simple systems thus provide a framework for understanding the productivity during upwelling of more complex natural systems. A key conclusion is going to be that the melting behaviour of these simple systems strongly suggests that the productivity functions generated by the MELTS calculations capture at least qualitatively the behaviour of the real mantle.

In the following discussions we consider both isentropic batch melting and fractional fusion; fractional fusion is envisioned as a sequence of infinitesimal isentropic melting steps, each followed by extraction of the melt phase, carrying its entropy out of the system with it. Both processes are defined by the restriction

$$d\boldsymbol{S} = S^l \, d\boldsymbol{M}; \quad (3.1)$$

i.e. the only changes we allow in the entropy of the system, $\boldsymbol{S}$ (extensive variables are boldface), are due to extraction of liquid and the resulting change is given by the specific entropy of the liquid, S^l, times the change in system mass, $\boldsymbol{M}$. The general forms we derive will apply to isentropic batch and incrementally isentropic fractional fusion as well as to any continuous melting or dynamic melting process (Langmuir et al. 1977) subject to the restriction that liquid mass is a function of no variables other than solid mass (e.g. $\boldsymbol{M}^l = 0$ for fractional fusion; $\boldsymbol{M}^l = \boldsymbol{M}^0 - \boldsymbol{M}^s$ where $\boldsymbol{M}^0$ is the initial system mass, a constant, for batch fusion; and

$$\boldsymbol{M}^l = \begin{cases} \boldsymbol{M}^0 - \boldsymbol{M}^s, & \text{for } \boldsymbol{M}^s \geqslant (1-f^*)\boldsymbol{M}^0, \\ \dfrac{f^*}{1-f^*}\boldsymbol{M}^s, & \text{for } \boldsymbol{M}^s \leqslant (1-f^*)\boldsymbol{M}^0, \end{cases} \quad (3.2)$$

for continuous fusion where f^* is a constant retained melt fraction). Note that $\boldsymbol{M}^l$ refers to the mass of liquid remaining in the system; extracted liquid is considered no further. We use the quantity F to refer to the melt fraction by mass normalized to original source mass for all melting processes:

$$F = 1 - \frac{\boldsymbol{M}^s}{\boldsymbol{M}^0}, \quad (3.3)$$

and the quantity f to refer to the mass fraction of liquid that remains in the source region,

$$f = \frac{\boldsymbol{M}^l}{\boldsymbol{M}^l + \boldsymbol{M}^s}, \quad (3.4)$$

i.e. for batch melting, $f = F$; for fractional melting, $f = 0$; and for continuous melting as defined in equation (3.2), $f = F$ until F reaches f^* and $f = f^*$ thereafter.

(a) *One-component systems*

The isentropic behaviour of a one-component system can be evaluated rigorously in a closed form. Taking P and T as the independent variables, for a single phase of one component we write

$$dS^\phi = \left(\frac{\partial S^\phi}{\partial T}\right)_P dT + \left(\frac{\partial S^\phi}{\partial P}\right)_T dP = \frac{C_p^\phi}{T} dT - V^\phi \alpha^\phi \, dP, \quad (3.5)$$

where S is specific entropy, C_p is isobaric heat capacity, V is specific volume, α is the isobaric coefficient of thermal expansion, and the superscript ϕ indicates the properties of a single phase. If we consider two phases, solid (s) and liquid (l), coexisting at equilibrium along a univariant curve (denoted 2ϕ), then for coexisting solid and liquid we have

$$\left(\frac{dS^s}{dP}\right)_{2\phi} = \frac{C_p^s}{T}\left(\frac{dT}{dP}\right)_{2\phi} - V^s\alpha^s, \qquad (3.6\,a)$$

and

$$\left(\frac{dS^l}{dP}\right)_{2\phi} = \frac{C_p^l}{T}\left(\frac{dT}{dP}\right)_{2\phi} - V^l\alpha^l. \qquad (3.6\,b)$$

For upwelling of a closed system at constant total specific entropy, S_0, the system must always satisfy

$$FS^l + (1-F)S^s = S_0. \qquad (3.7)$$

Hence for batch melting

$$F = \frac{S_0 - S^s}{S^l - S_s} = \frac{S_0 - S^s}{\Delta S_{\text{fus}}}, \qquad (3.8)$$

where ΔS_{fus} is the specific entropy of fusion for a one-component system, which is in general a function of P and T.

(i) *Constant coefficients*

In the special case that ΔS_{fus} is constant (equivalent to requiring $C_p^l = C_p^s$ and $V^l\alpha^l = V^s\alpha^s$), the isentropic productivity for batch melting can be obtained by differentiation of equation (3.8):

$$-\left(\frac{\partial F}{\partial P}\right)_S = \frac{1}{\Delta S_{\text{fus}}}\left(\frac{C_p^s}{T}\left(\frac{dT}{dP}\right)_{2\phi} - V^s\alpha^s\right) = \frac{1}{\Delta S_{\text{fus}}}\left(\frac{C_p^s}{T}\left(\frac{\Delta V_{\text{fus}}}{\Delta S_{\text{fus}}}\right) - V^s\alpha^s\right), \qquad (3.9)$$

where the second equality follows from the Clausius–Clapeyron equation for a one-component system. Since melting occurs over a range of pressure along a univariant curve with a finite slope, it also occurs over a range of temperatures. Hence equation (3.9) shows that even in the simplest possible case—a one-component system with constant ΔS_{fus}, ΔV_{fus}, C_p^s and $V^s\alpha^s$—isentropic productivity is not constant; i.e. it depends on temperature.

The magnitude of this effect can be estimated as follows. Neglecting the temperature difference due to the finite slope of the solid adiabat, the ratio of the temperature at the onset of isentropic melting to that at the completion of melting in a system with constant coefficients can be approximated by

$$\frac{T_0}{T_1} = \exp\left(\frac{\Delta S_{\text{fus}}}{C_p^l}\right) \qquad (3.10)$$

(Miller et al. 1991). If we take as typical values of the entropy of fusion ca. R per atom and of the liquid heat capacity ca. $3R$ per atom, we obtain $T_0/T_1 \sim 1.4$. For most silicates the $V^s\alpha^s$ term in equation (3.9) is very small, so melting at the completion of isentropic melting in a one-component system would typically be ca. 1.4 times more productive than at the onset of melting.

Although fractional melting is neither isentropic nor reversible, as indicated above we define an idealized adiabatic fractional melting process as a series of infinitesimal isentropic melting steps, each followed by complete extraction of the liquid, carrying its entropy out of the system. We note that each infinitesimal increment of melting is equivalent to the initial increment of batch melting of a system that has been reduced by a factor $(1-F)$ from the original mass of the system. Since F is defined as the mass fraction of the *original* system that is now liquid (equation (3.3)), we obtain

$$-\left(\frac{dF}{dP}\right)_{\text{fractional}} = \frac{1}{\Delta S_{\text{fus}}}\left(\frac{C_p^s}{T}\left(\frac{\Delta V_{\text{fus}}}{\Delta S_{\text{fus}}}\right) - V^s\alpha^s\right)(1-F) \qquad (3.11)$$

(for a more rigorous derivation of productivity for incrementally isentropic processes, see the Appendix). Thus, in a one-component system, productivity during fractional melting is initially the same as that of batch melting, but becomes steadily smaller (and asymptotically approaches zero as the system approaches 100% melt) as upwelling proceeds. We emphasize that in a one-component system with constant coefficients, productivity during fractional fusion defined this way differs from batch productivity solely because of decreasing source mass. Note that the productivity per unit mass of solid, another way to define productivity, is always equal for batch and fractional melting in a one-component system with constant ΔS_{fus} (except in the case of solid–solid phase changes; see Asimow et al. (1995a)).

We can obtain an expression suitable for batch and fractional melting, as well as intermediate processes, in which some but not all of the melt is left in the system (so-called continuous or dynamic melting; Langmuir et al. 1977) if we note that at any time the fraction of the original mass that remains in the system is $(1-F)/(1-f)$. For batch melting this term is equal to one, and for fractional melting it reduces to $(1-F)$. Thus, for the general process of isentropic melting steps possibly followed by some melt extraction in the constant-coefficient one-component case (see the Appendix for a more rigorous derivation),

$$-\left(\frac{dF}{dP}\right) = \frac{1}{\Delta S_{\text{fus}}}\left(\frac{C_p^s}{T}\left(\frac{\Delta V_{\text{fus}}}{\Delta S_{\text{fus}}}\right) - V^s\alpha^s\right)\frac{(1-F)}{(1-f)}. \qquad (3.12)$$

We will generally refer to $-(dF/dP)$ as the 'isentropic productivity' even though for fractional fusion the process is only isentropic in each infinitesimal melting step.

To illustrate melting behaviour in a one-component system, equations (3.9) and (3.11) have been applied to pure diopside (we neglect any incongruent melting behaviour; Biggar & O'Hara 1969; Kushiro 1972). Using the properties given in table 1 and taking the values of ΔS_{fus}, ΔV_{fus}, C_p^s and $V^s\alpha^s$ at the 1 bar melting point to apply at all P and T, we obtained a linear melting curve (figure 2a). The same phase relations are shown in S–P space in figure 2b, where it should be noted that although the entropy difference between coexisting solid and liquid is constant, the boundaries of the two-phase field are not linear, reflecting the $1/T$ dependence in equations (3.6) (Asimow et al. 1995a). Choosing an isentrope that intersects the melting curve at 7 GPa, we obtained the productivity curves shown in figure 2c for batch and fractional melting (for comparison, the hypothetical linear case is also shown), which have been integrated to yield the F versus P curves shown in figure 2d. Batch melting in this case yields a concave-up F versus P curve, while the curve for fractional melting is concave down. Productivity at the completion of batch melting in this case is a factor 1.6 higher than the initial productivity. The upward curvature of the batch melting curve seen here will be referred to below as the '$1/T$ effect'. This effect is of only secondary significance in multicomponent systems, but we draw attention to it here in order to illustrate the improbability of constant productivity for any isentropic melting process.

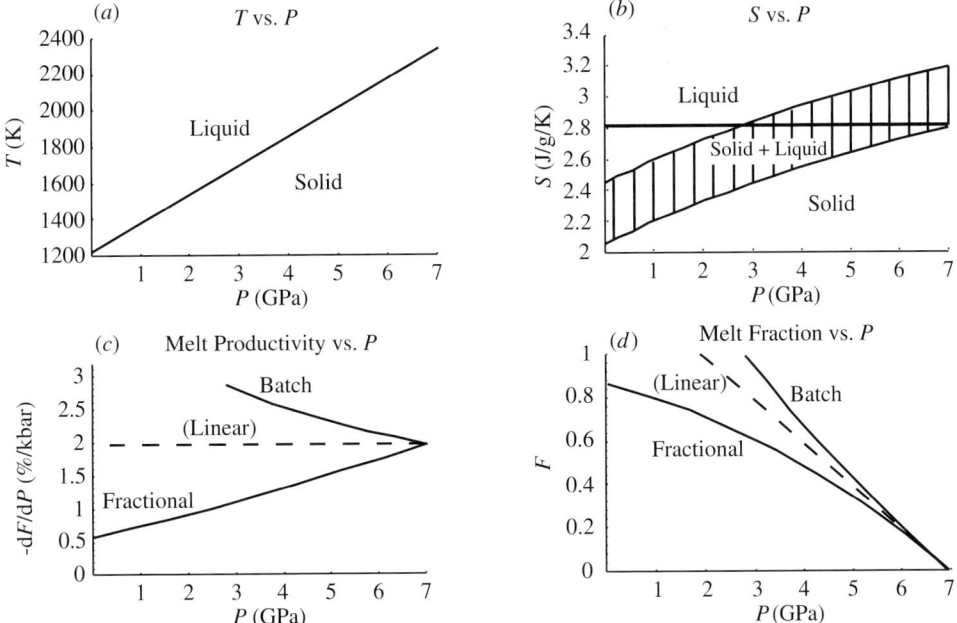

Figure 2. Isentropic melting behaviour of a hypothetical one-component system in which the heat capacity, $(dV/dT)_P$, ΔV_{fus} and ΔS_{fus} of diopside at 1 bar and 1664 K are taken to obtain over all P and T. (a) The solidus is linear. It has been chosen to go through 7 GPa and 2338 K, close to the actual diopside solidus (Rigden et al. 1989). (b) Isentropic melting is best illustrated with an S–P plot. Batch melting follows a horizontal line on this figure. The reference isentrope shown as a heavy horizontal line intersects the solidus at 7 GPa. (c) The isentropic productivity (expressed as percent melting per kbar pressure decrement) versus P for batch and fractional paths that intersect the solidus at 7 GPa. The dashed line is for comparative purposes only; it does not correspond to any isentropic path. (d) Melt fraction versus P for the same batch and fractional paths.

(ii) *Variable coefficients*

Figure 2c shows that for the case of constant coefficients, isentropic melting leads to increasing productivity in the batch case, reflecting the $1/T$ dependence, and decreasing productivity in the fractional case, reflecting decreasing source mass. We now apply the same analysis using more realistic variations in thermodynamic parameters as functions of P and T. In the case that ΔS_{fus} is not constant, equations (3.9) and (3.12) do not apply. Instead we begin from equation (3.1) and, as shown in the Appendix, we derive the following general expression for any process of isentropic melting steps possibly followed by melt extraction in a one-component system:

$$-\left(\frac{dF}{dP}\right) = \frac{1}{\Delta S_{\text{fus}}} \left(\frac{C_p^s + f(C_p^l - C_p^s)}{T} \left(\frac{dT}{dP}\right)_{2\phi} - [V^s \alpha^s + f(V^l \alpha^l - V^s \alpha^s)] \right) \frac{(1-F)}{(1-f)}. \quad (3.13)$$

For fractional melting in a one-component system with variable coefficients, f is zero, so equation (3.13) reduces to equation (3.11).

To examine the variable-coefficient case, we again used diopside as the example and treated it as a one-component system. The solidus curve predicted by the thermodynamic data given in table 1 is shown in figure 3a; the downward curvature results from the greater compressibility of the liquid relative to the solid. The same phase relations are shown in S–P space in figure 3b; the curvature of each boundary

Table 1. *Thermophysical properties of Di (figures 2 and 3) and a–b binary[a] (figures 4–6)*

parameter	diopside and a solids	b solid	di and a liqs	b liquid	units
S_0 (298 K)	142.5[b]	174.2			J mol^{-1} K^{-1}
H_0 (298 K)	-3200.583[b]	-2842.221			kJ mol^{-1}
V_0 (1664 K)	69.11[b]	same	82.34[c]	same	m^3 mol^{-1} × 10^{-6}
T_{fus} (1 bar)	1664 K	1164 K			
ΔS_{fus} (1 bar)	82.88[e]	same			J mol^{-1} K^{-1}
K_{T0}	90.7[c]	same	21.9[d]	24	GPa
K'_T	4.5[c]	same	6.9[d]	6.9	
α	3.2×10^{-5}[c]	same	6.5×10^{-5}[c]	same	K^{-1}
C_p	$305.41 - 160.49\, T^{-0.5}$ $-71.66 \times 10^5\, T^{-2}$ $+92.184 \times 10^7\, T^{-3}$[b]	same	353[c]	same	J mol^{-1} K^{-1}

[a] For binary example, end member a is identical to diopside, end member b is selected to have a 1 bar melting point 500 K lower. Unreferenced quantities for b are chosen arbitrarily to give well-behaved binary phase-loop up to 10 GPa.
[b] Berman (1988).
[c] Rigden *et al.* (1989) and sources therein.
[d] Lange & Carmichael (1990).
[e] Stebbins *et al.* (1983).

of the two-phase field is higher than in the constant coefficient case, reflecting the $(dT/dP)_{2\phi}$ term in equations (3.6). Using these coefficients, we computed $F(P)$ from equation (3.8) and $-(dF/dP)$ from equation (3.13) subject to the constraint $f = F$ along the batch adiabat that intersects the solidus at 7 GPa. For fractional melting, we computed $-(dF/dP)$ along the path that intersects the solidus at 7 GPa using equation (3.13) subject to the constraint $f = 0$, and integrated to obtain $F(P)$. The results are shown in figures 3c and 3d.

Comparison of figures 2c and 3c shows that in systems with variable coefficients, the increase in productivity with increasing melt fraction due to the curvature of the solidus can be substantial for batch melting; in the case shown it leads to large (e.g. a factor of 5.7 between $F = 0$ and $F = 0.9$) increases in productivity as melting proceeds. In fact, as shown by comparing the fractional fusion curves in figures 2c and 3c, the increase in productivity due to curvature of the solidus in the more realistic case overwhelms the tendency for productivity to decrease due to the decreasing mass of the source and results in increasing productivity with progressive melting even for fractional fusion (although there must be a maximum and productivity must eventually decrease at high melt fraction, since melt fraction normalized by original source mass must for fractional melting asymptotically approach $F = 1$ in equation (3.11)). We note that equation (3.13) also shows that a solidus slope that is negative or less than the adiabatic gradient will generally lead to crystallization rather than melting with decreasing pressure at constant entropy (Rumble 1976; Albarède 1983; Iwamori *et al.* 1995).

The strong increase in productivity in the variable-coefficient one-component case reflects the increase in the slope of the solidus with decreasing pressure. The same effect is present in multicomponent systems, where the analogous controlling variable

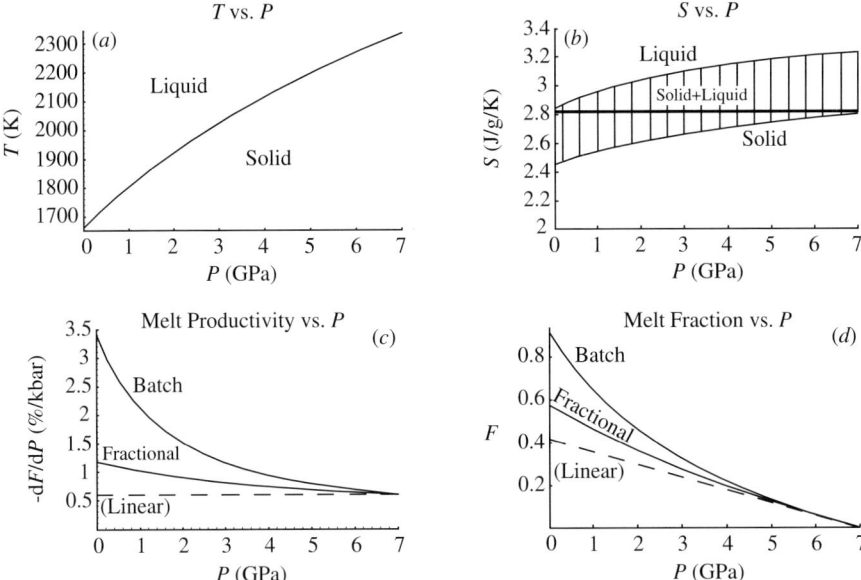

Figure 3. Isentropic melting behaviour of diopside using the thermodynamic data from table 1. (a) The solidus is concave down, due to greater compressibility of the liquid relative to the solid. (b) The curvature of the solidus translates into greater curvature of the edges of the two-phase field in S–P space, compared to figure 2b. The heavy horizontal line is an S–P path for batch melting. (c) The isentropic productivity versus P for batch and fractional paths that intersect the solidus at 7 GPa. The dashed line is for comparative purposes only; it does not correspond to any isentropic path. (d) Melt fraction versus P for the same batch and fractional paths.

is $(\partial T/\partial P)_F$, the slope of a constant melt fraction contour; note that all such contours are collapsed onto the univariant solidus in P–T space in a one-component system but are arrayed between the solidus and the liquidus (and are not, in general, parallel to either) in multicomponent systems. The influence of the slopes of these contours on isentropic productivity will be referred to below as the '$(\partial T/\partial P)_F$ effect'.

(b) Multicomponent systems

For the general case in a multicomponent system, the expression for $-(dF/dP)$ for isentropic or incrementally isentropic melting paths with possible melt extraction is derived in the Appendix:

$$-\frac{dF}{dP} = \frac{\left(\dfrac{C_p^s + f(C_p^l - C_p^s)}{T}\right)\left(\dfrac{\partial T}{\partial P}\right)_F - [V^s\alpha^s + f(V^l\alpha^l - V^s\alpha^s)] + \left(\dfrac{\partial S_X}{\partial P}\right)_F}{\left(C_p^s + f(C_p^l - C_p^s)\right)/T \left(\dfrac{\partial F}{\partial T}\right)_P + \dfrac{(1-f)}{(1-F)}(S^l - S^s) + \left(\dfrac{\partial S_X}{\partial F}\right)_P}, \quad (3.14)$$

where the superscript 's' now refers to the bulk properties of the residual (usually polymineralic) solid assemblage and $(\partial S_X/\partial P)_F$ and $(\partial S_X/\partial F)_P$ are shorthand notation for terms that reflect the effects on S^l and S^s of changes in liquid and mineral composition and of changes in the relative abundances of the minerals in the solid assemblage (see Appendix; for related equations, see Verhoogen 1965; McKenzie

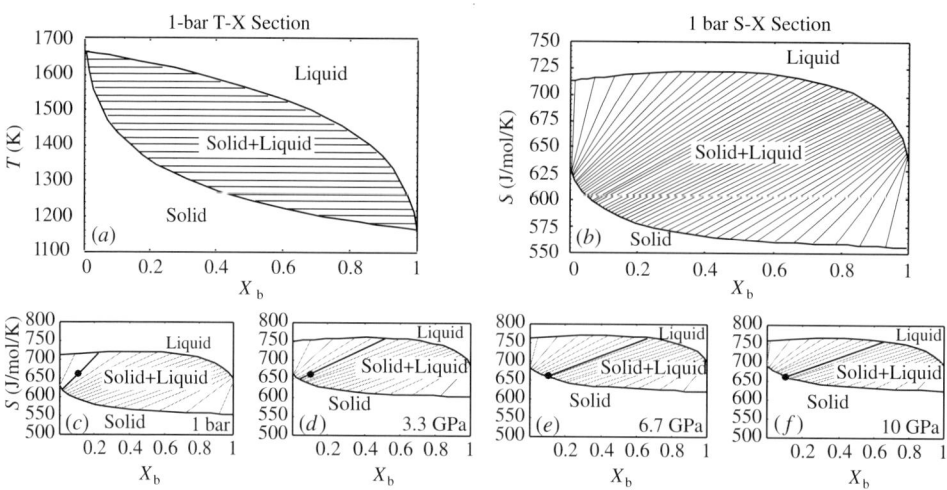

Figure 4. Phase diagram of a two-component model system a–b; both solid and liquid are ideal solutions. Model data are in table 1. (a) T versus X_b at 1 bar. (b) S versus X_b at 1 bar. Tie-lines in the two-phase field indicate entropy and composition of coexisting phases. (c)-(f) Analysis of isentropic melting is visualized by a series of S versus X_b sections at $P = 1$ bar, 3.3 GPa, 6.7 GPa and 10 GPa with the point $X_b = 0.1$, $S = 663$ J mol^{-1} K^{-1} and the tie-line that passes through it at each pressure highlighted. The position of the point along the highlighted tie-line gives the melt fraction by the lever rule; this sequence illustrates the importance, both of the movement of the phase loop as a function of pressure, and the rotation of the tie-lines towards the vertical near the end members in determining melt fraction and productivity.

1984; Iwamori et al. 1995). Note that ΔS_{fus}, defined as the entropy difference between a solid and liquid of the same composition, does not appear in this expression. Most earlier thermodynamic treatments of adiabatic melting in multicomponent systems have equated $S^l - S^s$ or $(\partial S/\partial F)_{P,T}$ (which is equivalent to the sum of the last two terms in the denominator of (3.14)) with ΔS_{fus} for the bulk peridotite, which has undoubtedly led to inaccuracies. The essential first-order change from equation (3.13) is the presence of a term in the denominator involving the partial derivative of melt fraction with respect to temperature at constant pressure, $(\partial F/\partial T)_P$, which we call the isobaric productivity. In a one-component system, isobaric melting occurs at a unique temperature, so this quantity is infinite. Hence the first term in the denominator of equation (3.14) as well as the terms due to compositional and modal changes in the entropy of the phases vanish for one-component systems and this expression reduces to equation (3.13). For multicomponent systems, $(\partial F/\partial T)_P$ thus joins $(\partial T/\partial P)_F$ as a key source of variability in isentropic productivity. Note that the $1/T$ effect will typically be of secondary importance in the multicomponent case, since it now contributes both to the numerator and denominator of the expression for $(\mathrm{d}F/\mathrm{d}P)$.

In this section, we explore the origins and importance for isentropic productivity of variations in $(\partial F/\partial T)_P$ in a simple two-component binary phase loop and then in MELTS simulations of peridotite melting. We will emphasize that $(\partial F/\partial T)_P$ reflects changes in liquid and solid composition during melting via conservation of mass as expressed in the lever rule. We then demonstrate that variations in both $(\partial F/\partial T)_P$ and $(\partial T/\partial P)_F$ are needed for a reasonably accurate understanding of the variations in $(\mathrm{d}F/\mathrm{d}P)$. We also show that variations of other parameters with melt fraction, including $S^l - S^s$, do not affect productivity variations by more than ca. 10%, even when phases are exhausted from the residue.

(c) Two-component systems

Before considering isentropic melting, we first evaluate isobaric productivity in two-component systems. We approach the problem in this way because isobaric melting can be treated using familiar phase diagrams from which the effect of composition on productivity can be deduced easily and because we wish to isolate isobaric productivity from the other terms in the general expression for incrementally isentropic productivity (equation (3.14)). In this exercise, we use as a model system a hypothetical binary a–b with complete solid solution in the solid and ideal mixing for both the liquid and solid solutions (figure 4). End member a has the thermodynamic properties of diopside; end member b has similar properties, except the melting point at 1 bar is arbitrarily chosen to be 500 °C lower than that of a. Model parameters for the end members are listed in table 1. We chose a complete solid solution model as our example, rather than a eutectic or peritectic involving solid phases of fixed composition, because all mantle phases are solid solutions and hence the phase loop captures the essential behaviour of the natural system (except when a phase is exhausted on melting, as discussed later); we chose this hypothetical binary rather than the actual diopside–hedenbergite system because the exaggerated difference in melting points of the two end members allows the effects of a finite melting interval to be more easily seen.

(i) Isobaric melting

In a two-component system, the productivity of isobaric melting with increasing temperature is simply a matter of conservation of mass. If we consider a system where the bulk composition is given by X_b, the mass fraction of component b, we can write

$$FX_b^l + (1 - F)X_b^s = X_b, \quad \text{or} \quad F = \frac{(X_b - X_b^s)}{(X_b^l - X_b^s)}, \qquad (3.15)$$

which is just a statement of mass balance (i.e. the familiar lever rule for graphical analysis of phase diagrams). For batch melting, differentiation of equation (3.15) leads to

$$\left(\frac{\partial F}{\partial T}\right)_P^{\text{batch}} = -\left(\frac{\partial X_b^s}{\partial T}\right)_P \left(\frac{1}{X_b^l - X_b^s}\right) - \frac{(X_b - X_b^s)\left(\partial(X_b^l - X_b^s)/\partial T\right)_P}{(X_b^l - X_b^s)^2}. \qquad (3.16)$$

For fractional melting the second term in equation (3.16) vanishes since the instantaneous solid composition is always equal to the bulk composition and source mass decreases as $(1 - F)$, which leads to

$$\left(\frac{\partial F}{\partial T}\right)_P^{\text{fractional}} = \left(\frac{\partial X_b^s}{\partial T}\right)_P \left(\frac{1}{X_b^l - X_b^s}\right)(1 - F). \qquad (3.17)$$

(a more rigorous derivation of the expression for isobaric fractional melting requires starting from extensive variables as in the derivation in the Appendix).

The inverse relationship between isobaric productivity and the compositional difference between coexisting liquid and solid (i.e. the $1/(X_b^l - X_b^s)$ factor) is generally the most important term in both equations (3.16) and (3.17) for the simple phase loop. The second term in (3.16) is important at high melt fraction (i.e. where $F = (X_b - X_b^s)/(X_b^l - X_b^s)$ is large) or when the solids are fixed or nearly fixed in composition (note that the first term goes to zero if the solid phases are fixed in composition); the derivative of solid composition in the first term is also important near

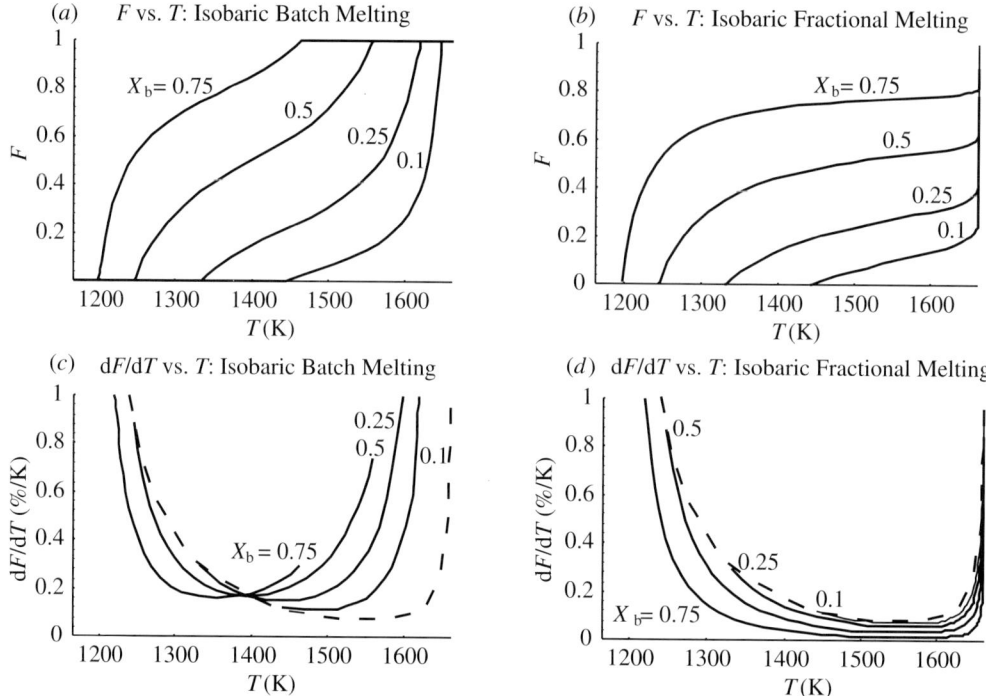

Figure 5. Isobaric batch and fractional melting of the model binary system a–b at 1 bar according to the phase relations shown in figure 4. X_b values indicate the bulk composition of the solid before the initiation of melting. (a) Batch melting, F versus T. (b) Fractional melting, F versus T. Note that all curves finish melting at $T = 1664$ K, the melting point of the a end member. (c) Isobaric productivity in percent melting per degree temperature increase for batch melting. The dashed curve shows the first term in equation (3.16) and is the locus of values of productivity on the solidus ($F = 0$) for various bulk compositions. (d) Isobaric productivity for fractional melting. The values differ from the dashed solidus productivity curve only by a factor $(1 - F)$.

the exhaustion of a phase from the residue when a multiphase residual assemblage is melting (see below). Equations (3.16) and (3.17) yield infinite $(\partial F/\partial T)_P$ during eutectic or peritectic melting.

The key effect of variable composition of the phases is to cause isobaric productivity to be small when the difference between liquid and solid compositions, $(X_b^l - X_b^s)$, is large. This is illustrated by the quantitative results (figure 5) based on the calculated phase relations for our model binary phase loop (figure 4a), in which the form of the melt fraction versus temperature curve varies with bulk composition mostly according to whether the compositional difference between the liquid and solid initially increases or always decreases with increasing melt fraction. Melt fraction versus temperature curves and isobaric productivity versus temperature curves for batch and fractional melting of the bulk compositions $X_b = 0.1, 0.25, 0.5$ and 0.75 are shown in figure 5. The dashed curves in figures 5c and 5d plot the first term in equation (3.16); this is the initial isobaric productivity (i.e. at the solidus) as a function of bulk composition. The fractional fusion curves differ from the dashed curve only by a factor $(1 - F)$; the batch melting curves differ from the dashed curve according to the second term in equation (3.16), which increases with F and changes sign at the widest point on the phase loop ($T = 1393$ K, $X_b^s = 0.15$, $X_b^l = 0.85$). Examination of the dashed curve in figures 5c and 5d shows that in this example, the multiplication by $(\partial X_b^s/\partial T)_P$ in the first term of equation (3.16) contributes a strong asymmetry to the productivity function, which is otherwise dominated by the (nearly symmetric about 1393 K) inverse compositional distance term. The net effect of all these terms is that compositions with $X_b < 0.06$ show a melt fraction versus temperature curve for batch melting that is always concave up (i.e. an isobaric productivity that always increases as melting proceeds). For fractional melting, productivity always increases for $X_b < 0.03$. For more b-rich bulk compositions, the melt fraction versus temperature curves (such as those illustrated in figures 5a and 5b for $X_b = 0.25, 0.5$ and 0.75) are initially concave down, but concave up at higher F. The critical bulk X_b below which the melt fraction versus temperature curve is everywhere concave up depends on the shape of the phase loop; in the diopside–hedenbergite system where the phase loop is much narrower, it occurs at $X_{Hd} \sim 0.4$ for batch melting and $X_{Hd} \sim 0.2$ for fractional melting. In the forsterite–fayalite system the corresponding values are $X_{Fa} \sim 0.14$ for batch melting and $X_{Fa} \sim 0.1$ for fractional melting. The location of this critical X_b cannot be read directly off the phase diagram; it depends on all the terms in equations (3.16) or (3.17) and does not correspond to the widest point on the phase loop. Note again that in the special case where the solid residue is fixed in composition (i.e. only batch melting is continuous in temperature), the first term in equation (3.16) vanishes and the difference in melt and solid composition always decreases with F, so the geometric effect leads to the melt fraction versus temperature curve being everywhere concave up.

(ii) *Isentropic melting*

There is no simple two-dimensional phase diagram with which to portray isentropic melting for a binary loop. Inspection of the general expression for isentropic melting (equation (3.14)) shows that the geometric effect related to the compositional distance between liquid and solid (i.e. the $(\partial F/\partial T)_P$ term discussed in the preceding paragraphs) is superimposed on the $1/T$ and $(\partial T/\partial P)_F$ effects that control isentropic productivity in one-component systems. The relationship among the terms in these equations can be visualized by examining figures 4b–f. Figure 4b shows S versus X_b at 1 bar for the model binary phase loop; figures 4c–f show a series of simplified S versus X_b sections on the same scale at successively higher pressures. With increasing pressure, the entire phase loop moves up (i.e. to higher values of specific entropy); in the one-component diopside-like end member a, this increase is illustrated in figure 3b. Isentropic melting of a particular bulk composition can then be visualized as the movement of the loop over a particular fixed point ($X_b = 0.1$, $S = 663$ J mol^{-1} K^{-1} in this example) as pressure decreases.

Although the shape of the phase loop is complex and changes with pressure, its overall downward movement with decreasing pressure tends to contribute, for any composition, to increasing productivity with decreasing P (i.e. a concave up F versus P diagram) just as it does in the one-component end members. This reflects the $1/T$ and $(\partial T/\partial P)_F$ dependencies described above. The melt composition effect is discernible in the rotation of the tie lines (in this case little influenced by decompression) towards the vertical from the center to the edges of the loop, which results in changes in the difference in composition between the solid and liquid with increased melting. Although it is not as easy to read as the diagrams for the isobaric melting case, examination of figure 4 shows that for a-rich compositions, the compositional difference between solid and melt decreases with decreasing pressure (and increasing melt fraction). Just as in the isobaric melting case, this leads to a purely geometric contribution tending to increase productivity as pressure decreases. However,

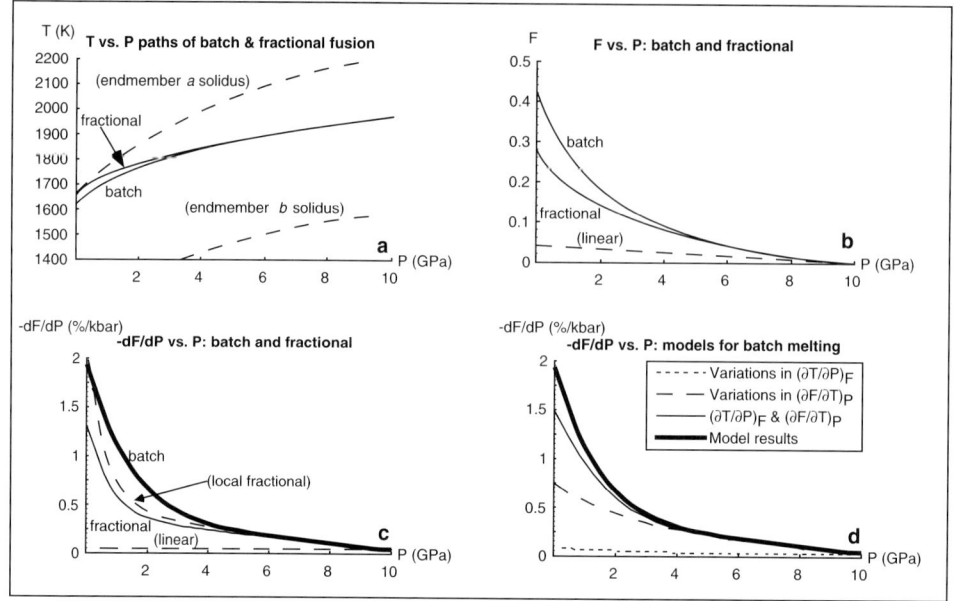

Figure 6. Isentropic batch and incrementally isentropic fractional melting of the model binary system a–b. (a) T versus P, showing the univariant melting curves for the end members and both batch and fractional isentropic melting paths for composition $a_{90}b_{10}$ that intersect the solidus at 10 GPa. (b) F versus P, showing the upward curvature characteristic of increasing productivity. The dashed line is a linear extrapolation of the productivity at the solidus; it does not correspond to any isentropic path. (c) Isentropic productivity versus P for batch and fractional melting. The dashed curve shows 'local' fractional productivity of a unit mass of solid at any pressure; the light solid curve shows $-dF/dP$ for incrementally isentropic fractional melting, where F is normalized to the original source mass (see text). (d) Isentropic productivity versus P for batch melting (heavy curve) compared to a calculation (light solid curve) of productivity based on equation (3.14) where all parameters except $(\partial T/\partial P)_F$ and $(\partial F/\partial T)_P$ were held constant at their values on the solidus at 10 GPa (see text for details). Also shown are calculations in which we allowed $(\partial T/\partial P)_F$ (dotted curve) or $(\partial F/\partial T)_P$ (dashed curve) to vary along the adiabat, holding the other quantity constant at its solidus value. Variations in $(\partial F/\partial T)_P$ capture the major qualitative features of the productivity, but $(\partial T/\partial P)_F$ variations are also required to get a good quantitative fit.

for more b-rich compositions, the compositional difference between solid and liquid initially increases with progressive isentropic melting; as a consequence, when this increase is quite pronounced, complex melt fraction versus P functions (including initially decreasing productivity) can result from the combination of this with the $1/T$ and $(\partial T/\partial P)_F$ effects.

Results for isentropic melting in the model binary system are shown in figure 6 for isentropic batch and fractional melting of the bulk composition $a_{90}b_{10}$ starting at 10 GPa. As anticipated in the above discussion of figures 4c–f, isentropic productivity increases with progressive melting. For fractional melting, figure 6c shows both the productivity normalized to original source mass, the usage we adopt, and the productivity relative to unit mass of solid present at any pressure (dotted line, labelled 'local fractional'). It is interesting to note that (except exactly at the solidus where they are identical) the local productivity of fractional melting in this system is lower than the productivity of batch melting at low melt fraction but slightly greater at high melt fraction (similar to models of peridotite melting; Hirschmann et al. 1994). Comparison of figures 5 and 6 shows that the contribution of the $(\partial T/\partial P)_F$ effect can lead to increasing isentropic productivity at all F even for compositions such as $X_b = 0.1$ that have initially decreasing isobaric productivity.

To demonstrate the contributions of the $(\partial T/\partial P)_F$ and $(\partial F/\partial T)_P$ terms to variations in isentropic productivity, we calculated productivity curves by substituting these quantities into equation (3.14), assuming all other parameters (C_p^s, C_p^l, T, f, $V^s\alpha^s$, $V^l\alpha^l$ and $(S^l - S^s)$) are constant at their values on the solidus at 10 GPa (although $(1-F)/(1-f)$ is also allowed to vary so that fractional melting is normalized properly and all terms due to compositional and modal changes in the entropy of the phases, i.e. $(\partial S_X/\partial P)_F$ and $(\partial S_X/\partial F)_P$, are set to zero). We also tried allowing only one of $(\partial T/\partial P)_F$ and $(\partial F/\partial T)_P$ to vary and holding the other constant along with the above list of parameters at its value on the solidus; the resulting three curves (labelled according to which quantity or quantities we allowed to vary) are shown in figure 6d. Only when we allow both $(\partial T/\partial P)_F$ and $(\partial F/\partial T)_P$ to vary do we reproduce the isentropic productivity function reasonably well, demonstrating that variations of *both* parameters control the detailed shape of the productivity function. Note that the chosen bulk composition in this model system has a total isobaric melting interval (i.e. an average $(\partial F/\partial T)_P$) for $0.1 < X_b < 0.9$ comparable to that of natural peridotite (Takahashi 1986); hence the quantitative importance of $(\partial F/\partial T)_P$ relative to other sources of productivity variation in equation (3.14) in this two-component system is comparable to its importance in peridotite melting.

(d) *Multicomponent systems*

Equation (3.14) shows how knowledge of $(\partial F/\partial T)_P$, $(\partial T/\partial P)_F$, and values of parameters such as $(S^l - S^s)$, C_p, T and $V\alpha$ can be translated into predictions of isentropic productivity and its variability for any system of arbitrary compositional complexity and variance. There are, however, factors other than those we have considered in the model systems treated above that contribute to the variability of these parameters and thus to variations in productivity during isentropic melting. For example, we have emphasized that the dominant term for isobaric melting along a binary phase loop is the inverse dependence on the compositional distance between liquid and solid. However, discontinuous reactions and phase exhaustion must also play important roles in productivity for polymineralic assemblages. We have not presented simple examples involving such phenomena, but they are readily treated in terms of the same parameters discussed above. For example, the exhaustion of a phase restricts the compositional variations available to the solid residue. This translates into a discontinuous drop in the rate of change of the composition of the solid with temperature (equivalent to the $(\partial X_b^s/\partial T)_P$ term in equations (3.16) and (3.17)) and therefore results in a discontinuous drop in productivity even though the compositional distance (equivalent to $X_b^l - X_b^s$ in equations (3.16) and (3.17)) is continuous. The effect of the derivative of bulk solid composition on productivity is also evident at the end of eutectic or peritectic melting in a simple system (where the change from infinite $(\partial X_b^s/\partial T)_P$ and $(\partial F/\partial T)_P$ to finite values results in a corresponding decrease in $-\mathrm{d}F/\mathrm{d}P$ in equation (3.14)), at the loss of a phase during cotectic melting in a ternary system, and at the exhaustion of clinopyroxene during melting of natural peridotite. Note that in none of our simulations of batch fusion have we observed a *drop* in productivity except on phase exhaustion (or addition, as in the case of the spinel-plagioclase transition (Asimow et al. 1995a)).

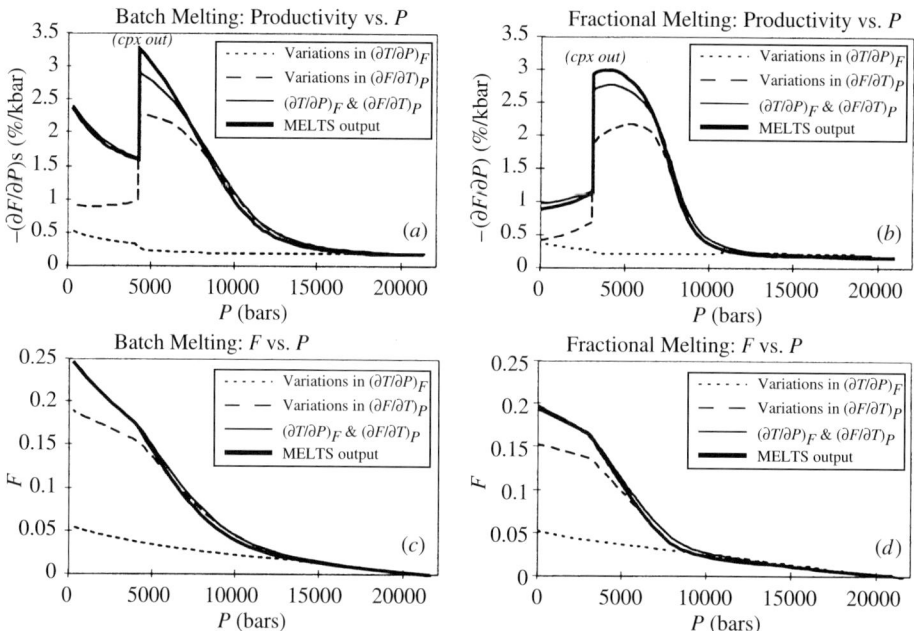

Figure 7. Isentropic melting of 9-component model fertile peridotite (Hart & Zindler 1986; Asimow et al. 1995a) based on calculations using MELTS. (a) Isentropic productivity versus P during batch melting (heavy curve) compared with productivity predicted by equation (3.14) with all parameters except (i) $(\partial T/\partial P)_F$ (dotted curve), (ii) $(\partial F/\partial T)_P$ (dashed curve), or (iii) $(\partial T/\partial P)_F$ and $(\partial F/\partial T)_P$ (light solid curve) held constant at their values near the middle of the melting paths at 11 kbar (see text for details). (b) Analogous to (a) for fractional melting. (c) Melt fraction F versus P for batch melting, compared to melt fraction expected by integrating the three curves in (a). (d) Analogous to (c) for fractional melting.

4. Model peridotite system

Given the simple rules developed above for one- and two-component systems and their generalization to multicomponent systems, we are now in a position to anticipate the productivity function of isentropically melting mantle peridotite during batch fusion and of incrementally isentropic fractional fusion. Although a rigorous analysis is needed to understand the interaction of all the variables controlling productivity in a complex multicomponent system, the simple arguments developed here give considerable insight into the overall behaviour. For example, our analysis makes clear that isentropic productivity is very unlikely ever to be even approximately constant. In addition, it suggests that the concave up melt fraction versus P functions predicted for peridotite melting by the MELTS algorithm (Hirschmann et al. 1994; Asimow et al. 1995a) are robust features of the behaviour of natural peridotite.

We have examined quantitatively controls on productivity using the results of isentropic batch and fractional MELTS calculations on a model peridotite. We used a nine-component model composition in the system SiO_2–TiO_2–Al_2O_3–Cr_2O_3–Fe_2O_3–FeO–MgO–CaO–Na_2O (composition from Hart & Zindler 1986). Choosing an adiabat that intersects the solidus at 22 kbar, we calculated batch melting by minimizing H at fixed S, P and bulk composition to obtain T, F and the compositions of coexisting liquid and solids. For fractional fusion, we searched in pressure for the point along the isentrope that has a fixed incremental melt fraction ($dF = 0.001$) and then took the entropy and composition of the residue as the reference for the next step. Batch isentropic productivity was calculated by differentiation of the F versus P results, and fractional productivity was determined by dividing dF by the pressure difference between successive melt extractions. Figure 7 shows the calculated productivity and melt fraction as functions of pressure for batch and fractional melting.

We have evaluated the extent to which the sources of variation isolated above (i.e. $(\partial T/\partial P)_F$ and $(\partial F/\partial T)_P$) combine to control variations in MELTS-predicted peridotite productivity via an exercise similar to the analysis of the binary case above. We assumed all other parameters (C_p^s, C_p^l, T, f, $V^s \alpha^s$, $V^l \alpha^l$ and ($S^l - S^s$) but not $(1-F)/(1-f)$, which normalizes fractional melting, or $(\partial S_X/\partial P)_F$ and $(\partial S_X/\partial F)_P$, which we neglect altogether by setting them to zero) in equation (3.14) to be constant (at their 11 kbar values, i.e. at the midpoint of the melting paths) and calculated the isentropic productivity along the melting path based on several different sets of values for $(\partial T/\partial P)_F$ and $(\partial F/\partial T)_P$: (i) the actual value of $(\partial T/\partial P)_F$ at each point on the melting path, with $(\partial F/\partial T)_P$ held constant at its 11 kbar value; (ii) the actual value of $(\partial F/\partial T)_P$ at each point on the melting path, with $(\partial T/\partial P)_F$ held constant at its 11 kbar value; and (iii) the actual values of both $(\partial T/\partial P)_F$ and $(\partial F/\partial T)_P$ along the melting path. The resulting curves are shown in figure 7, labelled by what was allowed to vary. For both batch and fractional fusion, shown in figures 7a and 7b, case (ii) captures the general form of the isentropic productivity function, including the rise to a peak at the exhaustion of clinopyroxene and the sharp drop-off. Case (iii), however, shows much better quantitative agreement (although differences are noticeable where productivity is large), demonstrating as for the binary case presented earlier that variations in *both* $(\partial T/\partial P)_F$ and $(\partial F/\partial T)_P$ must be taken into account to approximate accurately the productivity function. The fits to melt fraction for case (iii) shown in figures 7c and 7d are also very good, indicating that the overall amounts of melting during batch and fractional fusion of peridotite can be precisely modelled using equation (3.14) and that variations in T, $(S_l - S_s)$, C_p, $V\alpha$ and compositional derivatives are of secondary importance compared to $(\partial T/\partial P)_F$ and $(\partial F/\partial T)_P$. We emphasize that this exercise is entirely based on the internally consistent nature of the MELTS calculation; consequently, although it helps to isolate the key parameters in the peridotite productivity function as predicted by MELTS, it does not directly address the accuracy of the MELTS results for melt production in nature.

The good qualitative match to the isentropic productivity obtained solely by varying $(\partial F/\partial T)_P$ (figures 7a and 7b) implies that the source of much of the variation in isentropic productivity can be understood by examining the controls on $(\partial F/\partial T)_P$. Just as in the two-component system discussed above, the shape of the isobaric productivity function reflects the compositions of coexisting melt and residue and is dominated by the rate of change of the compositional difference between them (except near the exhaustion of a phase in the more complex system). Near-solidus melts of peridotite differ significantly in composition from the coexisting residue, and the composition of the liquid changes rapidly with increased melting at low melt fractions, becoming more similar to the composition of the residue with increased melting (e.g. in the sense that the melts become richer in normative olivine and poorer in normative plagioclase and incompatible elements, and thus more similar to peridotite; Takahashi & Kushiro 1983; Baker & Stolper 1994; Kushiro 1996). Our analysis indicates that this 'geometric effect', which influences the $(\partial F/\partial T)_P$ function and hence the isentropic productivity, is the main factor leading to low productivity near the solidus and the strongly concave up melt fraction versus pressure function predicted by the MELTS calculations. It must be emphasized that peridotite melting is neither invariant nor pseudo-invariant (there are many more components than phases and hence in both MELTS calculations and experiments the compositions of liquids vary continuously at all melt fractions) and thus eutectic and peritectic melting are very poor models for peridotite. The important point is that the increase in isentropic productivity with progressive melting in these calculations is dominated by the tendency for the liquids to be initially very distant from the source composition (due especially to high concentrations of incompatible elements like Na_2O) and to move towards the bulk composition with progressive melting.

As illustrated by figure 7, the discontinuous changes in isentropic productivity associated with phase exhaustion are also precisely mirrored by changes in isobaric productivity; in both cases this reflects the rate of change of the residual solid composition (analogous to the $(\partial X_b^s/\partial T)_P$ parameter in equations (3.16) and (3.17)). As the exhaustion of clinopyroxene is approached during batch melting, the bulk residual solid composition changes significantly and its temperature derivative is large; the result is a very high productivity in this region. At the actual disappearance of clinopyroxene from the residue, the derivative of bulk solid composition decreases discontinuously, resulting in a drop in productivity. In the fractional case, the shape of the productivity function just before clinopyroxene exhaustion is somewhat different (figure 7a versus 7b; note that this difference is not apparent in the F versus P figures, figure 7c versus 7d); the decrease in productivity in anticipation of clinopyroxene exhaustion during fractional fusion probably reflects the fact that the jadeite component of clinopyroxene is nearly exhausted a few kbar before the phase disappears, leading to a decrease in the temperature derivative of clinopyroxene composition and hence a decrease in both isobaric and isentropic productivity in advance of the much larger discontinuous drop at cpx-out. This example illustrates quite clearly a point made in the Introduction, that efforts to model or understand the productivity of peridotite melting, that do not include the effects of phase equilibria and changing solid and liquid compositions as melting progresses, are very unlikely to capture the essence of the isentropic melting process.

The predicted overall increase in isentropic productivity with melt fraction in the batch melting case, punctuated by drops in productivity upon exhaustion of phases from the residue, appears to be a general feature of simple systems with solid-solution, particularly when the solid solution(s) are close to the high-temperature end member(s). MELTS calculations suggest that it is also a robust feature of the more complex multicomponent peridotite system. The productivity function for fractional fusion can be more complex, but it is also likely to have a concave upward shape at low degrees of melting of relatively fertile peridotite. Note that productivity during fractional fusion of fertile peridotite, although lower than that of batch fusion at low melt fraction (but not exactly at the solidus, where they must be equal), is predicted to be comparable to that of batch fusion after several percent melting (figures 7a and 7b; see also Hirschmann et al. 1994). Although contrary to most previous speculations (Niu & Batiza 1991; Langmuir et al. 1992), recent experimental work appears to confirm this prediction (Hirose & Kawamura 1994).

5. Conclusions

There is no thermodynamic basis for assuming a constant rate of melt generation during isentropic depressurization. Even in a simple one-component system, the isentropic productivity depends on $1/T$ and the slope of the solidus, leading to increasing productivity with progressive melting (i.e. the melt fraction versus pressure function is concave up). Although other parameters appear in the general expression for isentropic productivity in multicomponent systems, the most important factors are the slopes of equal melt-fraction contours, $(\partial T/\partial P)_F$, and the isobaric productivity, $(\partial F/\partial T)_P$, both of which can be determined, in principle, from relatively straightforward phase equilibrium experiments. The isobaric productivity is the principal source of the variability of productivity during isentropic melting of peridotite, and it can be reduced to a simple statement of mass balance if the compositions of coexisting melt and residue are known. At low melt fractions, changes in the isobaric melt productivity are dominated by the decrease with progressive melting in the compositional difference between liquid and bulk residual solids.

Several authors have constructed models of peridotite melting in which $(\partial F/\partial T)_P$ is initially very high and decreases with progressive melting, based largely on analogy with low variance melting in simple systems (e.g. at a eutectic or peritectic). Thermodynamic modelling using MELTS, however, does not predict such behaviour. In contrast, initial liquids are predicted to differ significantly in composition from the coexisting bulk solid and to move closer in composition to the residue with progressive melting, leading to low productivity at the solidus and increases in productivity with increasing melt fraction (i.e. the same variations in productivity found for analogous simple systems). The same effects carry over into isentropic productivity. There are additional complexities related to phase changes and phase exhaustions, but their impact on isentropic productivity can also be understood by examining these same effects.

In summary, analysis of simple systems and thermodynamic calculations on complex peridotite compositions lead us to predict that isentropic melting of typical mantle peridotites will be characterized by an overall increase in isentropic productivity with melt fraction in the batch melting case, punctuated by drops in productivity upon exhaustion of each phase from the residue. The productivity function for fractional fusion can be more complex, but we predict that the concave upward shape of the melt fraction versus pressure curve predicted for the batch fusion case is also likely to be a characteristic of low degrees of fractional melting of relatively fertile peridotite.

The authors are grateful to Mark Ghiorso and Richard Sack, the authors of MELTS, for permission to play with their code and suit it to our needs. Mike O'Hara provided a helpful review and much important devil's advocacy. This work was supported by NSF grants OCE-9504517, EAR-9219899 and OCE-9314505. This is Division of Geological and Planetary Sciences contribution 5703.

Appendix A.

Here we derive a general expression for isentropic and for incrementally isentropic melt productivity in multicomponent systems. We consider only processes that obey the restriction

$$\mathrm{d}\boldsymbol{S} = S^{\mathrm{l}}\,\mathrm{d}\boldsymbol{M} \tag{A 1}$$

where $\boldsymbol{S}$ is the extensive entropy of the system, S^{l} is the specific entropy of the liquid phase, and $\boldsymbol{M}$ is the mass of the system. We also require that the mass of liquid in the source region, $\boldsymbol{M}^{\mathrm{l}}$, be a function only of the mass of solid in the source region,

$\boldsymbol{M}^s$, and constants such that

$$\frac{\mathrm{d}\boldsymbol{M}^1}{\mathrm{d}\boldsymbol{M}^s} = \left(\frac{\partial \boldsymbol{M}^1}{\partial \boldsymbol{M}^s}\right)_Y \quad \text{and} \quad \left(\frac{\partial \boldsymbol{M}^1}{\partial Y}\right)_{\boldsymbol{M}^s} = 0 \tag{A 2}$$

for any variable Y. For multicomponent systems, the superscript s refers to bulk properties of the polymineralic solid assemblage. These constraints limit the processes to those for which entropy change of the system only occurs by extraction of melt and for which there is a strict coupling between melt production and melt extraction. For example, for batch fusion, $\mathrm{d}\boldsymbol{M} = 0$ (i.e. $\mathrm{d}\boldsymbol{M}^1 = -\mathrm{d}\boldsymbol{M}^s$), so equation (A 1) means that the process is isentropic. For fractional fusion, the mass of the system decreases due to removal of liquid (i.e. $\mathrm{d}\boldsymbol{M} = \mathrm{d}\boldsymbol{M}^s$), and the entropy of the system decreases by the amount carried away by the liquid; since this is the only way in which the entropy of the system changes, fractional fusion subject to the constraint of equation (A 1) can be envisioned as a series of infinitesimal increments of isentropic fusion followed by complete melt removal.

Given equations (A 1) and (A 2), the changes in the state of the system are entirely determined by two variables, so we can write the total differential of $\boldsymbol{S}$ in terms of P and $\boldsymbol{M}^s$:

$$\mathrm{d}\boldsymbol{S} = \left(\frac{\partial \boldsymbol{S}}{\partial P}\right)_{\boldsymbol{M}^s} \mathrm{d}P + \left(\frac{\partial \boldsymbol{S}}{\partial \boldsymbol{M}^s}\right)_P \mathrm{d}\boldsymbol{M}^s. \tag{A 3}$$

Since $\boldsymbol{M} = \boldsymbol{M}^1 + \boldsymbol{M}^s$, equations (A 1) and (A 3) lead to

$$\left(\frac{\partial \boldsymbol{S}}{\partial P}\right)_{\boldsymbol{M}^s} \mathrm{d}P + \left(\frac{\partial \boldsymbol{S}}{\partial \boldsymbol{M}^s}\right)_P \mathrm{d}\boldsymbol{M}^s = S^1 \mathrm{d}\boldsymbol{M}^1 + S^1 \mathrm{d}\boldsymbol{M}^s. \tag{A 4}$$

Dividing equation (A 4) by $\mathrm{d}P$, rearranging, and applying the chain rule

$$\frac{\mathrm{d}\boldsymbol{M}^1}{\mathrm{d}P} = \frac{\mathrm{d}\boldsymbol{M}^1}{\mathrm{d}\boldsymbol{M}^s} \frac{\mathrm{d}\boldsymbol{M}^s}{\mathrm{d}P} \tag{A 5}$$

leads to an expression for the change in solid mass with pressure:

$$\frac{\mathrm{d}\boldsymbol{M}^s}{\mathrm{d}P} = \left(\frac{\partial \boldsymbol{S}}{\partial P}\right)_{\boldsymbol{M}^s} \bigg/ \left[S^1 + S^1 \frac{\mathrm{d}\boldsymbol{M}^1}{\mathrm{d}\boldsymbol{M}^s} - \left(\frac{\partial \boldsymbol{S}}{\partial \boldsymbol{M}^s}\right)_P\right]. \tag{A 6}$$

We now evaluate the partial derivatives that appear in equation (A 6). We differentiate

$$\boldsymbol{S} = M^s S^s + M^1 S^1 \tag{A 7}$$

to obtain

$$\left(\frac{\partial \boldsymbol{S}}{\partial P}\right)_{\boldsymbol{M}^s} = M^s \left(\frac{\partial S^s}{\partial P}\right)_{\boldsymbol{M}^s} + M^1 \left(\frac{\partial S^1}{\partial P}\right)_{\boldsymbol{M}^s} + S^1 \left(\frac{\partial \boldsymbol{M}^1}{\partial P}\right)_{\boldsymbol{M}^s}. \tag{A 8}$$

The restriction, equation (A 2), causes the last term in equation (A 8) to vanish. The evaluation of the remaining terms in equation (A 8) is related to equations (3.6 a) and (3.6 b) in the text, except that partial derivatives at constant $\boldsymbol{M}^s$ appear in place of total derivatives along the two-phase boundary, and more importantly we must now include derivatives that describe changes in the compositions of the phases:

$$\mathrm{d}S^1 = \left(\frac{\partial S^1}{\partial T}\right)_{P,X^1} \mathrm{d}T + \left(\frac{\partial S^1}{\partial P}\right)_{T,X^1} \mathrm{d}P + \sum_{i=1}^{n^1} \left(\frac{\partial S^1}{\partial X_i^1}\right)_{T,P,X_{j \neq i}^1} \mathrm{d}X_i^1, \tag{A 9 a}$$

$$dS^s = \left(\frac{\partial S^s}{\partial T}\right)_{P,X^s} dT + \left(\frac{\partial S^s}{\partial P}\right)_{T,X^s} dP + \sum_{k=1}^{n^s}\left[S^k\, d\gamma^k + \gamma^k \sum_{i=1}^{n^k} \left(\frac{\partial S^k}{\partial X_i^k}\right)_{T,P,X_{j\neq i}^k} dX_i^k\right], \quad \text{(A 9 b)}$$

where X_i^l is the mass fraction of component i in the n^l component liquid phase and

$$\left(\frac{\partial S^l}{\partial X_i^l}\right)_{T,P,X_{j\neq i}^l}$$

should be recognized as the partial specific entropy of component i in the liquid. In (A 9 b), S^s represents a weighted sum over n^s solid phases, γ^k is the mass fraction of the kth solid phase in the bulk assemblage, S^k is the specific entropy of the kth phase, X_i^k is the mass fraction of component i in the kth solid phase of n^k components, and

$$\left(\frac{\partial S^k}{\partial X_i^k}\right)_{T,P,X_{j\neq i}^k}$$

is the partial specific entropy of component i in solid phase k. All these new quantities are, in general, functions of temperature, pressure and bulk composition. For brevity in what follows we define the entire last term in (A 9 a) as dS_X^l and the entire last term in (A 9 b) as dS_X^s, since they represent the changes in S^l and S^s due to compositional and/or modal changes in the liquid and bulk solid, respectively. We note that these terms have been neglected without comments in all previous treatments of isentropic melting of which we are aware. In most cases, first-order approximations of productivity remain reasonable when dS_X^l and dS_X^s are neglected.

From (A 9 a) and (A 9 b) we can obtain

$$\left(\frac{\partial S^l}{\partial P}\right)_{M^s} = \frac{C_p^l}{T}\left(\frac{\partial T}{\partial P}\right)_{M^s} - V^l\alpha^l + \left(\frac{\partial S_X^l}{\partial P}\right)_{M^s}, \quad \text{(A 9 c)}$$

$$\left(\frac{\partial S^s}{\partial P}\right)_{M^s} = \frac{C_p^s}{T}\left(\frac{\partial T}{\partial P}\right)_{M^s} - V^s\alpha^s + \left(\frac{\partial S_X^s}{\partial P}\right)_{M^s}, \quad \text{(A 9 d)}$$

where the last terms are abbreviations for sums similar to those in (A 9 a) and (A 9 b) except $(\partial X_i^l/\partial P)_{M^s}$ is substituted for dX_i^l, $(\partial \gamma^k/\partial P)_{M^s}$ for $d\gamma^k$, and $(\partial X_i^k/\partial P)_{M^s}$ for dX_i^k. Using the intensive variable $f = M^l/(M^l + M^s)$ for the mass fraction of liquid in the system and combining equations (A 8), (A 9 c) and (A 9 d) leads to

$$\left(\frac{\partial S}{\partial P}\right)_{M^s} = (M^s + M^l)\left(\frac{C_p^s + f(C_p^l - C_p^s)}{T}\right)\left(\frac{\partial T}{\partial P}\right)_{M^s}$$
$$-[V^s\alpha^s + f(V^l\alpha^l - V^s\alpha^s)] + f\left(\frac{\partial S_X^l}{\partial P}\right)_{M^s} + (1-f)\left(\frac{\partial S_X^s}{\partial P}\right)_{M^s}, \quad \text{(A 10)}$$

and below for brevity we will use the definition

$$\left(\frac{\partial S_X}{\partial P}\right)_{M^s} \equiv f\left(\frac{\partial S_X^l}{\partial P}\right)_{M^s} + (1-f)\left(\frac{\partial S_X^s}{\partial P}\right)_{M^s}.$$

Next we take the partial derivative of equation (A 7) with respect to M^s at constant P:

$$\left(\frac{\partial S}{\partial M^s}\right)_P = S^s + S^l\left(\frac{\partial M^l}{\partial M^s}\right)_P + M^s\left(\frac{\partial S^s}{\partial M^s}\right)_P + M^l\left(\frac{\partial S^l}{\partial M^s}\right)_P. \quad \text{(A 11)}$$

To simplify equation (A 11) we can obtain from (A 9a) and (A 9b):

$$\left.\begin{array}{l}\left(\dfrac{\partial S^{\rm l}}{\partial \boldsymbol{M}^{\rm s}}\right)_P = \dfrac{C_p^{\rm l}}{T}\left(\dfrac{\partial T}{\partial \boldsymbol{M}^{\rm s}}\right)_P + \left(\dfrac{\partial S_X^{\rm l}}{\partial \boldsymbol{M}^{\rm s}}\right)_P \\[2ex] \left(\dfrac{\partial S^{\rm s}}{\partial \boldsymbol{M}^{\rm s}}\right)_P - \dfrac{C_p^{\rm s}}{T}\left(\dfrac{\partial T}{\partial \boldsymbol{M}^{\rm s}}\right)_P + \left(\dfrac{\partial S_X^{\rm s}}{\partial \boldsymbol{M}^{\rm s}}\right)_P,\end{array}\right\} \qquad (\text{A}\,12)$$

where again the last terms in each expression are defined similarly to the last terms in (A 9a) and (A 9b) except that partial derivatives with respect to $\boldsymbol{M}^{\rm s}$ at constant P replace all the total differentials, which together with the definition of f and (A 2) lead to

$$\left(\dfrac{\partial S}{\partial \boldsymbol{M}^{\rm s}}\right)_P = S^{\rm s} + S^{\rm l}\dfrac{\partial \boldsymbol{M}^{\rm l}}{\partial \boldsymbol{M}^{\rm s}}$$

$$+ (\boldsymbol{M}^{\rm s} + \boldsymbol{M}^{\rm l})\left(\dfrac{C_p^{\rm s} + f(C_p^{\rm l} - C_p^{\rm s})}{T\,(\partial \boldsymbol{M}^{\rm s}/\partial T)_P} + f\left(\dfrac{\partial S_X^{\rm l}}{\partial \boldsymbol{M}^{\rm s}}\right)_P + (1-f)\left(\dfrac{\partial S_X^{\rm s}}{\partial \boldsymbol{M}^{\rm s}}\right)_P\right), \qquad (\text{A}\,13)$$

and for further brevity we define

$$\left(\dfrac{\partial S_X}{\partial \boldsymbol{M}^{\rm s}}\right)_P \equiv f\left(\dfrac{\partial S_X^{\rm l}}{\partial \boldsymbol{M}^{\rm s}}\right)_P + (1-f)\left(\dfrac{\partial S_X^{\rm s}}{\partial \boldsymbol{M}^{\rm s}}\right)_P.$$

We now substitute equations (A 10) and (A 13) into (A 6) to obtain our final expression in terms of extensive mass variables:

$$-\dfrac{\mathrm{d}\boldsymbol{M}^{\rm s}}{\mathrm{d}P} = \dfrac{\dfrac{C_p^{\rm s} + f(C_p^{\rm l} - C_p^{\rm s})}{T}\left(\dfrac{\partial T}{\partial P}\right)_{\boldsymbol{M}^{\rm s}} - [V^{\rm s}\alpha^{\rm s} + f(V^{\rm l}\alpha^{\rm l} - V^{\rm s}\alpha^{\rm s})] + \left(\dfrac{\partial S_X}{\partial P}\right)_{\boldsymbol{M}^{\rm s}}}{\left(\dfrac{C_p^{\rm s} + f(C_p^{\rm l} - C_p^{\rm s})}{T\,(\partial \boldsymbol{M}^{\rm s}/\partial T)_P}\right) - \dfrac{(S^{\rm l} - S^{\rm s})}{(\boldsymbol{M}^{\rm s} + \boldsymbol{M}^{\rm l})} + \left(\dfrac{\partial S_X}{\partial \boldsymbol{M}^{\rm s}}\right)_P}, \qquad (\text{A}\,14)$$

where we retain $S^{\rm l} - S^{\rm s}$ and $(\partial S_x/\partial \boldsymbol{M}^{\rm s})_P$ instead of attempting to define the last two terms in the denominator as $\Delta S_{\rm fus}$ since this does not correspond to the common understanding of the meaning of $\Delta S_{\rm fus}$. Now the definition of $F = 1 - (\boldsymbol{M}^{\rm s}/\boldsymbol{M}^0)$, the fraction of the initial solid mass that has been melted, gives

$$\dfrac{\mathrm{d}F}{\mathrm{d}P} = -\dfrac{1}{\boldsymbol{M}^0}\dfrac{\mathrm{d}\boldsymbol{M}^{\rm s}}{\mathrm{d}P}, \quad \left(\dfrac{\partial F}{\partial Y}\right)_P = -\dfrac{1}{\boldsymbol{M}^0}\left(\dfrac{\partial \boldsymbol{M}^{\rm s}}{\partial Y}\right)_P \quad \text{and} \quad \left(\dfrac{\partial Y}{\partial P}\right)_F = \left(\dfrac{\partial Y}{\partial P}\right)_{\boldsymbol{M}^{\rm s}}, \qquad (\text{A}\,15)$$

for any variable Y (including notably T and S_X, which can be shown using the definitions above), which relations together with the equation $(1-F)/(1-f) = (\boldsymbol{M}^{\rm s} + \boldsymbol{M}^{\rm l})/\boldsymbol{M}^0$ allow us to eliminate all extensive variables from equation (A 14) and produce our final productivity equation:

$$-\dfrac{\mathrm{d}F}{\mathrm{d}P} = \dfrac{\left(\dfrac{C_p^{\rm s} + f(C_p^{\rm l} - C_p^{\rm s})}{T}\left(\dfrac{\partial T}{\partial P}\right)_F - [V^{\rm s}\alpha^{\rm s} + f(V^{\rm l}\alpha^{\rm l} - V^{\rm s}\alpha^{\rm s})] + \left(\dfrac{\partial S_X}{\partial P}\right)_F\right)}{\left(\dfrac{C_p^{\rm s} + f(C_p^{\rm l} - C_p^{\rm s})}{T\,(\partial F/\partial T)_P}\right) + \dfrac{(1-f)}{(1-F)}(S^{\rm l} - S^{\rm s}) + \left(\dfrac{\partial S_X}{\partial F}\right)_P}. \qquad (\text{A}\,16)$$

Equation (A 16) can be simplified directly for batch and fractional melting by taking $F = f$ and $f = 0$, respectively. Furthermore, equation (A 16) reduces to the correct form (equation (3.13)) for one-component systems, where (i) it is straightforward to set $\Delta S_{\text{fus}} = (S^l - S^s)$ and there are no S_X terms since composition is constant and only one solid phase can participate in melting except at an invariant point, (ii) the last two terms in equation (A 11) and hence the first term in the denominator of equation (A 16) vanish since $(\partial T/\partial F)_P = 0$ when isobaric melting takes place at a unique temperature, and (iii) $(\partial T/\partial P)_F = (dT/dP)_{2\phi}$ since all melting is restricted to the univariant two-phase curve. Finally, equation (A 16) can be reduced to the constant coefficient one-component case (equation (3.12)) since

$$\Delta S_{\text{fus}}(P,T) = \Delta S_{\text{fus}}(P_0,T_0) + \int_{T_0}^{T} \frac{(C_p^l - C_p^s)}{T} dT + \int_{P_0}^{P} (V^l\alpha^l - V^s\alpha^s) dP \quad \text{(A 17)}$$

means that constant ΔS_{fus} also requires $(C_p^l - C_p^s) = 0$ and $(V^l\alpha^l - V^s\alpha^s) = 0$.

We also include here the expression for calculating the P–T path of upwelling material undergoing any of the melting processes described by equations (A 1) and (A 2). For batch melting it is simple to show

$$\left(\frac{\partial T}{\partial P}\right)_S = \left(\frac{\partial T}{\partial P}\right)_F + \left(\frac{\partial F}{\partial P}\right)_S \bigg/ \left(\frac{\partial F}{\partial T}\right)_P \quad \text{(A 18)}$$

(e.g. Albarède 1992). A simple derivation beginning from the total differential of M^s expressed in terms of P and T followed by substitution of F for M^s leads to the corresponding result for (dT/dP) subject to equation (A 1) rather than constant S:

$$\frac{dT}{dP} = \left(\frac{\partial T}{\partial P}\right)_F + \frac{dF}{dP} \bigg/ \left(\frac{\partial F}{\partial T}\right)_P. \quad \text{(A 19)}$$

References

Albarède, F. 1983 Limitations thermiques à l'ascension des magmas hydratés. *C. R. Acad. Sci. Paris* **296**, 1441–1444.

Albarède, F. 1992 How deep do common basaltic magmas form and differentiate? *J. Geophys. Res.* **97**, 10 997–11 009.

Asimow, P. D., Hirschmann, M. M., Ghiorso, M. S., O'Hara, M. J. & Stolper, E. M. 1995a The effect of pressure-induced solid–solid phase transitions on decompression melting of the mantle. *Geochim. Cosmochim. Acta* **59**, 4489–4506.

Asimow, P. D., Hirschmann, M. M., Ghiorso, M. S. & Stolper, E. M. 1995b Isentropic melting processes in the mantle. In *Plume 2 Conf.* (ed. D. L. Anderson et al.), pp. 12–14. Terra Nostra 3/1995. Bonn: Alfred–Wegener–Stiftung.

Baker, M. B., Hirschmann, M. M., Ghiorso, M. S. & Stolper, E. M. 1995 Compositions of low-degree partial melts of peridotite: Results from experiments and thermodynamic calculations. *Nature* **375**, 308–311.

Baker, M. B. & Stolper, E. M. 1994 Determining the composition of high-pressure mantle melts using diamond aggregates. *Geochim. Cosmochim. Acta* **58**, 2811–2827.

Berman, R. G. 1988 Internally-consistent thermodynamic data for minerals in the system Na_2O–K_2O–CaO–MgO–FeO–Fe_2O_3–Al_2O_3–SiO_2–TiO_2–H_2O–CO_2: representation, estimation, and high temperature extrapolations. *J. Petrol.* **89**, 168–183.

Biggar, G. M. & O'Hara, M. J. 1969 Solid solutions at atmospheric pressure in the system CaO–MgO–SiO_2 with special reference to the instabilities of diopside, akermanite, and monticellite. *Prog. Exp. Pet.* pp. 89–96. London: NERC.

Cawthorn, R. G. 1975 Degrees of melting in mantle diapirs and the origin of ultrabasic liquids. *Earth Planet. Sci. Lett.* **27**, 113–120.

Ghiorso, M. S. & Sack, R. O. 1995 Chemical mass transfer in magmatic processes. IV. A revised and internally consistent thermodynamic model for the interpolation and extrapolation of liquid-solid equilibria in magmatic systems at elevated temperatures and pressures. *Contrib. Mineral. Petrol.* **119**, 197–212.

Hart, S. R. 1993 Equilibration during mantle melting: A fractal tree model. *Proc. Natl. Acad. Sci.* **90**, 11 914–11 918.

Hart, S. R. & Zindler, A. 1986 In search of a bulk-earth composition. *Chem. Geol.* **57**, 247–267.

Hess, P. C. 1992 Phase equilibria constraints on the origin of ocean floor basalt. In *Mantle flow and melt generation at mid-ocean ridges* (ed. J. Phipps Morgan *et al.*), pp. 67-102. (Geophysical Monograph 71.) Washington, DC: AGU.

Hirose, K. & Kawamura, K. 1994 A new experimental approach for incremental batch melting of peridotite at 1.5 GPa. *Geophys. Res. Lett.* **21**, 2139–2142.

Hirschmann, M. M., Stolper, E. M. & Ghiorso, M. S. 1994 Perspectives on shallow mantle melting from thermodynamic calculations. *Min. Mag.* A **58**, 418–419.

Iwamori, H., McKenzie, D. P. & Takahashi, E. 1995 Melt generation by isentropic mantle upwelling. *Earth Planet. Sci. Lett.* **134**, 253–266.

Kinzler, R. J. & Grove, T. L. 1992 Primary magmas of mid-ocean ridge basalts. 2. Applications. *J. Geophys. Res.* **97**, 6907–6926.

Klein, E. M. & Langmuir, C. H. 1987 Global correlations of ocean ridge basalt chemistry with axial depth and crustal thickness. *J. Geophys. Res.* **92**, 8089–8115.

Kushiro, I. 1972 Determination of liquidus relations in synthetic silicate systems with electron probe analyses: The system forsterite–diopside–silica at 1 atm. *Am. Mineral.* **57**, 51–74.

Kushiro, I. 1996 Partial melting of a fertile mantle peridotite at high pressure: and experimental study using aggregates of diamond. In *Earth processes: reading the isotopic clock* (ed. A. Basu & S. Hart), pp. 109–122. (Geophysical Monograph 95.) Washington, DC: AGU.

Lange, R. L. & Carmichael, I. S. E. 1990 Thermodynamic properties of silicate liquids with emphasis on density, thermal expansion and compressibility. In *Modern methods of igneous petrology: understanding magmatic processes* (ed. J. Nicholls & J. K. Russell), Rev., Mineral., no. 24, pp. 25–64. Washington, DC: Mineralogical Society of America.

Langmuir, C. H., Bender, J. F., Bence, A. E., Hanson, G. N. & Taylor S. R. 1977 Petrogenesis of basalts from the FAMOUS area: Mid-Atlantic Ridge. *Earth Planet. Sci. Lett.* **36**, 133–156.

Langmuir, C. H., Klein, E. M. & Plank, T. 1992 Petrological systematics of mid-ocean ridge basalts: constraints on melt generation beneath ocean ridges. In *Mantle flow and melt generation at mid-ocean ridges* (ed. J. Phipps Morgan *et al.*), pp. 183–280. (Geophysical Monograph 71.) Washington, DC: AGU.

Longhi, J. 1992 Origin of green glass magmas by polybaric fractional fusion. *Proc. Lunar Planet. Sci.* **22**, 343–353.

McKenzie, D. P. 1984 The generation and compaction of partial melts. *J. Petrol.* **25**, 713–765.

McKenzie, D. P. & Bickle, M. J. 1988 The volume and composition of melt generated by extension of the lithosphere. *J. Petrol.* **29**, 625–679.

McKenzie, D. P. & Bickle, M. J. 1990 A eutectic parameterization of mantle melting. *J. Phys. Earth* **38**, 511–515.

McKenzie, D. P. & O'Nions, R. K. 1991 Partial melt distributions from inversion of rare earth element concentrations. *J. Petrol.* **32**, 1021–1091.

Miller, G. H., Stolper, E. M. & Ahrens, T. H. 1991 The equation of state of a molten komatiite. 2. Applications to komatiite petrogenesis and the Hadean mantle. *J. Geophys. Res.* **96**, 11 849–11 864.

Niu, Y. & Batiza, R. 1991 An empirical method for calculating melt compositions produced beneath mid-ocean ridges: application to axis and off-axis (seamounts) melting. *J. Geophys. Res.* **96**, 21 753–21 777.

Ramberg, H. 1972 Mantle diapirism and its tectonic and magmagenic consequences. *Phys. Earth. Planet. Interiors* **5**, 45–60.

Rigden, S. M., Ahrens, T. J. & Stolper, E. M. 1989 High-pressure equation of state of molten anorthite and diopside. *J. Geophys. Res.* **94**, 9508–9522.

Rumble, D. 1976 The adiabatic gradient and adiabatic compressibility. *Carn. Inst. Wash. Yb.* **75**, 651–655.

Scott, D. R. & Stevenson, D. J. 1989 A self-consistent model of melting, magma migration and buoyancy-driven circulation beneath mid-ocean ridges. *J. Geophys. Res.* **94**, 2973–2988.

Solomatov, V. S. & Stevenson, D. J. 1994 Can sharp seismic discontinuities be caused by non-equilibrium phase transformations? *Earth Planet. Sci. Lett.* **125**, 267–279.

Sparks, D. W. & Parmentier, E. M. 1991 Melt extraction from the mantle beneath spreading centers. *Earth Planet. Sci. Lett.* **105**, 368–378.

Spiegelman, M. 1993 Flow in deformable porous media. 2. Numerical analysis—the relationship between shock waves and solitary waves. *J. Fluid Mech.* **247**, 39–63.

Stebbins, J. F., Carmichael, I. S. E. & Weill, D. J. 1983 The high-temperature liquid and glass heat contents and heats of fusion of diopside, albite, sanidine, and nepheline. *Am. Mineral.* **68**, 717–730.

Takahashi, E. 1986 Melting of a dry peridotite KLB-1 up to 14 GPa: implications on the origin of peridotitic upper mantle. *J. Geophys. Res.* **91**, 9367–9380.

Takahashi, E. & Kushiro, I. 1983 Melting of a dry peridotite at high pressures and basalt magma genesis. *Am. Mineral.* **68**, 859–879.

Turcotte, D. L. & Ahern, J. L. 1978 A porous flow model for magma migration in the asthenosphere. *J. Geophys. Res.* **83**, 767–772.

Verhoogen, J. 1965 Phase changes and convection in the Earth's mantle. *Phil. Trans. R. Soc. Lond.* A **258**, 276–283.

A review of melt migration processes in the adiabatically upwelling mantle beneath oceanic spreading ridges

By P. B. Kelemen[1], G. Hirth[1], N. Shimizu[1], M. Spiegelman[2] and H. J. B. Dick[1]

[1] Woods Hole Oceanographic Institution, Woods Hole, MA 02543, USA
[2] Lamont Doherty Earth Observatory, Palisades, NY 10964, USA

We review physical and chemical constraints on the mechanisms of melt extraction from the mantle beneath mid-ocean ridges. Compositional constraints from MORB and abyssal peridotite are summarized, followed by observations of melt extraction features in the mantle, and constraints from the physical properties of partially molten peridotite. We address two main issues. (1) To what extent is melting 'near-fractional', with low porosities in the source and chemical isolation of ascending melt? To what extent are other processes, loosely termed reactive flow, important in MORB genesis? (2) Where chemically isolated melt extraction is required, does this occur mainly in melt-filled fractures or in conduits of focused porous flow?

Reactive flow plays an important role, but somewhere in the upwelling mantle melting must be 'near fractional', with intergranular porosities less than 1%, and most melt extraction must be in isolated conduits. Two porosity models provide the best paradigm for this type of process. Field relationships and geochemical data show that replacive dunites mark conduits for focused, chemically isolated, porous flow of mid-ocean ridge basalt (MORB) in the upwelling mantle. By contrast, pyroxenite and gabbro dikes are lithospheric features; they do not represent conduits for melt extraction from the upwelling mantle. Thus, preserved melt extraction features do not require hydrofracture in the melting region. However, field evidence does not rule out hydrofracture.

Predicted porous flow velocities satisfy ^{230}Th excess constraints ($ca.$ 1 m yr^{-1}), provided melt extraction occurs in porous conduits rather than by diffuse flow, and melt-free, solid viscosity is less than $ca.$ 10^{20} Pa s. Melt velocities of $ca.$ 50 m yr^{-1} are inferred from patterns of post-glacial volcanism in Iceland and from ^{226}Ra excess. If these inferences are correct, minimum conditions for hydrofracture may be reached in the shallowest part of melting region beneath ridges. However, necessary high porosities can only be attained within pre-existing conduits for focused porous flow. Alternatively, the requirement for high melt velocity could be satisfied in melt-filled tubes formed by dissolution or mechanical instabilities.

Melt-filled fractures or tubes, if they form, are probably closed at the top and bottom, limited in size by the supply of melt. Therefore, to satisfy the requirements for geochemical isolation from surrounding peridotite, melt-filled conduits may be surrounded by a dunite zone. Furthermore, individual melt-filled voids probably contain too little melt to form sufficient dunite by reaction, suggesting that dunite zones must be present before melt extraction in fractures or tubes.

1. Introduction

(a) Residual peridotites are not in equilibrium with MORB

MORB forms by partial melting of adiabatically decompressing mantle peridotite beneath spreading ridges (Allègre et al. 1973; McKenzie & Bickle 1988). Nevertheless, MORB is not in equilibrium with residual peridotite (harzburgite and lherzolite). For example, MORB is not saturated in orthopyroxene (Opx) at Moho pressures, whereas Opx is a major constituent of mantle peridotites dredged and drilled from the mid-ocean ridges (hereafter, 'abyssal peridotites') and mantle peridotites in ophiolites. Liquids parental to MORB are saturated with Opx at pressures greater than 8 kbar (O'Hara 1965; Stolper 1980; Elthon & Scarfe 1980), greater than 15 km below the Moho. In addition, most abyssal peridotites and mantle peridotites in ophiolites are strongly depleted in light rare earth elements (REE) and other highly incompatible elements, and are therefore far from trace element equilibrium with MORB (Johnson et al. 1990; Johnson & Dick 1992; Kelemen et al. 1995b; Dick & Natland 1996; Ross & Elthon 1997a,b). These observations indicate that MORB is a mixture that preserves evidence for partial melting of mantle peridotite over a range of pressures, much higher than the pressure at the base of the crust (Klein & Langmuir 1987; Salters & Hart 1989; Kinzler & Grove 1992).

(b) MOST melt extraction is in chemically isolated conduits of focused flow

Polybaric, near-fractional melting models closely approximate the composition of MORB and residual peridotites (Johnson et al. 1990). To produce large fractionations between light and heavy REE, and between elements such as Ti/Zr and Lu/Hf in residual peridotites, small melt fractions must be segregated from their sources and mixed to form MORB. To preserve these fractionations, and to produce disequilibrium between melt and Opx at low pressure, most melts must be transported to the crust without re-equilibration with surrounding peridotite. Diffuse porous flow of large amounts of melt through residual peridotite from the source to the crust would lead to extensive chemical reaction, decreasing or erasing trace element fractionation (Spiegelman & Kenyon 1992; Hart 1993; Iwamori 1993). Reaction would also bring derivative liquids close to low pressure Opx saturation (Kelemen 1990); such liquids cannot comprise a significant fraction of MORB. Thus, focused flow of melt in spatially restricted conduits is required for MORB extraction from the mantle. However, diffuse, reactive flow of small amounts of melt, at low melt/rock ratios, cannot be ruled out.

(c) Melt inclusions and magma conduits

Shimizu and co-workers discovered extensive trace element heterogeneity in glass inclusions hosted in magnesian olivine crystals within MORB. These include 'ultra-depleted melts' (UDM), silica-rich, Opx-saturated liquids that are strongly light REE depleted, and have high Ti/Zr, consistent with derivation by low pressure partial melting of the most depleted abyssal peridotite (Sobolev & Shimizu 1993). This observation substantiates the hypothesis that MORB is a mixture of melts derived from variably depleted sources at different depths beneath ridges.

Light REE enriched melts are also found in olivine-hosted glass inclusion suites (Sobolev & Shimizu 1994). These were originally believed to represent small degree melts of garnet lherzolite. Such an interpretation would place strong constraints on the nature of magma conduits beneath ridges. They would, in some instances, be required to retain unmixed melt fractions derived from the greatest depths. However, heavy REE ratios in the inclusions are inconsistent with derivation of melts directly from garnet lherzolite. Instead, Shimizu (1997) proposed that melts of garnet peridotite reacted with spinel peridotite to produce the observed liquid compositions, and Sobolev (1997) proposed that observed liquids are mixtures with an unusually large proportion of garnet lherzolite melts.

Alternatively, light REE enriched melt inclusions at slow spreading ridges can be explained by melt/rock reaction in the shallow mantle, without any involvement of garnet. Rising magma encounters the base of the conductive geotherm at about 20 km depth (Sleep 1975). Reaction between primitive MORB and depleted harzburgite under conditions of decreasing liquid mass can produce light REE enriched liquids with a flat heavy REE slope, the same as melt inclusions from the Mid-Atlantic Ridge (figure 1). This model is not unique, but we present it to emphasize that the presence of light REE enriched melt inclusions places few constraints on melt extraction beneath ridges.

(d) 'Near-fractional' melting models

As noted in §1b, geochemical data on abyssal peridotite have been explained as the result of fractional melting. However, there are significant discrepancies between clinopyroxene (Cpx) compositions predicted for pure fractional melting and observed incompatible element ratios in Cpx in abyssal peridotite (Johnson & Dick 1992; Sobolev & Shimizu 1993; Iwamori 1993). For instance, La/Sm, Ce/Sm and Zr/Ti ratios are higher, at a given Nd/Sm, than predicted by purely fractional melting. This discrepancy has been explained by 'near-fractional' melting processes characterized by a constant intergranular porosity in the rock during melting, and/or reactive porous flow during melt extraction. For example, in 'incremental melting', fixed fractions of melt are periodically extracted and melt extraction is chemically isolated. Johnson et al. (1990) estimated that melt fractions in this model must be less than 1%, assuming that the increment is the same throughout the melting region. In 'continuous' or 'dynamic melting', a threshold intergranular porosity must be achieved before melt extraction. With continued melting, this threshold porosity is retained in the rock, and melt extraction is chemically isolated (Langmuir et al. 1977; Maaloe & Scheie 1982; Johnson & Dick 1992; Sobolev & Shimizu 1993). Johnson & Dick (1992) found that a retained porosity of 0.1–0.5% fits the abyssal peridotite data best, assuming that the intergranular porosity is the same throughout the melting region. Models in which melt reacts or mixes with residual peridotite have also been proposed (Elthon 1992; Iwamori 1993) – again assuming that this amount is constant throughout the melting region. These models provide a close fit to the data provided that the amount of reaction or mixing is small.

(e) Not all melting is necessarily 'near-fractional'

To explore the flexibility of constraints from the abyssal peridotite data on melting and melt extraction, we investigated two models with non-constant intergranular porosity. In the first, melting occurs in a closed system (batch melting) until a melt fraction (porosity) of 7 wt% is reached; then the intergranular porosity drops to zero, and melting is fractional. This is an example of a model in which a percolation threshold must be achieved before melt is extracted. The threshold is higher in pyroxene-rich lherzolite in the initial stages of melting, and drops to zero in depleted

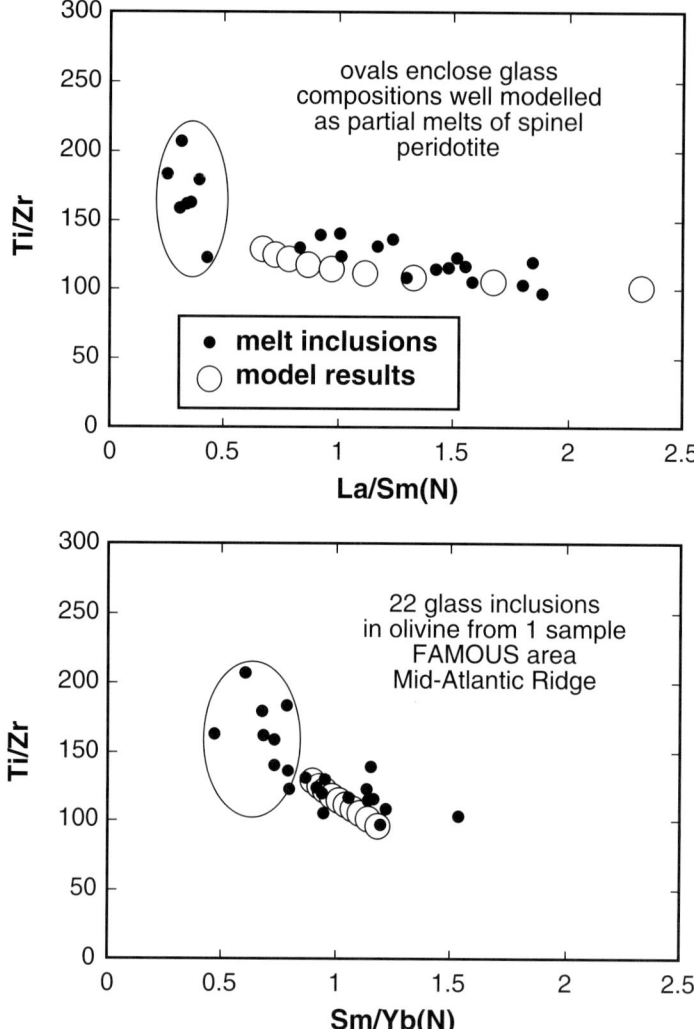

Figure 1. Melt/rock reaction between MORB and shallow mantle peridotite can form light rare earth element (REE) enriched liquids such as those in melt inclusions in olivine along the Mid-Atlantic Ridge (Shimizu 1997). To model melt/rock reaction, we used the DePaolo (1981) formalism, with a MORB initial liquid (chondrite normalized La, Ce, Nd, Sm, Eu, Gd, Dy, Er, Yb: 6, 7, 8, 9, 9.5, 10, 10, 10, 10), and a depleted harzburgite solid reactant (2×10^{-13}, 2×10^{-4}, 0.019, 0.18, 0.25, 0.27, 0.34, 0.39, 0.51) with 63 olivine: 23 orthopyroxene: 9 clinopyroxene: 5 spinel, produced by 15% fractional melting of MORB source mantle (melting model described in caption for figure 2). The mass of solid reactants/solid products, Ma/Mc, was set at 0.97, for liquid mass is decreasing slightly as a result of conductive cooling near the base of the lithosphere. Solid phase proportions were held constant. Results are plotted for values of F (fraction of initial liquid mass) from 1 to 0.1. Ovals enclose glass compositions well modelled as partial melts of spinel peridotite. The melt/rock reaction model accounts well for all other glass compositions.

harzburgites near the top of the melting column (Nicolas 1986). The correspondence of model results to the abyssal peridotite data is remarkably close (figure 2).

We were surprised by this result, and so tested an equally extreme model in which melting is fractional until 7 wt% melt is extracted, after which an additional 13%

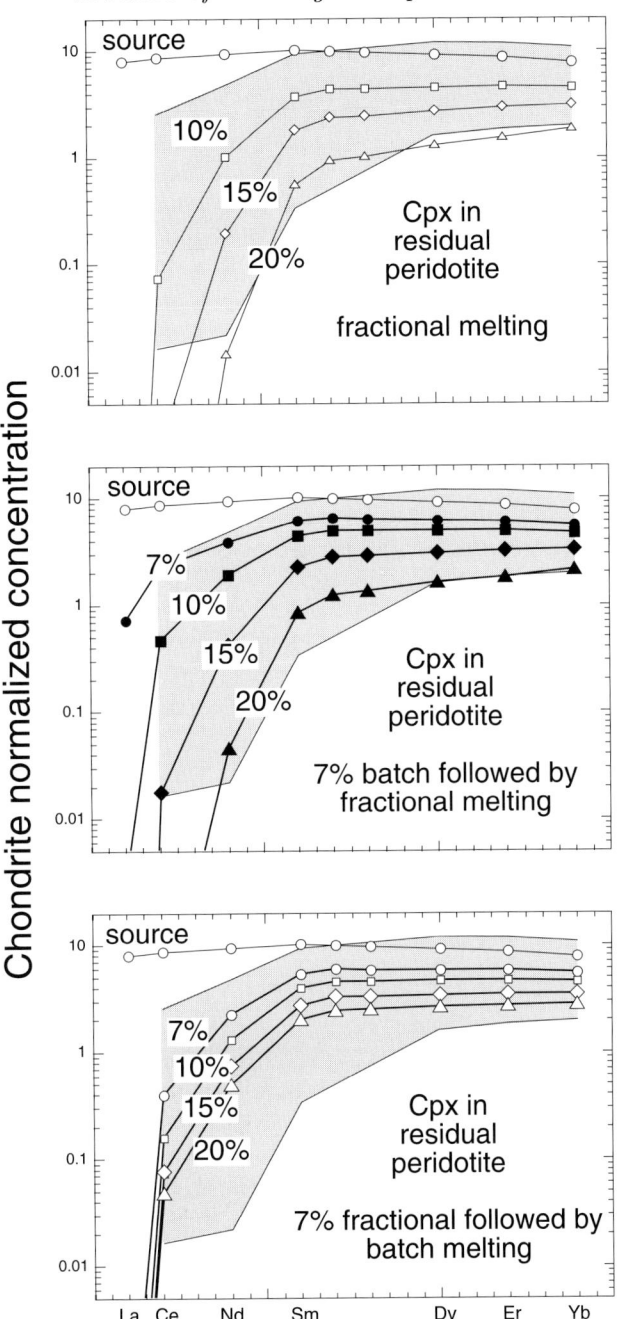

Figure 2. Calculated REE in clinopyroxene (Cpx) in residues of partial melting of mantle peridotite in spinel lherzolite facies, with various different combinations of batch and fractional melting. Grey shaded area illustrates the range of REE in Cpx in abyssal peridotites (Johnson et al. 1990; Johnson & Dick 1992). No unique combination of batch and fractional melting processes is required to account for the abyssal peridotite data. However, an interval of near-fractional melting, on the order of 3 wt%, is required to produce large light/heavy REE fractionation. Calculations used the methods of Gast (1968) and Shaw (1970), melt and source modes from Kelemen et al. (1992), partition coefficients compiled by Kelemen et al. (1993), and initial peridotite composition from Shimizu (1997), with $Yb = 2\times$ chondritic concentration.

batch melting occurs. This model also provides a good fit to the peridotite data. Our point is not that either model is correct, but to emphasize that if the porosity (or the amount of reactive porous flow) is variable, rather than constant over the entire melting column, there are a rich variety of melt extraction mechanisms that produce residues similar to abyssal peridotite. Therefore, the *maximum* intergranular porosity and the *integrated* amount of reactive flow during partial melting and melt transport is poorly constrained. However, Cpx compositions in abyssal peridotites require that, *at some point in the melting process, near-fractional melting did occur*. As a rule of thumb, the minimum amount of near-fractional melting, F, must be greater than or equal to the smallest crystal/liquid distribution coefficient for elements which are fractionated in residual peridotites during the melting process, such as Nd and Sm. Thus, ⩾3% near-fractional melting is apparently required to account for the abyssal peridotite data.

(f) U/Th isotopic disequilibrium and reactive flow at the base of the melting regime

U/Th isotopic disequilibrium in MORB indicates that melt extraction in the presence of garnet involves porous flow at intergranular porosities less than 0.2% (McKenzie 1985; Williams & Gill 1989; Beattie 1993*a*, *b*; Spiegelman & Elliot 1993; Iwamori 1994; Richardson & McKenzie 1994; Lundstrom *et al.* 1995). Since fractionation of highly incompatible elements such as U/Th requires very small intergranular porosities, melt must leave its source almost as fast as it forms. In near-fractional melting models, this constraint reduces the solid residence time for radioactive U. Since garnet is apparently required to hold more U than Th in the source, and virtually all the U is in the liquid after 1–2% melting, even in the presence of garnet, excesses of ^{230}Th are only produced in the initial stages of fractional melting. U/Th isotopic disequilibrium has also been explained by reactive porous flow over some portion of the melt extraction region (Spiegelman & Elliott 1993; Lundstrom *et al.* 1995). Reactive porous flow incorporates both small intergranular porosities and long ingrowth times in the source region. Reactive porous flow could produce ^{230}Th excesses anywhere that garnet is present in the melting region.

Hirth & Kohlstedt (1996) concluded that small amounts of H_2O-rich melt begin to form at depths of 120–70 km beneath mid-ocean ridges with 'normal' mantle potential temperatures (1300 °C). This is in addition to the possible formation of carbonate-rich melt at greater depth (Plank & Langmuir 1992, and references therein). The proportions of volatile-rich melts will be small, buffered by the limited available H and CO_2. Thus, small amounts of melt will be produced over a large depth interval, at *ca.* 0.1% kbar^{-1}. This hypothesized deep melting regime must be strikingly different from the anhdyrous melting regime that produces the bulk of MORB. It is plausible that reactive porous flow in the deep, volatile-rich melting regime produces the observed excesses of ^{230}Th in MORB.

(g) Reactive flow also occurs in the shallow mantle

Based on U/Th isotopic disequilibrium, reactive porous flow of melt may be important near the base of the melting regime (§ 1 *f*). In a similar way, Th/Ra disequilibrium data suggest that reactive flow may occur near the top of the melting regime. ^{226}Ra has a half life of about 1600 years. If ^{226}Ra disequilibrium were produced by melting of garnet lherzolite, rapid ascent of magma would be required, providing a strong constraint on transport models. However, reactive porous flow could produce ^{226}Ra isotopic disequilibrium in the absence of garnet, even at shallow depths, provid-

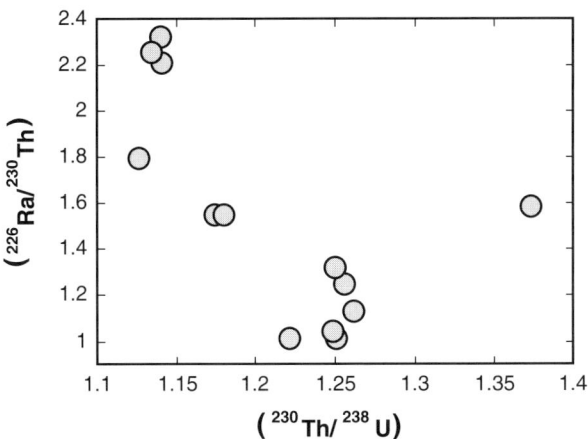

Figure 3. Plot of data for ^{226}Ra and ^{230}Th excesses in MORB. There is a negative correlation, which suggests that the excesses are produced by different processes and/or at different depths. Data are those compiled by Lundstrom et al. (1995) and Sims (1995).

ed that Ra is more incompatible than Th for spinel lherzolite/liquid (Ra partitioning data are not available). Thus, ^{226}Ra excess excess may be created in a different part of the melting regime than ^{230}Th excess. This inference is supported by data that show a negative correlation between ^{230}Th/^{238}U excess and ^{226}Ra/^{230}Th excess in MORB, implying that two separate geochemical regimes produce the two different excesses (figure 3).

An unequivocal example of reactive porous flow is the formation of plagioclase lherzolites sampled by dredging along mid-ocean ridges. Many plagioclase lherzolites cannot be entirely residual in origin, and can be interpreted as residual peridotites that were refertilized by crystallization of plagioclase, Cpx, and olivine from melt migrating by porous flow (Dick 1989; Elthon 1992). Similar 'impregnated peridotites' are common near the crust–mantle transition zone in the Oman ophiolite (Rabinowicz et al. 1987; Ceuleneer & Rabinowicz 1992; Boudier & Nicolas 1995).

Insight into the scale of reactive flow is provided by geochemical variability in abyssal and ophiolite peridotites. Average peridotite compositions on the 1000 km^2 scale follow predicted melting trends (figure 4). However, abyssal and ophiolite peridotites from on the 100 km^2 scale show 'decoupled' major and trace elements; locally, there is no correlation between trace element ratios and Cpx abundance. In general, decoupled major and trace element variation can result from (1) redistribution of major elements by dissolution and precipitation reactions (Quick 1981; Kelemen 1990; Kelemen et al. 1992), combined with (2) 'cryptic' metasomatism in which trace elements are redistributed by liquid in major element equilibrium with host peridotite (Menzies et al. 1985; Bodinier et al. 1990). Both of these processes involve liquid migrating by reactive porous flow. Additional hypotheses and evidence for reactive porous flow in ophiolite peridotites have been presented by Gregory (1984), Takazawa et al. (1992), Ozawa (1994), Ozawa & Shimizu (1995) and Quick & Gregory (1995).

(h) Geophysical data indicate that average porosity is low

The conclusion that intergranular porosity must be small over an interval of ⩾3% melting (§ 1 e) is supported by geophysical data. There are anomalously slow shear

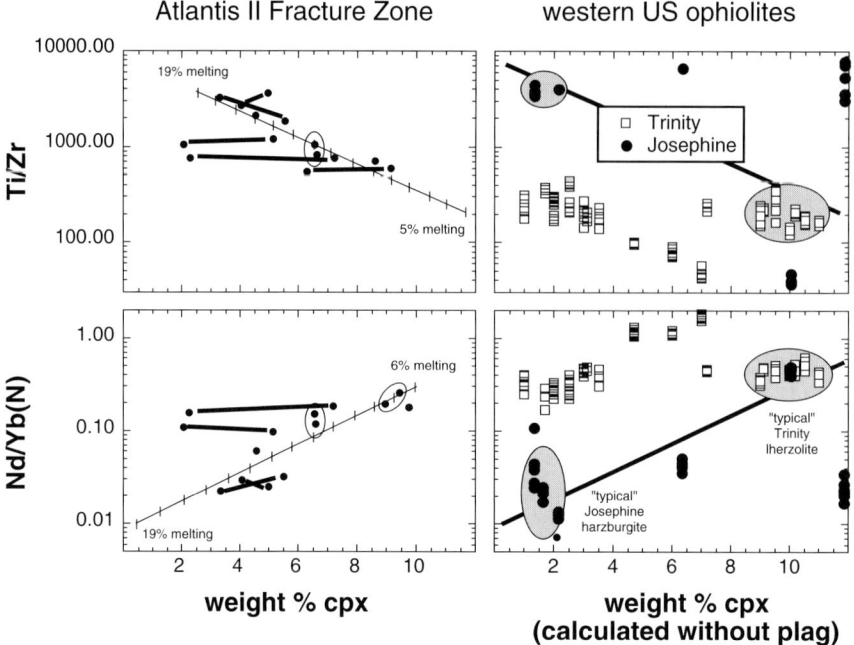

Figure 4. Trace element contents and proportions of Cpx in abyssal peridotites (Johnson & Dick 1992) and in mantle peridotites from the China Mountain area in the Trinity peridotite (Quick 1981a, b; Kelemen et al. 1992) and the western half of the Josephine peridotite north of the Oregon border (Dick 1976 and our unpublished data). For the abyssal peridotite data, samples obtained from the same dredge haul are connected by bold, straight lines or enclosed within ovals. For Trinity data, Cpx mode was calculated for plagioclase-free peridotite using plagioclase = Cpx + Opx + spinel − olivine. Curves on each diagram are residues of melting calculated by Johnson & Dick (1992). For both abyssal peridotites and ophiolites, the mode varies dramatically within a single dredge haul or peridotite massif (ca. 1 km^2 for dredge hauls, 100 km^2 for Trinity samples, 200 km^2 for Josephine samples). In contrast, trace element ratios are relatively constant on this scale. The wide spread in Cpx mode for the Trinity and Josephine reflects ubiquitous but volumetrically minor, centimetre- to metre-scale variation; 'typical' peridotites are relatively uniform within each massif, and the two massifs are distinctly different. If one were to 'dredge' the Trinity, one would likely obtain a lherzolite with mild light REE depletion. In contrast, dredging of the Josephine would likely yield a harzburgite with extreme light REE depletion.

wave velocities, attributable to the presence of melt, beneath the East Pacific Rise and the Mid-Atlantic Ridge (Forsyth 1992; Forsyth et al. 1996). However, the intergranular porosity is small, between 3 and 6.5% at 45 km if melt is in grain-boundary tubes, and less than 1% if melt is in grain boundary films. Since a substantial proportion of melt is present in films rather than tubes (Waff & Faul 1992; Jin et al. 1994; Hirth & Kohlstedt 1995a; Faul 1997), the interconnected, intergranular porosity is probably less than 3%. The shear wave data do not rule out the presence of additional melt in isolated pockets. However, analysis of gravity from the southern East Pacific Rise indicates that little melt is present in interconnected pores or isolated pockets (Magde et al. 1995). Thus, the geophysical data support the inference from geochemical observations that the average intergranular porosity in the melting column is small, and that melt extraction by porous flow on some length scale must be chemically isolated.

(i) Conclusion: two porosity models are needed

The foregoing sections present evidence for (1) chemically isolated melt transport by focused flow of melt in spatially restricted conduits (§ 1 b), and (2) diffuse, reactive porous flow (§§ 1 f, g) beneath spreading ridges. This apparent paradox can be understood in the context of two porosity models (Wang 1993). In these models, the bulk of the liquid flux is in highly permeable conduits, but a significant quantity of geologically important trace elements are derived from processes involving a much smaller flux of liquid moving by diffuse porous flow in the interstices between these conduits.

2. Melt migration features in the mantle section of ophiolites

Localized melt migration is essential in extraction of MORB from the mantle. In this section, we review data on localized melt migration features at the outcrop scale in ophiolites and their analogues in abyssal peridotite. We concentrate on data from the Oman ophiolite, because several observations demonstrate that it formed at an oceanic spreading centre. The crustal section includes a continuous layer of sheeted dikes. The main lava series and the dikes are tholeiitic lavas with REE and trace element contents similar to MORB (Alabaster *et al.* 1982). In detail, REE and other incompatible trace elements in the Oman lavas are lower, at a given Cr concentration, than in 'normal' MORB. However, the similar compositions of Oman lavas and MORB indicate that the processes that formed the ophiolite were similar to processes operating at mid-ocean ridges. On the basis of radiometric age data, the lack of crustal thickness variation, and other geological observations, it is probable that the ophiolite formed at a fast-spreading ridge (Tilton *et al.* 1981; Nicolas 1989).

(a) Mantle dunites

(i) Dunites are conduits for MORB transport in the asthenosphere

Dunites are rocks composed of more than 90% olivine, with lesser amounts of spinel and, locally, pyroxene. For brevity, we use only the term dunite to refer to such rocks, and the term peridotite to refer only to rocks with less than 90% olivine. 'Discordant' dunites are the most common magmatic feature in the mantle section of ophiolites. We distinguish between 'mantle dunites' and dunites in the crust–mantle transition zone of ophiolites, which may have a different origin. Mantle dunites comprise 5–15% of the exposed mantle section in Oman (Boudier & Coleman 1981; Lippard *et al.* 1986, table 3.6). They preserve sharp, irregular contacts that are locally discordant to banding and crystallographic lineation in the harzburgite (Lippard *et al.* 1986, p. 53). Because harzburgites and most dunites have a foliation parallel to the palaeo-Moho, it is inferred that both were transposed by corner flow beneath a spreading ridge. Thus, transposed dunites were formed in upwelling mantle beneath a ridge.

Geochemical data suggest that dunites mark conduits for focused flow of MORB through the adiabatically upwelling mantle. In Oman, a substantial proportion of dunites contain minor Cpx. Kelemen *et al.* (1995b) demonstrated that these Cpx have REE in equilibrium with liquids similar to MORB and to the dikes and lavas that form the upper crust of the ophiolite. In addition, spinel in dunites has high molar $Cr/(Cr+Al)$, denoted 'Cr#' and high Ti, like most spinels in primitive MORB. As in Oman, Cpx in medial gabbro selvages within dunites formed in shallow mantle at the East Pacific Rise (sampled at Hess Deep) are in trace element equilibrium

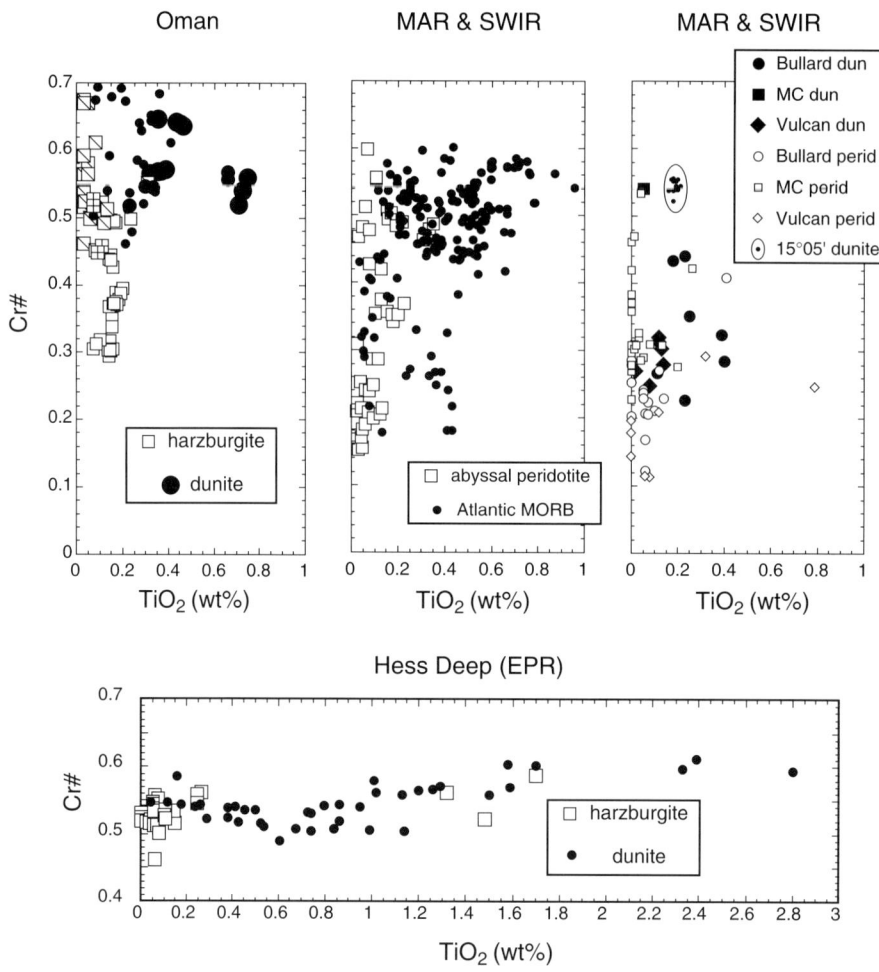

Figure 5. Compositions of spinel in dunites and associated peridotites from the Oman ophiolite (Pallister & Hopson 1981; Augé 1987; Kelemen et al. 1995b), and from abyssal peridotites dredged and drilled from the Mid-Atlantic Ridge (MAR), Southwest Indian Ridge (SWIR), and East Pacific Rise (Hess Deep – EPR) (Dick & Bullen 1984; Allan & Dick 1996; Arai & Matsukage 1996; Dick & Natland 1996; and our unpublished data; 'MC' is an abbreviation for Marie Celeste Fracture Zone). Also shown for comparison are compositions of spinel in MORB from the Atlantic (compiled by Dick & Bullen 1984). Dunites generally have Cr/(Cr+Al), denoted 'Cr#', as high as the most depleted peridotites, but higher TiO_2 than any associated peridotites. Spinels in MORB are compositionally similar to those in dunite, and different from those in residual peridotite.

with MORB (Dick & Natland 1996). Spinels in these dunites have high Cr# and high Ti, similar to spinel in Oman dunites and in MORB (Allan & Dick 1996; Arai & Matsukage 1996; Dick & Natland 1996). Our unpublished data show that high Cr# and high Ti are also typical of abyssal dunite from the Mid-Atlantic and SW Indian Ridges (figure 5). Thus, dunites from Oman and mid-ocean ridges have spinel compositions similar to those found in MORB, supporting the notion that dunites are conduits for MORB transport. An essential caveat is that some ophiolites are polygenetic, and not all dunites in ophiolites reflect equilibration with melts similar to MORB (Kelemen et al. 1992; Takazawa et al. 1992; Ozawa 1994; Ozawa & Shimizu 1995; Quick & Gregory 1995; Gruau et al. 1995; Takazawa 1996).

(ii) *Dunite dimensions*

The size distribution and proportion of dunites can be used to estimate melt/rock ratios within melt extraction conduits. The largest mantle dunites in Oman are in the Wadi Tayin massif (Hopson et al. 1981). We find that these are tabular, in two cases more than 200 m thick and up to 10 km long. They are locally massive, but elsewhere are anastamosing, with elongate harzburgite inclusions up to 10 m thick and 100 m long. Another large dunite, in the Hilti massif (Al Wasit geological map, Oman Ministry of Petroleum & Minerals 1987) is more than 100 m thick and more than 10 km long, sub-parallel to the Moho.

The resolution at which smaller dunites in Oman have been mapped varies. In the well-mapped Wadi Tayin massif (Fanjah, Samad and Quryat geological maps 1986), there are more than 35 dunites more than 1 km long in *ca.* 800 km². Most are transposed, sub-parallel to the Moho, indicating that they have undergone corner flow. Their map widths generally exceed 100 m (their true thickness ranges from 20 to 100 m). Thus, 'large' dunites comprise *ca.* 0.5% of the exposed mantle section. Smaller dunites are ubiquitous. If the estimate that dunite comprises 5–15% of the mantle section in Oman is correct (§ 2 a (i)), then 'small' dunite bodies must be more than 90% of this total.

Because the proportion of dunite may increase upward and toward the centre of the melting region, transposed dunites in the 10 km thick mantle section of the ophiolite may represent most or all of the dunite formed in the upwelling column. If so, and if the melting region was *ca.* 100 km deep, then dunite comprises 0.5–1.5% of the melting region. 'Large' dunites comprise *ca.* 10% of this total. If most melt flowed through dunite conduits, and the mean extent of melting was 10%, then the average melt/rock ratio in all dunites would be 20:1 to 8:1. If most melt flowed in 'large' dunites, the melt/rock ratio would be 200:1 to 80:1.

(iii) *Most mantle dunites are replacive features*

Hypotheses for the origin of mantle dunites are reviewed by Kelemen et al. (1995a, b), who conclude on the basis of geological and geochemical data that most dunites are replacive features, formed by dissolution of pyroxene and concomitant precipitation of olivine in magma migrating by porous flow. Please see Dick (1976, 1977), Quick (1981), Berger & Vannier (1984), Bodinier (1988), Kelemen (1990), Kelemen et al. (1992), and Quick & Gregory (1995) for important contributions, and Kelemen et al. (1995a, b) for additional references and discussion. Here, we add previously unpublished ideas.

In addition to the hypothesis that they are replacive, mantle dunites have been interpreted as fractures filled with olivine precipitated from migrating melt (i.e. 'cumulates'). However, data on $Mg/(Mg+Fe^{2+})$, denoted 'Mg#', suggests that few if any large dunites are cumulates. Compared to olivine in mantle peridotite, olivine in cumulate dunite should have lower Mg#. In fact, Mg#'s are similar in olivine from dunites and associated mantle peridotites (Kelemen 1990; Kelemen et al. 1995b). If dunites are cumulate, their limited range of Mg# indicates that they represent a very small fraction of the parental magma mass. For example, Takahashi (1992) reports less than 0.5% variation of Mg# in a tabular dunite *ca.* 10 m wide and more than 5 km long. This requires that the dunite, if cumulate, represents less than 2% of initial liquid mass, provided the initial liquid had less than 15 wt% MgO. A restricted range of Mg# is the norm for mantle dunite; it is unlikely that the combination of cooling rate and melt flux would always produce such small degrees of crystallization. Thus, high and relatively constant Mg#'s in large mantle dunites indicate that most are not cumulate. Since melt/rock reaction maintains nearly constant Mg# (Kelemen 1986, 1990), a replacive origin can explain the similar olivine Mg#'s in dunites and host peridotite, as well as their mono-mineralic composition and contact relationships.

(iv) *How much dunite can be formed by reaction?* < 5% *of the melting region*

The solubility of pyroxene in ascending melt is about 1 wt% kbar^{-1} of decompression (figure 3 in Kelemen *et al.* 1995*a*). The maximum amount of dunite that can be formed is 3 wt% of the mantle in the melting region, assuming the average degree of melting is 10 wt%, the average pressure of melting is 13 kbar, the pressure at the base of the crust is 3 kbar, and average peridotite has 30 wt% pyroxene. If this were the case, the average melt/rock ratio in these dunites would be 2:1 and all magmas would be saturated with Opx. However, primitive MORB is far from Opx saturation, as noted in § 1 *a*. Therefore, the melt/rock ratio must be more than 2:1. If the average melt/rock ratio were more than 100:1 (i.e. dunite proportion less than *ca.* 0.1%), the resulting magmas would be far from equilibrium with Opx and the dunite formation process would have little geochemical effect on the composition of MORB. In this case, the aggregate liquids that comprise MORB could be regarded as a mixture of polybaric partial melts. If the melt/rock ratio has an intermediate value, *ca.* 10:1 (i.e. 1% dunite), then MORB would be far from Opx saturation, but dunite formation might have other effects on the composition of MORB.

(v) *Is there a 'dunite signature' in MORB? Maybe*

As noted in § 2 *a* (iv), if the melt/rock ratio in dunite conduits is *ca.* 10, then dunite formation should have a compositional effect on MORB. In fact, MORB has characteristics unlike aggregated, polybaric partial melts that could be produced by dunite formation. Figure 6 is a plot of Ce/Yb(N) in liquids against Cr# in spinel. MORB, and liquids calculated to be in equilibrium with Oman dunites, have Ce/Yb(N) near 1, as high as in liquids in equilibrium with the least depleted abyssal peridotites. Spinel in MORB and in dunites has Cr# of about 0.6, as high as spinel in the most depleted abyssal peridotites. This reflects the fact that REE ratios in MORB record the average degree of melting of the source, while Cr# apparently records the highest degree of melting.

The apparent discrepancy in degrees of melting recorded by Cr# versus Ce/Yb(N) can be explained if liquids involved in dunite formation constitute a moderate proportion of MORB (i.e. dunites comprise *ca.* 1% of the rock). The dunite forming reaction, pyroxene + liquid → olivine + spinel + liquid, can increase the Cr-content of derivative liquids, but has little effect on REE ratios (figure 6). Pending improved models for spinel/liquid equilibria, this hypothesis must be regarded as conjectural. Alternatively, mixing of Ce-rich, Cr-poor liquid produced at small degrees of melting with Cr-rich, Ce-poor liquid produced by large degrees of melting could produce the observed MORB composition, assuming that variability of Al and Yb in mantle-derived magmas is negligible. However, what would become of magmas produced by intermediate degrees of melting? If they were added, the mixture would be in equilibrium with spinel with intermediate Cr#. Thus, the dunite hypothesis may be

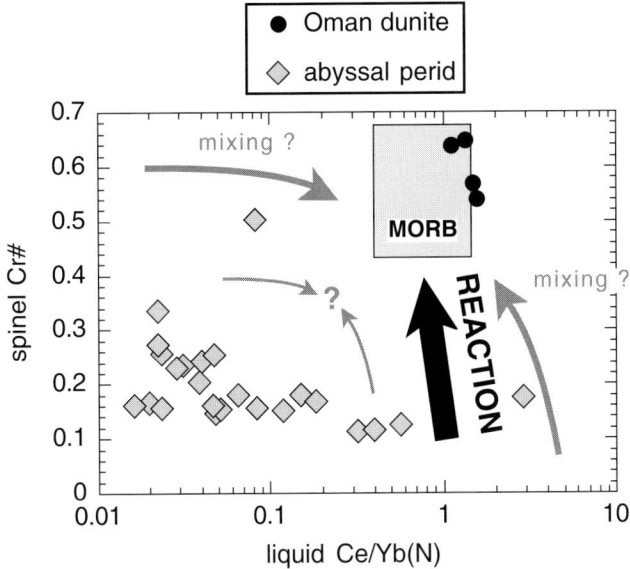

Figure 6. Calculated Cr# in spinel and chondrite normalized Ce/Yb, denoted 'Ce/Yb(N)', in liquid, illustrating that MORB may include a compositional 'signature' of dunite formation. Field for MORB is based on compiled compositions of spinels (Dick & Bullen 1984) and liquids (Langmuir et al. 1992). Dunite and peridotite Ce/Yb(N) were calculated from Cpx compositions and Cpx/liquid distribution coefficients of Hart & Dunn (1993). Data for Cpx in abyssal peridotites are from Johnson et al. (1990) and Johnson & Dick (1992), and data for Cpx in dunites are from Kelemen et al. (1995b), for Oman ophiolite samples. Mass balance calculations for dunite forming reactions yield the trend labeled 'REACTION'. For example, this trend results if peridotite and liquid have REE contents as for figure 1, peridotite has 0.5 wt% Cr_2O_3 and 3–10 wt% Al_2O_3, initial liquid has 100 ppm Cr and 15 wt% Al_2O_3, dunite has 1% spinel, melt/rock ratio is 10, ratio of peridotite reactant to dunite product is 1.1, olivine/liquid = 2 for Cr and 0.0001 for Al, and spinel/liquid = 200 for Cr and 2 for Al.

preferable. A signature of dunite formation in MORB has also been suggested based on spinel compositions (Arai & Matsukage 1996), and major element trends (Niu & Batiza 1993; Gaetani et al. 1995).

(vi) *The reactive infiltration instability (RII)*

Daines & Kohlstedt (1994) and Kelemen et al. (1995a) proposed that dunites originate as dissolution channels. Since pyroxene dissolution combined with olivine precipitation increases liquid mass under conditions of constant temperature or constant enthalpy (Kelemen 1990; Daines & Kohlstedt 1994), it follows that such reactions will increase magma mass – and porosity – under adiabatic conditions. Dissolution of pyroxene in olivine-saturated magma is analogous to porous flow of fluid through a partially soluble rock. Theoretical and experimental studies show that diffuse porous flow in these systems is unstable. Because initial perturbations in permeability lead to enhanced fluid flow, which in turn leads to more rapid dissolution, diffuse flow breaks down into focused porous flow through high porosity dissolution channels, elongate in the direction of fluid flow (Chadam et al. 1986; Hoefner & Fogler 1988; Daccord 1987). This is termed the 'reactive infiltration instability' (RII).

The RII was first considered for rigid porous media, with an initial, planar solution front normal to the direction of fluid flow. With numerical experiments (Kelemen

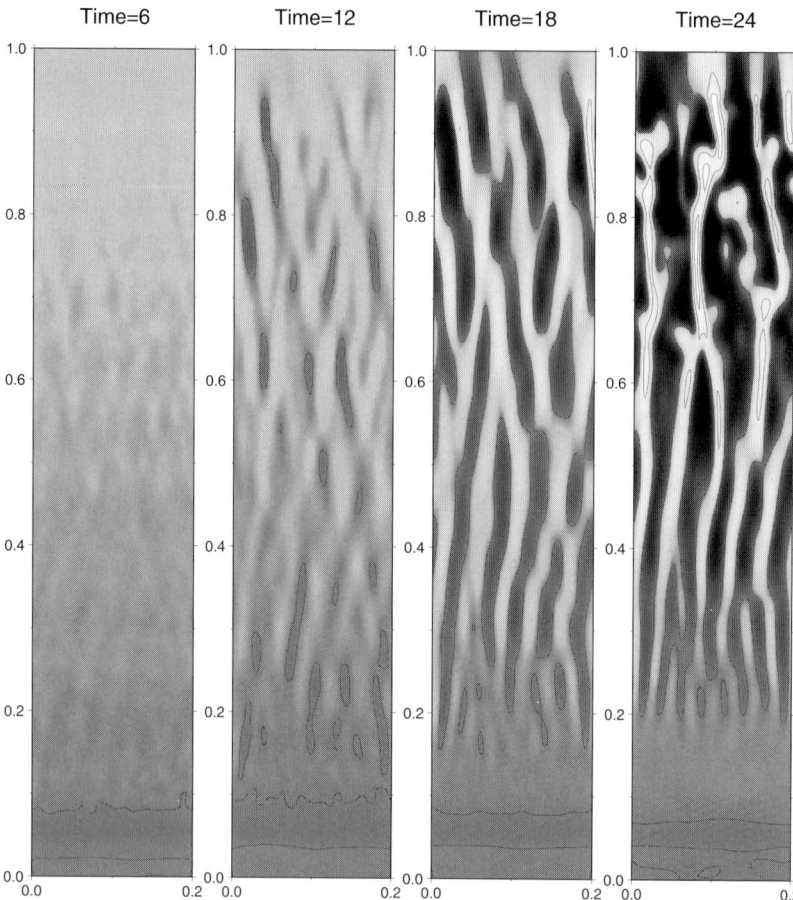

Figure 7. Evolution of porosity from a fully nonlinear calculation for porous flow in a viscously compacting, soluble medium, illustrating how high porosity dissolution channels form and coalesce. All times have the same grey scale map. Contours of 1% porosity are shown. Final frame has a minimum porosity near 0.1%, and a maximum near 3%. The minimum porosity is limited by a non-constant bulk viscosity that makes it difficult to compact very small porosities (Sleep 1988). Numerical parameters are 65×321 grid points, 36276 time steps. Boundary conditions are constant flux bottom and free-flux top with periodic (wrap-around) sides. These are preliminary results from Spiegelman et al. (1996).

et al. 1995a) and a stability analysis (Aharonov et al. 1995), we have shown that the RII occurs in a solubility gradient with no initial solution front, and in viscously compacting porous media. The RII is continuously present in time and space in a solubility gradient; as long as soluble solids remain in the system, dissolution channels grow unstably. Because the solubility of peridotite in adiabatically decompressing melts increases with decreasing pressure, the RII can occur throughout the melting region.

The most obvious manifestation of dissolution reactions in the mantle is the formation of dunite by dissolution of pyroxene. However, even olivine has a finite solubility in adiabatically ascending melt, so that the RII could form high porosity channels *within* dunite. If melt supply is sufficient, this might ultimately form 'open', melt-filled tubes. Ongoing research on the RII includes investigation of branching and coalescence of channels and partitioning of flow between dissolution channels and the surrounding matrix. Aharonov et al. (1995) predicted, and preliminary numerical experiments confirm, that dissolution channels coalesce downstream (upward, in the mantle), provided that solubility increases downstream (e.g. figure 7). In these calculations, where the permeability contrast between channels and matrix exceeds 1000, more than 95% of the flux is focused in dissolution channels that comprise less than 10% of the volume (Spiegelman et al. 1996).

(vii) *Mechanical instabilities*

Stevenson (1989) noted that in viscously deforming media, weak regions may have a slightly lower pressure than surrounding, stronger regions. Therefore, he proposed that melt will tend to flow from strong regions into weaker regions. Because melt fraction is negatively correlated with the strength of partially molten peridotite (Hirth & Kohlstedt 1995a, b), this can unstably form melt-rich shear zones within a melt-poor matrix (the 'Stevenson instability'). Similar hypotheses have been proposed by Sleep (1988). A role for mechanical focusing of melt flow is demonstrated by dunites in the Josephine peridotite that formed syn-kinematically along ductile shear zones in conductively cooled, shallow mantle at ca. 1100 °C (Kelemen & Dick 1995). Factors that localized melt flow probably included the Stevenson instability, anisotropic permeability due to vertical foliation in shear zones, and the RII.

(b) *Dikes: lithospheric features, and – in Oman – generally not formed by MORB*

Pyroxenite and gabbroic dikes are observed in the mantle section of many ophiolites. The dikes are generally discordant to the high temperature foliation in residual peridotites (Boudier & Coleman 1981; Lippard et al. 1986; Nicolas 1989). Formation of pyroxene and plagioclase is indicative of crystallization from evolved, low temperature liquid. Such liquids form by crystal fractionation due to conductive cooling. We infer that the dikes, as well as tabular dunites with medial pyroxenites (Trinity peridotite, Quick 1981b), formed off-axis, in lithosphere with a conductive geotherm.

In Oman, compositional data demonstrate that many gabbronorite and pyroxenite dikes in the mantle section were not conduits for MORB extraction. Cpx in 6 of 8 dikes from the Wadi Tayin massif are strongly light REE depleted (figure 8), and have high Ti/Zr (Farrier et al. 1997). In these respects they are identical to Cpx in peridotites from the same massif (Kelemen et al. 1995b) and abyssal peridotites (Johnson et al. 1990). The dikes represent small amounts of crystallization of melt derived from a source similar to the Oman harzburgites. These melts had REE contents very different from MORB. Instead, the melts were similar to 'ultra-depleted' melts (UDM, §1c) observed in glass inclusions within some olivine phenocrysts in MORB (Sobolev & Shimizu 1993).

A 'cumulate' origin for Wadi Tayin dikes is indicated by high Mg# in Cpx (83–95 mol%) and high anorthite in plagioclase (85–95 mol%), demonstrating that all dikes contained little or no 'trapped liquid' (Farrier et al. 1997). Thus they are not asthenospheric melt pockets that cooled to form gabbroic rocks in the lithosphere. Most of the dikes are websterites or gabbronorites (Opx-rich). Thus, the liquids were saturated in Opx, similar to UDM and very different from MORB. We conclude that melts which formed the Wadi Tayin dikes were derived from a source similar to the surrounding harzburgite, and that the dikes are shallow, lithospheric features. Further, their similarity to UDM – the lowest pressure melt in the polybaric mixture comprising MORB – indicates that UDM may be transported from the shallow

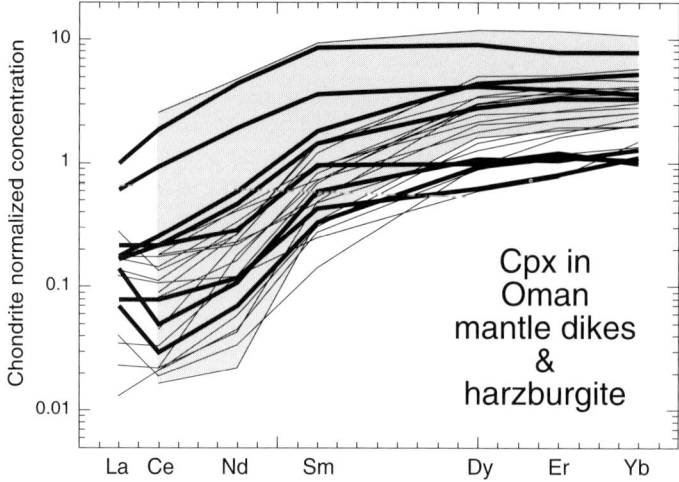

Figure 8. REE contents of Cpx in gabbronorite and websterite dikes from the Wadi Tayin massif of the Oman (bold lines), compared to Cpx in harzburgite from the same massif (light lines) and in abyssal peridotite (shaded field). Six of eight dikes analysed had Cpx REE contents different from Cpx in equilibrium with MORB, and very similar to Cpx in residual peridotites, suggesting that the dikes are formed by crystal fractionation from liquids derived by low pressure partial melting of highly depleted peridotite. Data for dikes from Farrier et al. (1997), for Oman peridotite from Kelemen et al. (1995b), and for abyssal peridotite from Johnson et al. (1990) and Johnson & Dick (1992).

mantle in dikes. If UDM have a transport mechanism different from other melts, this could explain why UDM are occasionally found as glass inclusions within olivine in MORB (Sobolev & Shimizu 1993), and why some cumulate gabbros from mid-ocean ridges have strongly light REE depleted compositions (Ross & Elthon 1993, 1997a, b).

There are mantle dikes that did form in equilibrium with MORB-like liquids. As noted in §2a(i), Cpx in medial gabbro within dunite from the shallow mantle lithosphere at Hess Deep are in trace element exchange equilibrium with normal MORB (Dick & Natland 1996). Similarly, gabbroic dikes and sills that formed in equilibrium with melts similar to MORB are found in the upper 1–2 km of the mantle section below the crust in the Maqsad region of the Oman ophiolite (Benoit et al. 1997; Kelemen et al. 1997). Thus, on the basis of limited data, it seems that most gabbroic dikes more than 2 km below the base of the crust in the Oman ophiolite formed off-axis, from depleted liquids similar to UDM, whereas most gabbroic dikes and sills close to the base of the crust form from liquids similar to MORB.

Dikes in peridotite from the Mid-Atlantic Ridge have been recovered by dredging, submersible sampling and drilling (Cannat 1996; Cannat & Casey 1995). Most of these are different from gabbroic dikes of the Oman ophiolite, extending to more evolved compositions with enriched minor and trace element compositions. Gabbroic dikes from the Mid-Atlantic Ridge formed in equilibrium with liquids more enriched in incompatible elements than MORB, by a combination of melt/rock reaction and extensive crystal fractionation within cold, lithospheric mantle peridotite.

(c) Are dunites associated with cracks? Some are. (Most are not?)

Most mantle dunites have a replacive origin. However, there might have been 'cryptic fractures' in the centre of most or all dunites during their formation, resulting in a kind of 'uncertainty principle' for geologists seeking to determine whether fractures play an important role in melt extraction from the asthenosphere. Melts rising adiabatically in a crack will not become saturated in any solid phase. Such a crack might close, leaving no trace, when liquid is no longer present. In this sense, field observations do not require asthenospheric melt flow in cracks, but they cannot be used to rule it out.

The tabular nature of many dunites has been used to infer that they are related to fractures. Indeed, tabular dunites with a medial pyroxenite, common in the Lanzo peridotite (Boudier & Nicolas 1972), the Horoman peridotite (Takahashi 1992) and the Trinity peridotite (Quick 1981b), may be porous reaction zones around melt-filled fractures (Nicolas 1986). Similar features have also been recovered at Hess Deep, in drill core from the crust–mantle transition zone (Dick & Natland 1996). However, mass balance constraints (§ 3 f) suggest that it is unlikely that tabular dunites form around individual fractures. While the tabular geometry of many dunites probably does not arise from reactive porous flow alone (though Tait et al. (1992) report tabular dissolution channels), tabular dunites could form as a result of the RII combined with anisotropic permeability in peridotite and/or an anisotropic stress distribution (see § 2 a (vii) and § 4 d). Cracks may form within dunite, due to their high intergranular porosity and large grain size (§§ 3 b, f). This is most likely near the base of the lithosphere.

(d) Chromitites: focused flow, but not necessarily in asthenospheric fractures

Chromitites are composed of more than 50% massive chromian spinel. In ophiolites, they are enclosed within dunite, rather than peridotite. Chromitites are indicative of focused melt flow (Leblanc & Ceuleneer 1992). Because the solubility of Cr-spinel is low in silicate melts, chromitites must have scavenged Cr from 300 to 400 times their mass in liquid. Thus, the integrated melt/rock ratio for both chromitites and some of the surrounding dunites is more than 300. The fact that chromite crystallizes in massive concentrations, despite its low concentration in magmas, reflects sluggish kinetics for chromite nucleation compared to crystal growth. Basaltic magmas must commonly be supersaturated in chromite, but still able to transport Cr until they come into contact with chromite that has already nucleated.

Chromitites have been interpreted to form in melt-filled voids by magmatic sedimentation (Lago et al. 1982; Nicolas 1986, 1990). Alternatively, they may form in porous media. Sedimentation was suggested by 'nodular' chromitites (ellipsoid chromite grains a few centimetres in diameter), resembling a grain-supported conglomerate. However, in 'orbicular' chromitites, also common, chromite ovoids are not in contact and are 'hollow', with olivine cores. Some show alternating rings of chromite and olivine (Nicolas 1989, fig. 10.12c). These are similar to orbicular structures in granites (Wang & Merino 1993, fig. 1) and concretions formed during diagenesis in porous rocks (Wang & Merino 1995).

Lago et al. (1982) and Nicolas (1986) found that flux calculations based on suspension of chromite crystals by magmatic flow were consistent with the rate of magmatic accretion of the crust, and suggested that this supported the hypothesis that melt extraction occurs in hydrofractures extending from near the base of the melting

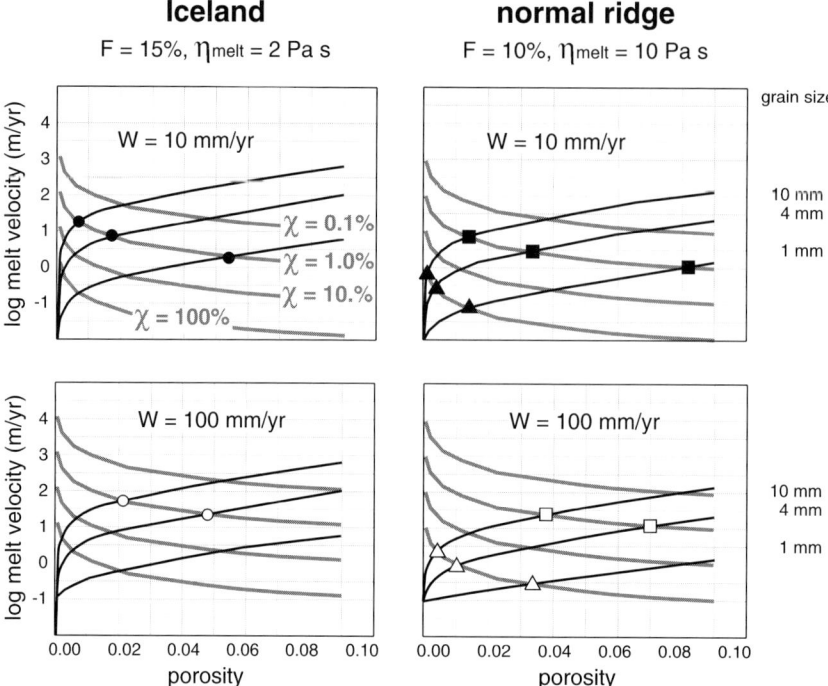

Figure 9. Velocities, relative to the Earth's surface, for melts migrating by porous flow in the mantle. In general, porous flow velocities in the mantle can be high enough to satisfy melt velocity constraints for mid-ocean ridges based on U/Th isotopic disequilibrium, the magmatic response to Icelandic deglaciation, and the steady state melt flux. Velocities calculated using Darcy's law with permeability from Cheadle (1993) are shown as black, solid lines for solid grain sizes of 10, 4 and 1 mm. Values are shown for two melt viscosities, 2 Pa s ('Iceland') and 10 Pa s ('normal ridge'), and two solid upwelling velocities, 10 and 100 mm yr^{-1}. Also shown as light grey lines are solutions to a one-dimensional melt flux equation for mid-ocean ridges, $w = WF/(\phi\chi)$, in which w and W are melt and solid velocities, ϕ is the porosity, F is the average fraction of melting (taken to be 15% for Iceland and 10% for normal ridges) and χ is the proportion of melt conduits. Intersections between the curves specify melt velocity and porosity of steady-state flow for a given grain size and proportion of melt conduits. Circles, squares and triangles highlight specific intersection points; these points are also shown in figure 10.

regime to the Moho. However, we show in §3a that porous flow can also supply magma at the rate of magmatic accretion of the crust (7000 m² per 10–100 years per unit length of ridge). Even accepting that chromitites are magmatic sediments, it does not follow that they form in fractures in the asthenosphere. They could form in shallow mantle. Most chromitites are found within a few kilometres of the Moho. High melt velocities do not require flow in a pipe or a tabular opening with a large vertical dimension (Spera 1980; Lago et al. 1982). In addition, even if chromitites form in melt-filled voids spanning tens of kilometres in the asthenosphere, these voids could be formed by the RII (§2a(vi)), mechanical instabilities (§2a(vii)), or a combination of these processes, as well as by fracture.

(e) Summary

Dunites and chromitites mark conduits for chemically isolated extraction of MORB within the ascending mantle. Dunites are replacive features formed by dissolution of

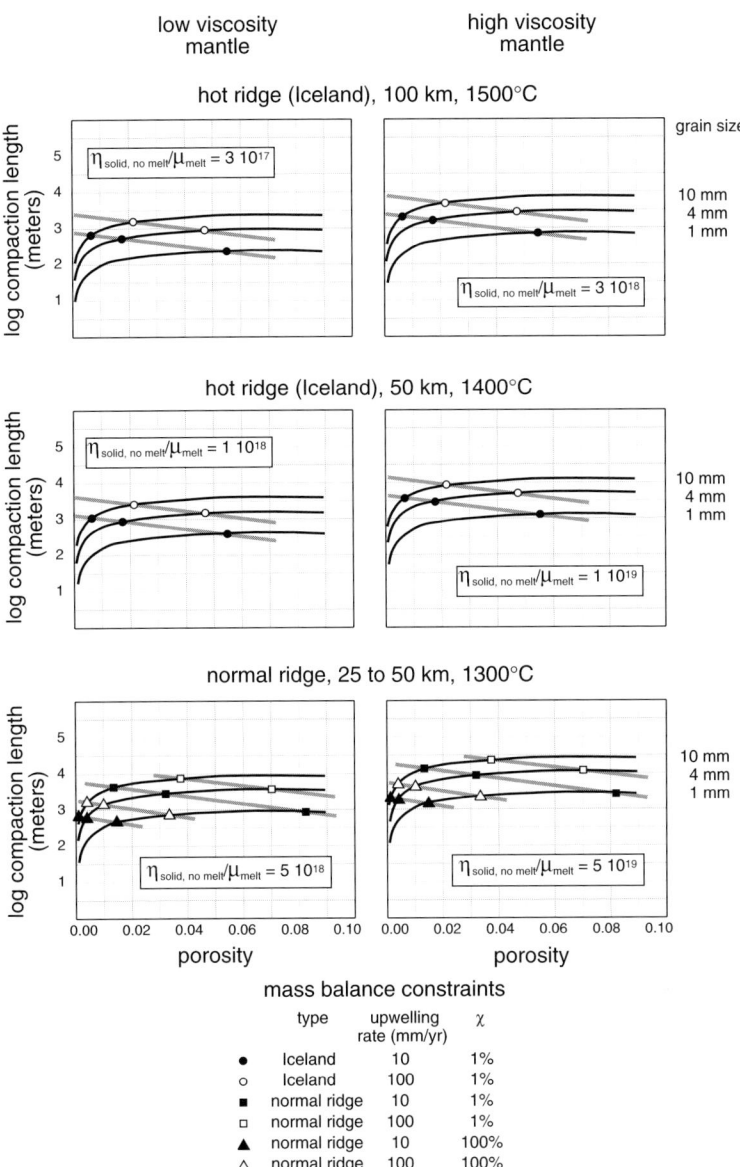

Figure 10. Calculated compaction lengths for a variety of mantle temperatures, solid and melt viscosities, and solid grain sizes, shown as black, solid lines. Also shown as light grey lines are porosities and compaction lengths consistent with the one-dimensional, steady state melt flux equation for mid-ocean ridges (from figure 9). In general, these results suggest that hydrofracture is unlikely in the region of adiabatic upwelling beneath mid-ocean ridges. The likelihood of hydrofracture is enhanced if melt flow is restricted to a few, high porosity conduits within coarse-grained rocks. If mantle viscosities are as high as allowable given current geophysical data, then the minimum compaction length for hydrofracture, 10^4 m, is exceeded for conduits of focused flow ($\chi \leqslant 1\%$) within rocks with an average grain size more than 4 mm.

pyroxene in melt ascending along conduits of focused porous flow. Fractures within dunites in the asthenosphere are not required, but cannot be ruled out on the basis of geological data. In contrast, pyroxenite and gabbro dikes are lithospheric features and do not provide information about melt transport in the asthenosphere.

3. Constraints from physics

(a) *Is porous flow fast enough to account for melt velocity estimates? Yes.*

Four types of melt velocity estimates are available for oceanic spreading ridges. Two come from U-series disequilibria, a third is based on geological observations in Iceland, and a fourth is based on simple mass balance constraints for melt extraction from a one-dimensional column.

Excesses of ^{230}Th (half life of *ca.* 75 000 years) and ^{226}Ra (*ca.* 1600 years) in fresh, zero age MORB indicate that melt transport from the point where these excesses grew to the surface was rapid compared to the half life of ^{230}Th and ^{226}Ra. Since garnet is apparently required to hold more U than Th in a solid phase (Beattie 1993a, b), excesses of radiogenic ^{230}Th are produced only in the deepest part of the melting region, below the garnet–spinel peridotite transition, perhaps 75 km beneath the crust (Spiegelman & Elliot 1993; Iwamori 1994; Richardson & McKenzie 1994; Lundstrom et al. 1995). It has been proposed that garnet pyroxenite contributes significantly to the composition of MORB (Chabaux & Allègre 1994; Hirschmann & Stolper 1996). Since garnet is stable to lower pressures in pyroxenite bulk compositions than in peridotites, ^{230}Th excesses could theoretically be produced by reactive porous flow in garnet pyroxenites at depths less than 75 km. However, pyroxenite bodies in mantle peridotite samples always form tabular bands. Reactive porous flow is unlikely to be restricted to garnet pyroxenite bands within spinel peridotite.

Thus, ^{230}Th disequilibrium requires melt transport rates *ca.* 75 km per 75 000 years, or 1 m yr^{-1}. If ^{226}Ra disequilibrium must form in the presence of garnet, then melt velocities must be *ca.* 50 m yr^{-1}. However, as noted in §1g, ^{226}Ra disequilibrium may be formed in the shallow mantle; in this case it is not a strong constraint on mantle melt transport velocities. Jull & McKenzie (1996) report an outpouring of basaltic magma associated with deglaciation in Iceland. They infer that melt was produced by decompression due to deglaciation, throughout the melting region extending to perhaps 100 km depth, and reached the surface within 2000 years. This requires melt velocities *ca.* 50 m yr^{-1}. However, other explanations are possible for the post-glacial volcanism; e.g. long term storage of magma in a lower crustal reservoir.

In a one-dimensional, upwelling column, steady state melt velocity is related to fraction of melting and solid upwelling by $w = WF/\phi$, where w and W are melt and solid velocities and F and ϕ are the fraction of melting and the porosity (Spiegelman 1993c). If virtually all melt flow is focused in conduits, this may be modified to $w = WF/(\phi\chi)$, where χ is the proportion of conduits. As noted in §2a(iv), if melt flow conduits are dunite dissolution channels, then chemical data require that χ be $\leqslant 3\%$. Geochemical data (§2a(v)) suggest that χ is *ca.* 1%. A value of *ca.* 1% may also be inferred from the proportion of dunite in the Oman ophiolite (§2a(ii)). However, note that if more than 90% of the melt is transported in 10% of dunite conduits, and dunite conduits comprise 1% of the melting region, $\chi \approx 0.1\%$.

Solutions for $F = 15\%$ (Iceland) and 10% (normal ridges), and $\chi = 100, 10, 1$ and 0.1% are plotted as wide, shaded curves in figure 9. These can be compared to melt velocities inferred from post-glacial volcanism in Iceland and from U/Th disequilibrium at normal ridges. If mantle upwelling beneath Iceland is comparable to the spreading rate and $\phi > 0.1\%$, then melt velocities of order 50 m yr^{-1} are only achieved for $\chi \leqslant 1\%$. Beneath a normal, slow spreading ridge (upwelling rate of 10 mm yr^{-1}, $\phi > 0.1\%$), χ must be less than 100% to reach a melt velocity of 1 m yr^{-1}. Thus melt extraction must occur in spatially restricted channels ($\chi < 1$) in order for steady state flow velocities to satisfy velocity constraints.

The mass balance constraints on melt velocity can also be compared to buoyancy-driven, porous flow velocities (solid lines in figure 9), calculated using Darcy's law with permeability estimated by Cheadle (1993), as reported in McKenzie (1989). We use the Cheadle estimates (similar to Von Bargen & Waff 1986) because of uncertainty in extracting permeability from experiments by Riley & Kohlstedt (e.g. 1991), Daines & Kohlstedt (1993), and Faul (1997). Permeability varies with grain size as well as porosity. Olivine and pyroxene grain size in the mantle is generally 3–10 mm (Mercier 1980). Mantle dunites are coarser-grained than surrounding peridotites. Dunites commonly have an average grain size ca. 10 mm. In syn- and post-kinematic dunites, olivine has irregular shapes, interfingering with adjacent crystals (Nicolas 1989, fig. 2.7a; Kelemen & Dick 1995, fig. 7B). The irregular grain shapes preserved in the dunites are not consistent with static, sub-solidus grain growth. We therefore infer that the grain size in dunites is larger than that in surrounding peridotites during melt flow in the mantle.

Buoyancy-driven, permeability-based melt velocities beneath Iceland and normal mid-ocean ridges are shown as black solid lines in figure 9 for solid upwelling at 10 and 100 mm yr^{-1}. Melt viscosity is inversely proportional to the average pressure of melting (Kushiro 1986). Beneath Iceland, with melt viscosity of 2 Pa s and mantle upwelling similar to the spreading rate, melt velocities of 1 m yr^{-1} (fast enough to preserve ^{230}Th excess) can be attained at ϕ of 0.2 and less than 0.1% for grain sizes of 4 and 10 mm. To achieve melt velocities of 50 m yr^{-1}, inferred from Icelandic deglaciation volcanism, ϕ of 7 and 2% are required for the same grain sizes. Beneath a normal, slow-spreading ridge, with melt viscosity of 10 Pa s and passive upwelling, melt velocities of 1 m yr^{-1} are attained at ϕ of 1.5 and 0.2% for 4 and 10 mm grains.

In the previous paragraph we have not incorporated mass balance. Since the total melt flux is limited by the rate of decompression melting, high liquid flow velocities at high porosity cannot be sustained at steady state because they require more liquid than is produced. Therefore, assuming *both* (1) that the Cheadle (1993) permeability estimates are accurate and (2) that the one-dimensional, steady state flux calculation is a good approximation, steady-state melt velocities and porosities are defined by the intersections of the mass balance curves and the permeability-based curves in figure 9. These intersections specify melt velocity and porosity given values of grain size, W, F and χ. With $F = 10\%$ and $\chi = 1\%$, the requirement that melt moves at more than 1 m yr^{-1} beneath Iceland is easily satisfied; for example if the grain size is 4 mm, ϕ must be ca. 1.7%. However, with $F = 10\%$ and $\chi = 1\%$, melt velocities from the deglaciation constraint ($\geqslant 50 \text{ m yr}^{-1}$) can only be satisfied if channels have a grain size more than 20 mm and $\phi \lesssim 0.2\%$. These grain sizes are larger than generally observed, even in dunite. Alternatively, if $\chi = 0.1\%$ (e.g. conduits are 1% of the melting region but a subset of 10% of conduits carries 90% of the melt flux) then melt velocities of 50 m yr^{-1} require grain size of 10 mm and $\phi \approx 2\%$.

Beneath a normal, slow spreading ridge, with $\chi = 1\%$, the requirement from ^{230}Th excess that melt moves at more than 1 m yr^{-1} is satisfied, for example, with grain size of 4 mm and $\phi \approx 3.3\%$. If $\chi = 10\%$, grain size must be greater than ca. 4 mm and

$\phi \lesssim 1\%$. However, if ^{226}Ra disequilibrium forms in the presence of garnet, required melt velocities of 50 m yr^{-1} beneath a slow spreading ridge with $\chi = 1\%$ are only attained at unreasonably large grain sizes. For a reasonable grain size of 10 mm, χ must be less than 0.02% and $\phi \gtrsim 6\%$.

Note that, in all cases discussed in this section, most conduit porosities required to satisfy melt velocity constraints are higher than source porosities required to produce U-series disequilibrium (ca. 0.1%), and higher than source porosities required for the interval of 'near-fractional' melting necessary to account for abyssal peridotite REE compositions. However, there is no reason to assume that conduit porosities and source porosities should be equal.

(b) Can hydrofracture occur in adiabatically ascending mantle? Maybe, but if so, then only in pre-existing conduits.

It has been argued that hydrofracture is unlikely in the viscously deforming, partially molten upper mantle (Stevenson 1989). The reasoning is that the mantle can deform viscously at differential stresses considerably lower than the tensile strength of the solid matrix. However, hydrofractures may also form if the melt pressure exceeds the confining pressure (σ_3) by a value equal to the tensile strength, even when the differential stress is low. The *maximum possible* ΔP (melt pressure, σ_3) due to melt buoyancy in a viscously compacting solid matrix is of the order $\Delta \rho g \delta_c$, where $\Delta \rho$ is the difference in density between melt and solid, g is the gravitational constant, and δ_c is the compaction length (McKenzie 1984). The compaction length is defined as

$$\sqrt{k_\phi(\zeta + \tfrac{4}{3}\mu_s)/\mu_1},$$

where ζ and μ_s are the bulk viscosity (deformation due to ΔP) and shear viscosity (deformation due to $\sigma_1 - \sigma_3$) of the partially molten matrix, k_ϕ is the permeability, and μ_1 is the melt viscosity. The values of ζ and μ_s are difficult to determine from experimental deformation data. Note that the larger of the two will dominate in determining the compaction length. We denote this larger value as $\eta_{\text{solid w/melt}}$, assume that experimental constraints on the influence of melt on viscosity determine this quantity, and rewrite a simplified expression for compaction length as

$$\sqrt{k_\phi(\eta_{\text{solid w/melt}})/\mu_1}.$$

The maximum value of ΔP would be attained in a porous system that suddenly started to compact, or in a compacting system beneath an overlying obstruction in the permeability. Because the compaction process acts to minimize pressure differences between solid and liquid, steady state stresses in continuously compacting systems without an overlying obstruction are smaller than this value (Spiegelman 1993a, b).

These calculations are for buoyancy-driven flow. Where forced flow drives melt from a broad region into a narrow zone, as may occur at the base of the crust beneath a spreading ridge, melt pressures may be higher. However, note that some coalescence of melt flow beneath ridges is driven by 'suction', i.e. low melt pressure (§ 4d). Thus the effect of forced flow on melt pressure is uncertain, and remains a topic of continuing research.

Unfortunately, our understanding of processes that control magma fracturing in viscous materials is limited. However, a combination of constraints derived from the rock mechanics literature and from deformation experiments on peridotite provides an estimate for the tensile strength of partially molten mantle rocks. In most experiments, samples are enclosed in metal capsules, which prevents melt from migrating out of the aggregate. The melt pressure in this case is roughly equal to the confining pressure, although it may be as high as the mean pressure. Thus, the effective confining pressure is very low, which enhances the formation of fractures.

Experiments show evidence for melt-enhanced brittle deformation of peridotite at differential stresses greater than 500 MPa (Bussod & Christie 1991). Together with an assumption that the effective confining pressure in these experiments was ca. 0, this result can be used to estimate the uniaxial compressive strength of partially molten peridotite. For comparison, numerous experiments conducted on similar compositions at stresses as high as 250 MPa show no evidence for brittle failure (Beeman & Kohlstedt 1993; Hirth & Kohlstedt 1995b). Griffith failure criteria indicate that the tensile strength of a brittle rock is 1/8 of the uniaxial compressive strength, in general agreement with experimental observations (Scholz 1990), although values between 1/10 and 1/20 are observed (Paterson 1978). We use a value of 1/10 as a minimum estimate for the tensile strength of partially molten mantle (i.e. $500/10 = 50$ MPa). The value of 1/10 is justified because the fracture energy of viscous materials is generally high as a result of crack-tip blunting (i.e. crack-tip stresses are reduced by plastic deformation). With a *minimum* strength of 50 MPa, and of 500 kg m^{-3}, a compaction length *greater than* 10^4 m is required for hydrofracture.

Estimates for the compaction length as a function of porosity are shown for different grain sizes and values of the viscosity ratio $\eta_{\text{solid w/melt}}/\mu_1$ in figure 10. We have included the influence of melt on the viscosity of the solid using experimental data on partially molten dunites (Hirth & Kohlstedt 1995a,b). Data on the viscosity of the partially molten aggregate from both the dislocation and diffusion creep regimes are well fit by the relationship

$$\eta_{\text{solid w/melt}} = \eta_{\text{no melt}} \exp(-45\phi),$$

where $\eta_{\text{no melt}}$ is the viscosity of the melt free aggregate, and ϕ is porosity (Hirth & Kohlstedt 1997). As in §3a, permeabilities were calculated using Cheadle (1993). As in figure 9, we also show constraints imposed by mass balance for a one-dimensional melting column, which allow us to specify grain size and porosity for a given proportion of melt flow conduits, denoted χ.

Beneath Iceland, we estimate that the melt-free mantle viscosity is between 6×10^{17} and 6×10^{18} at a depth of 100 km and between 2×10^{18} and 2×10^{19} at 50 km. These values were calculated using experimental flow laws (Chopra & Paterson 1984; Hirth & Kohlstedt 1996), assuming that the melting region is H$_2$O-free and that deformation is accommodated by dislocation creep. As in §3a, melt viscosity beneath Iceland is taken to be 2 Pa s. The minimum compaction length for hydrofracture, on the order of 10^4 m, is only approached beneath Iceland if the higher mantle viscosity estimate is used, grain size is on the order of 10 mm, and porosities are greater than 3% (figure 10). Recall that for Iceland, melt velocities implied by ^{230}Th excess, 1 m yr^{-1}, can be attained with $F = 15\%$, $W = 10$ mm yr^{-1}, $\chi = 1\%$, grain size of 4 mm, and $\phi = 1.7\%$ (§3a). At these conditions compaction lengths will be $< 10^4$ m beneath Iceland, even in the high viscosity mantle case. This result is fairly general for $\chi \geqslant 1\%$. However, melt velocities inferred from post-glacial volcanism, 50 m yr^{-1}, can be attained with $F = 15\%$, $W = 10$ mm yr^{-1}, $\chi = 0.1\%$, grain size of 10 mm, and $\phi = 2\%$. In the high viscosity case, these conditions imply a compaction length close to 10^4 in the shallow part of the upwelling mantle. Thus, hydrofracture might be possible in the upper part of the melting regime in Iceland, provided that (1) melting is already focused in high ϕ channels with coarse grains, (2) mantle viscosity is high, and (3) the post-glacial burst of volcanism was produced by melting due to deglaciation.

Beneath a normal mid-ocean ridge, where the mantle is cooler, the melt-free solid viscosity is estimated to be between 5×10^{19} and 5×10^{20} over the interval from 25 to 50 km. Note that the commonly used value of around 1×10^{19} is estimated for an asthenospheric channel beneath a rigid plate (Hager 1991) where viscosities may be lower than in the melting region under ridges, due to the presence of dissolved hydrogen in mantle minerals outside the melting region (Hirth & Kohlstedt 1996). The melt viscosity beneath a normal ridge will be close to 10 Pa s.

For normal mid-ocean ridges, we show solutions to the mass balance equations, with $\chi = 100$ and 1%, and mantle upwelling of 10 and 100 mm yr^{-1}. Diffuse porous flow ($\chi = 100\%$) is restricted to very low porosities, permeabilities and compaction lengths, and therefore cannot satisfy melt velocity constraints or produce hydrofracture. Larger grain size and/or higher porosity can satisfy velocity constraints, but lead to compaction lengths > 10^4 m with the higher mantle viscosity estimates. For example, the melt velocity requirement from ^{230}Th excess is satisfied for $F = 10\%$, $W = 10$ mm yr^{-1}, $\chi = 1\%$, grain size of 4 mm, and $\phi = 3\%$, corresponding to a compaction length of ca. 10^4 m in the high viscosity case. For melt velocities ca. 50 m yr^{-1}, inferred from Th/Ra isotopes if ^{226}Ra excess is produced at the base of the melting column, we found that for a grain size of 10 mm, χ must be less than 0.02% and $\phi \gtrsim 6\%$ (§ 3 a). Even for low estimates of mantle viscosity, these values lead to a compaction length approaching 10^4 m.

In summary, the conditions for hydrofracture may be approached beneath Iceland and normal mid-ocean ridges, but only if the pressure difference between liquid and solid is close to the maximum possible value, and if flow is previously focused into high porosity channels composed of coarse-grained rock. Thus, coarse-grained dunite conduits with high porosities could undergo hydrofracture. However, recall that (1) 10^4 m is the estimate of the *minimum* length of a porous column sufficient for hydrofracture, (2) the *maximum possible* ΔP in a compacting system is $\Delta \rho g \delta_c$, and (3) the maximum ΔP will be approached only if compaction starts suddenly in a previously static system or if there is an overlying drop in permeability. An overlying drop in the permeability is likely only in the shallowest parts of the melting region beneath mid-ocean ridges, at the spinel–plagioclase peridotite transition (Asimow et al. 1995) and/or at the base of the conductively cooled lithosphere (Kelemen et al. 1996).

(c) Would a percolation threshold increase melt pressure? Probably not.

Nicolas (1986, 1990) has proposed that permeability in fertile lherzolite is limited (see Toramaru & Fujii 1986; Faul 1997), so that a large intergranular porosity is created in ascending mantle until a 'percolation threshold' is suddenly exceeded over a large vertical distance. This scenario would serve to minimize the effect of compaction, which tends to decrease solid–liquid pressure differences to less than $\Delta \rho g \delta_c$ over time. It is unclear why a percolation threshold should be attained suddenly over a large vertical distance. One can imagine an arbitrary initial condition in which an upwelling column above the solidus is devoid of melt. Over time, decompression melting could produce a large vertical distance with constant porosity in this column. However, beneath the region of constant porosity and above the solidus, this process must form a region with non-constant porosity, decreasing gradually with depth to a porosity of zero at the solidus. After the overlying column with constant porosity passed through the percolation threshold, the mantle beneath the interconnected column, with porosities smaller than the percolation threshold, would still retain melt. This leads to a steady state in which partially molten peridotite becomes interconnected over a small vertical interval at a fixed height above the solidus, the point where the melt fraction exceeds the percolation threshold. Thus, it is unclear how a large column of partially molten peridotite can suddenly become interconnected in an actively upwelling mantle system. However, as outlined in §3b, increasing grain size and melt flux in porous flow channels as they develop might gradually cause liquid overpressure to approach the tensile strength of partially molten peridotite. Whereas Nicolas (1986) has proposed that dunites form around fractures, in the scenario envisioned in §3b, fractures would form within porous dunite conduits. And, although this process does not include any role for a percolation threshold, it could give rise to a periodic hydrofracture process similar in other respects to that envisioned by Nicolas (1986, 1990).

(d) Cracks from above?

Fowler & Scott (1997) have proposed that fractures originating in the lithosphere may propagate downward into the viscous, partially molten mantle, initiating melt-filled cracks. The central point is that, even when the mantle is 'too weak' for hydrofractures to form in situ, brittle fractures nucleated in the lithosphere may propagate too fast for viscous dissipation to relieve stress concentrations, and therefore may penetrate to arbitrary depths. It is unclear whether fractures initiated in this way can tap a large enough area of partially molten peridotite by suction to form melt-filled conduits.

(e) Closed, melt-filled conduits in porous media are not chemically isolated

The past few sections have concentrated on physical mechanisms for melt extraction in the upwelling asthenosphere beneath ridges. Hydrofracture seems unlikely to arise if porous melt transport is diffuse, but hydrofracture may occur under some circumstances within conduits for focused porous flow. In addition, as noted in §2a (vi), the RII may be capable of producing melt-filled conduits without hydrofracture. Formation of liquid-filled voids as a result of dissolution has been demonstrated for rigid, soluble porous media in a variety of analogue experiments (Daccord 1987; Daccord & Lenormand 1987), and is likely to occur in viscous media as well (Kelemen et al. 1995a; Aharonov et al. 1995). Mechanical instabilities (§2a (vii)) could also give rise to melt-filled voids.

Regardless of how they form, it is possible to constrain the extent of melt-filled conduits. The predicted flux in planar or cylindrical conduits is huge compared to melting and crustal production rates at ridges. Poiseuille flow in a melt-filled channel with a radius or half width of 0.5 m will have a cross-sectional flux of ca. 10^6 m^2 yr^{-1}, at least 1000 times greater than cross-sectional rates of oceanic crustal production (10^2–10^3 m^2 yr^{-1}). Thus, melt-filled conduits must be transient and rare. Furthermore, since the flux greatly exceeds the melt production rate, and because a conduit cannot instantaneously 'suck' all the melt from the entire melting region, it is unlikely that a melt-filled conduit could extend from the base of the melting region to the base of the crust. Therefore, transient melt-filled conduits – if they form – are likely to be closed at both ends. For this reason, melt in such a conduit must flow outward, into the surrounding peridotite, at the top and inward, from the peridotite, at the base. Illustrations of this flow pattern, for static conduits created by dissolution in rigid porous media, are given by Kelemen *et al.* (1995*a*). Similar flow patterns can be inferred from the results of Richardson *et al.* (1996) for moving, melt-filled conduits, closed at both ends, in viscous porous media. Thus, because it intrinsically involves porous flow through surrounding rocks, magma transport in melt-filled conduits probably cannot account for the requirement of chemical isolation of melt from residual mantle peridotite (§ 1 *b*).

(*f*) *Pre-existing dunite zones around melt-filled conduits*

If melt-filled conduits pass through dunite channels, magma within them will be isolated from reaction with residual mantle peridotite (figure 11). How might this be explained? Small dunite reaction zones might form around melt-filled conduits as a result of divergent porous flow from the top of the conduit. The solubility of Opx in adiabatically ascending basaltic liquid is about 1% kbar^{-1} of decompression (§ 2 *a* (iv)). Thus a given mass of liquid rising from 90 km along an adiabatic PT gradient can convert an equal mass of peridotite containing 30% pyroxene to dunite before it becomes pyroxene saturated. To preserve liquid compositions far from Opx saturation, the mass of dunite formed/mass of liquid in the conduit must be less than 1. However, Richardson *et al.* (1996) find that porous flow from a conduit into a viscous porous wall rock penetrates the wall rock to a radial distance of about one compaction length. Compaction lengths in mantle peridotite with $\phi > 0.01\%$ may vary from 10 to 10^4 m (figure 10). Thus, liquid within a melt-filled conduit ascending over a depth interval of 50 km would undergo porous flow through a cross-sectional area of 10–500 km^2 of peridotite host rock. Either (1) conduits must have cross-sectional areas much larger than this, which is unlikely, or (2) melt-filled conduits, closed at the top and bottom, must traverse pre-existing dunite channels in order to retain liquids far from Opx saturation. Perhaps the latter scenario is plausible. In § 3 *b*, we showed how hydrofractures, if they arise, would form within pre-existing conduits for focused porous flow of melt (dunites). This is equally likely for melt-filled tubes arising from the RII and/or mechanical instabilities. Wiggins & Spiegelman (1995) have shown a tendency for three-dimensional solitary waves in viscous peridotite to repeatedly follow specific pathways of high permeability. Closed, melt-filled fractures or tubes may be very similar in their behaviour to solitary waves. A third possibility, of course, is that melt-filled fractures or tubes never form, and *all* melt is transported by focused porous flow in dunite channels.

4. Plate scale melt transport: how is melt flow focused to the ridge?

A variety of models have been proposed to explain focusing of magmatic accretion to within a few kilometres of the ridge axis at both fast and slow spreading ridges (Sleep 1984; Whitehead *et al.* 1984; Crane 1985; Rabinowicz *et al.* 1987; Phipps Morgan 1987; Spiegelman & McKenzie 1987; Scott & Stevenson 1989; Sparks & Parmentier 1991; Scott 1992; Spiegelman 1993*c*). Some aspects of these models are relevant to the physical mechanisms of melt transport.

(*a*) *Fracture is not an effective mechanism for focusing to the ridge*

Sleep (1984) and Phipps Morgan (1987) proposed that fractures parallel to the maximum compressive stress, in upwelling mantle undergoing corner flow, will not focus

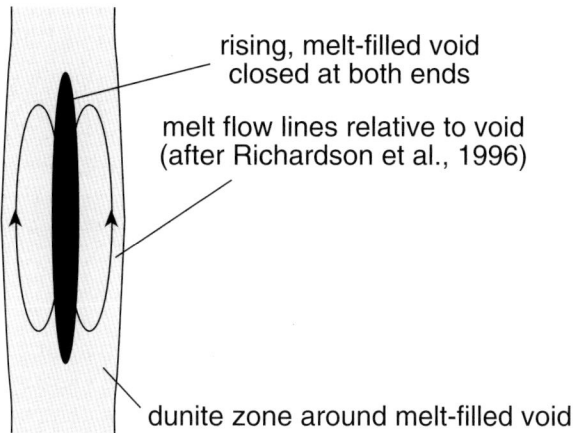

Figure 11. Schematic illustration of a melt-filled void (e.g. a fracture or dissolution channel) closed at both top and bottom, rising through partially molten, porous, viscously compacting mantle peridotite. As shown by Richardson et al. (1996), liquid will flow outward from the top and inward at the bottom of the void, with flow lines extending radially outward for about one compaction length. Thus, in order to maintain observed disequilibrium between melts and mantle peridotite, such melt-filled voids must move through a pre-existing dunite zone.

melt flow, and instead will be distributed over a region 40 km wide on either side of the ridge. Nicolas (1990) questioned this, pointing out that if solid mantle upwelling itself is focused beneath the ridge due to buoyancy-driven flow, the stress field will be different from that in passive flow models. He calculated that the maximum compressive stress would be sub-horizontal (dip less than 30°) from 35 to 15 km depth, trapping melt in sills. These sills would steepen to become subvertical dikes at less than 15 km. Transport of melt in sills rising with the solid velocity over 20 km would result in net transport times longer than 10^5 years. This is too slow to account to preserve ^{230}Th excess in MORB, and therefore seems unlikely. However, if hydrofractures follow pre-existing conduits for focused porous flow, and the pre-existing conduits coalesce toward the ridge, this would produce both flow in fractures and focused crustal accretion.

(b) Focused solid flow plus channels?

Whitehead et al. (1984) and Crane (1985) proposed that buoyancy due to melt and to chemical depletion in decompressing mantle would give rise to focused upwelling of peridotite beneath ridges. This influential idea spawned a series of models for focused, buoyancy-driven upwelling (Rabinowicz et al. 1987; Scott & Stevenson 1989; Turcotte & Phipps Morgan 1992; Scott 1992). Diapiric structures are observed in mantle from beneath a spreading ridge in the Oman ophiolite (Nicolas 1989; Ceuleneer & Rabinowicz 1992; Nicolas & Boudier 1995). However, some of types of focused solid flow are not viable in the sense that they produce liquids very different from MORB (Spiegelman 1996). In addition, Hirth & Kohlstedt (1996) suggest that – as a consequence of dehydration due to partial melting – mantle viscosities may be too high to produce buoyancy-driven flow strong enough to account for focusing of melt to the ridge axis. In any case, melt flow in chemically isolated conduits is required to preserve disequilibrium between MORB and mantle peridotite (§ 1 b). Focused solid upwelling, combined with formation of flow channels, might account for melt focusing to ridge axes.

(c) Coalescing dissolution channels? Maybe.

As noted in §2a(vi), dissolution channels formed as a consequence of the RII may coalesce downstream where solubility of the solid matrix increases downstream. Aharonov et al. (1995) showed that, in the initial stages of the instability, the preferred wavelength in channel spacing increases with increasing permeability. Channels on a small scale may be regarded as 'pores' on a larger scale. Thus, the larger scale permeability is higher, and the preferred wavelength for channels may be larger. We inferred that this leads to coalescence of channels in time and in space. Initial results of numerical modelling of the RII, illustrated in figure 7, show coalescence in time and space. In these preliminary results, the spacing of the largest channels depends mainly on the compaction length (Spiegelman et al. 1996).

Can coalescence of channels formed by the RII lead to focused melt extraction beneath mid-ocean ridges? Maybe. At first, it seems unlikely, since the aspect ratio (height/width) for coalescence of channels in figure 7 is larger than 1, whereas the aspect ratio of the melting region must be close to 1 (Plank & Langmuir 1992). However, our preliminary numerical modelling has been in a static solid reference frame, whereas beneath a mid-ocean ridge, melt generated near the solidus rises into mantle in which a conduit geometry has already been established. There may be large, pre-existing dissolution channels near the centre of the upwelling system that 'suck' melt inward. In addition, numerical modelling has been for dissolution reactions alone, so that porosities between channels become very small. With decompression melting as well as dissolution, lateral flow from one conduit into another may be facilitated by larger inter-channel porosities. For these reasons, lateral pressure gradients due to dissolution channels beneath mid-ocean ridges might lead to coalescence with smaller aspect ratios than in figure 7. Also, as noted in §4d, there may be several other causes of lateral pressure gradients which would enhance coalescence of a dissolution channel network.

Isotopic heterogeneity in MORB within a single ridge segment (Langmuir et al. 1992) indicates that a single network of coalescing conduits may not be sufficient to explain melt extraction. Instead, several networks might converge at the base of a ridge segment, each tapping spatially and chemically distinct domains within the upwelling mantle. Therefore, the aspect ratio of individual channel networks need not be (must not be?) as small as the aspect ratio of the entire melting regime, and additional factors must combine to ensure that all channel networks converge near the ridge axis.

(d) Other mechanisms for lateral focusing of porous flow

Sparks & Parmentier (1991), and Spiegelman (1993c) proposed that partial crystallization of melt rising into the conductively cooled lithosphere on either side of a ridge may create a permeability barrier that impedes vertical melt flow. Because the base of the lithosphere slopes upward toward the ridge axis, like a tent, ascending melt would migrate toward the ridge below the permeability barrier in a high porosity band. In its simplest form, this hypothesis is inconsistent with the geochemical constraints reviewed in §§1a, b. All abyssal peridotite and ophiolite mantle must pass through the roof of the tent. MORB and mantle peridotite would not preserve evidence for disequilibrium if solid flow lines all passed through a horizon in which aggregated, ascending melt was transported to the ridge axis. However, if combined with the presence of channels for chemically isolated melt flow within the roof, the tent mechanism could play a role in focusing melt to the ridge axis.

Phipps Morgan (1987) and Spiegelman & McKenzie (1987) showed that pressure gradients in viscously deforming peridotite undergoing corner flow might focus porous flow of melt toward a ridge axis. Required mantle viscosities, *ca.* 10^{20} Pa s, were higher than contemporary estimates of the viscosity of the mantle beneath ridge axes. However, a more recent viscosity estimate suggests that this process may be viable (Hirth & Kohlstedt 1996). Phipps Morgan (1987) also suggested that anisotropy in permeability might focus melt to the ridge axis. Field data from the Josephine peridotite confirm that melt does preferentially flow along the foliation in peridotites with a strong lattice preferred orientation (Kelemen & Dick 1995). Since diffuse porous flow cannot be the primary mode of MORB extraction from the mantle, this mechanism must be combined with the formation of chemically isolated conduits to explain magma transport beneath ridges.

5. Conclusion

For the moment, a clear picture has emerged for many aspects of melt extraction beneath ridges. Most melt is extracted from the adiabatically upwelling mantle by focused flow in dunite conduits; around these conduits, small amounts of melt migrate by reactive porous flow. The lowest pressure melts, 'ultra-depleted' liquids from the shallow mantle just beneath the conductive lithosphere, may be extracted primarily in fractures unrelated to dunite conduits.

Melt velocity constraints from U/Th isotope systematics can be satisfied by porous flow of melt from the base of the melting regime to the base of the crust. Provided that conduit porosities are larger than source porosities and melt-free mantle viscosities are less than *ca.* 10^{20} Pa s, flow in fractures is not required. Melt velocities of *ca.* 50 m yr^{-1} have been inferred from patterns of post-glacial volcanism and from ^{226}Ra excess. If these inferences are correct, the minimum conditions for hydrofracture may be reached in the upwelling mantle beneath ridges, even if mantle viscosities are less than *ca.* 10^{20} Pa s. Hydrofracture could arise periodically within coarse, high porosity dunite conduits, especially in the shallowest part of the melting region near the transition to conductively cooled lithosphere. Melt-filled tubes might also form in the asthenosphere by dissolution and/or mechanical instabilities within dunite conduits. If they form, fractures or tubes are probably closed at the top and bottom, and follow pre-existing dunites in order to maintain observed disequilibrium between MORB and shallow mantle peridotites. On a plate scale, focusing of melt flow toward ridges is best explained as a combination of porous flow mechanisms including the coalescence of dissolution channels, 'suction' of melt toward the ridge resulting from corner flow, flow along high permeability planes in foliated peridotite, and, perhaps, flow along a porous conduit beneath a permeability barrier near the base of the lithosphere.

There are several key areas of uncertainty regarding mechanisms of melt extraction. These include the relationship between permeability, porosity, and grain size in partially molten mantle peridotite, the tensile fracture strength of partially molten peridotite, the extent and nature of 'near-fractional' melting processes, the depth of formation of ^{226}Ra isotopic disequilibrium, the proportion of melt extraction due to hydrofracture in the asthenosphere, and the factors governing the size, shape and fluid mechanics of melt-filled cracks or tubes in viscous, porous media. In addition, substantial insight can be expected as the grain scale kinetics of melt/rock interaction are more fully explored.

Our thanks to Joe Cann, who suggested that we write this paper, and to Adolphe Nicolas, Benoit Ildefonse, Françoise Boudier, Einat Aharonov, Ken Sims, Stan Hart, Lucy Slater, Dan McKenzie, Chris Richardson, Matthew Jull, Karl Gronvold, Uli Faul, Marc Hirschmann, Dave Scott, Jack Whitehead, Dave Kohlstedt, Martha Daines, Brian Evans, Ed Stolper, and Kent Ross for advice, encouragement, reviews, and access to unpublished manuscripts. Work on this paper and collection of previously unpublished data were supported by US National Science Foundation grants OCE-9314013, OCE-9416616 and EAR-9418228 (P.B.K.), and EAR-9405845 (G.H.).

References

Aharonov, E., Whitehead, J. A., Kelemen, P. B. & Spiegelman, M. 1995 Channeling instability of upwelling melt in the mantle. *J. Geophys. Res.* **100**, 20 433–20 450.

Alabaster, T., Pearce, J. A. & Malpas, J. 1982 The volcanic stratigraphy and petrogenesis of the Oman ophiolite complex. *Contrib. Mineral. Petrol.* **81**, 168–183.

Allan, J. F. & Dick, H. J. B 1996 Cr-rich spinel as a tracer for melt migration and melt-wall rock interaction in the mantle: Hess Deep, Leg 147. *Sci. Res. Ocean Drill. Prog.* **147**, 157–172.

Allègre, C. J., Montigny, R. & Bottinga, Y. 1973 Cortège ophiolitique et cortège océanique, géochimie comparée et mode de genèse. *Bull. Soc. géol. France* **15**, 461–477.

Arai, S. & Matsukage, K. 1996 Petrology of gabbro-troctolite-peridotite complex from Hess Deep, Equatorial Pacific: implications for mantle-melt interaction within the oceanic lithosphere. *Sci. Res. Ocean Drill. Prog.* **147**, 135–155.

Asimow, P. D., Hirschmann, M. M., Ghiorso, M. S., O'Hara, M. J. & Stolper, E. M. 1995 The effect of pressure-induced solid–solid phase transitions on decompression melting of the mantle. *Geochim. Cosmochim. Acta* **59**, 4489–4506.

Augé, T. 1987 Chromite deposits in the northern Oman ophiolite: mineralogical constraints. *Mineral. Deposita* **22**, 1–10.

Beattie, P. 1993a The generation of uranium series disequilibria by partial melting of spinel peridotite: constraints from partitioning studies. *Earth Planet. Sci. Lett.* **117**, 379–391.

Beattie, P. 1993b Uranium–thorium disequilibria and partitioning on melting of garnet peridotite. *Nature* **363**, 63–65.

Benoit, M., Polvé, M. & Ceuleneer, G. 1997 Trace element and isotopic characterization of mafic cumulates in a fossil mantle diapir (Oman ophiolite). *Chem. Geol.* (In the press.)

Berger, E. T. & Vannier, M. 1984 Les dunites en enclaves dans les basaltes alcalins des iles océaniques: approche pétrologique. *Bull. Miner.* **107**, 649–663.

Beeman, M. L. & Kohlstedt, D. L. 1992 Deformation of fine-grained aggregates of olivine plus melt at high temperatures and pressures. *J. Geophys. Res.* **98**, 6443–6452.

Bodinier, J. L. 1988 Geochemistry and petrogenesis of the Lanzo peridotite body, Western Alps. *Tectonophysics* **149**, 67–88.

Bodinier, J. L., Vasseur, G., Vernieres, J., Dupuy, C. & Fabries, J. 1990 Mechanisms of mantle metasomatism: geochemical evidence from the Lherz orogenic peridotite. *J. Petrol.* **31**, 597–628.

Boudier, F. & Coleman, R. G. 1981 Cross section through the peridotite in the Samail Ophiolite, southeastern Oman Mountains. *J. Geophys. Res.* **86**, 2573–2592.

Boudier, F. & Nicolas, A. 1972 Fusion partielle gabbroique dans la Iherzolite de Lanzo. *Bull. Suisse Min. Pet.* **52**, 39–56.

Boudier, F. & Nicolas, A. 1995 Nature of the Moho Transition Zone in the Oman ophiolite. *J. Petrol.* **36**, 777–796.

Bussod, G. Y. & Christie, J. M. 1991 Textural development and melt topology in spinel lherzolite experimentally deformed at hypersolidus conditions. *J. Petrol.* (Special Lherzolite Issue), pp. 17–39.

Cannat, M. 1996 How thick is the magmatic crust at slow spreading oceanic ridges? *J. Geophys. Res.* **101**, 2847–2857.

Cannat, M. & Casey, J. F. 1995 An ultramafic lift at the Mid-Atlantic Ridge: successive stages of magmatism in serpentinized peridotites from the 15° N region. In *Mantle and lower crust exposed in oceanic ridges and in ophiolites* (ed. R. L. M. Vissers & A. Nicolas), pp. 5–34. Amsterdam: Kluwer.

Ceuleneer, G. & Rabinowicz, M. 1992 Mantle flow and melt migration beneath oceanic ridges: models derived from observations in ophiolites. *Geophys. Monograph* **71**, 123–154.

Chadam, J., Hoff, D., Merino, E., Ortoleva, P. & Sen, A. 1986 Reactive infiltration instabilities. *J. Appl. Math.* **36**, 207–221.

Cheadle, M. J. 1993 The physical properties of texturally equilibrated partially molten rocks. *Eos* **74**, 283.

Chopra, P. N. & Paterson, M. S. 1984 The role of water in the deformation of dunite. *J. Geophys. Res.* **89**, 7861–7876.

Crane, K. 1985 The spacing of rift axis highs: dependence upon diapiric processes in the underlying asthenosphere? *Earth Planet. Sci. Lett.* **72**, 405–414.

Daccord, G. 1987 Chemical dissolution of a porous medium by a reactive fluid. *Phys. Rev. Lett.* **58**, 479–482.

Daccord, G. & Lenormand, R. 1987 Fractal patterns from chemical dissolution. *Nature* **325**, 41–43.

Daines, M. J. & Kohlstedt, D. L. 1993 A laboratory study of melt migration. *Phil. Trans. R. Soc. Lond.* A **342**, 43–52.

Daines, M. J. & Kohlstedt, D. L. 1994 The transition from porous to channelized flow due to melt/rock reaction during melt migration. *Geophys. Res. Lett.* **21**, 145–148.

DePaolo, D. J. 1981 Trace element and isotopic effects of combined wallrock assimilation and fractional crystallization. *Earth Planet. Sci. Lett.* **53**, 189–202.

Dick, H. J. B. 1976 The origin and emplacement of the Josephine peridotite of Southwestern Oregon. Ph.D. thesis, Yale University, New Haven, NJ, USA.

Dick, H. J. B. 1977 Evidence for partial melting in the Josephine peridotite. *Bull. Oregon St. Dept. Geol. Miner. Ind.* **96**, 63–78.

Dick, H. J. B. 1989 Abyssal peridotites, very slow spreading ridges and ocean ridge magmatism. In *Magmatism in the ocean basins* (ed. A. D. Saunder & M. J. Norry), pp. 71–105. Geological Society Special Publications.

Dick, H. J. B. & Bullen, T. 1984 Chromian spinel as a petrogenetic indicator in abyssal and alpine-type peridotites and spatially associated lavas. *Contrib. Mineral. Petrol.* **86**, 54–76.

Dick, H. J. B. & Natland, J. H. 1996 Late stage melt evolution and transport in the shallow mantle beneath the East Pacific Rise. *Sci. Res. Ocean Drill. Prog.* **147**, 103–134.

Elthon, D. & Scarfe, C. M. 1980 High-pressure phase equilibria of a high-magnesia basalt: implications for the origin of mid-ocean ridge basalts. *Carnegie Inst. Wa. Yrbk*, pp. 277–281.

Elthon, D. 1992 Chemical trends in abyssal peridotites: refertilization of depleted suboceanic mantle. *J. Geophys. Res.* **97**, 9015–9025.

Farrier, K., Kelemen, P. B. & Shimizu, N. 1997 Off-axis gabbro and pyroxenite dikes in the Oman ophiolite: transport of ultra-depleted melts in lithospheric fractures. (In preparation.)

Faul, U. H. 1997 The permeability of partially molten upper mantle rocks from experiments and percolation theory. *J. Geophys. Res.* (In the press.)

Forsyth, D. W. 1992 Geophysical constraints on mantle flow and melt generation beneath mid-ocean ridges. *Geophys. Monograph* **71**, 1–65.

Forsyth, D. W. 1996 Partial melting beneath a Mid-Atlantic Ridge segment detected by teleseismic PKP delays. *Geophys. Res. Lett.* **23**, 463–466.

Fowler, A. C. & Scott, D. R. 1997 Hydraulic crack propagation in a porous medium. *Geophys. J. Int.* (In the press.)

Gaetani, G. A., DeLong, S. E. & Wark, D. A. 1995 Petrogenesis of basalts from the Blanco Trough, northeast Pacific: inferences for off-axis melt generation. *J. Geophys. Res.* **100**, 4197–4214.

Gast, P. W. 1968 Trace element fractionation and the origin of tholeiitic and alkaline magma types. *Geochim. Cosmochim. Acta* **32**, 1057–1089.

Gruau, G., Bernard-Griffiths, J., Lecuyer, C., Henin, O., Mace, J. & Cannat, M. 1995 Extreme Nd isotopic variation in the Trinity ophiolite complex and the role of melt/rock reactions in the oceanic lithosphere. *Contrib. Mineral. Petrol.* **121**, 337–350.

Gregory, R. T. 1984 Melt percolation beneath a spreading ridge: evidence from the Semail peridotite, Oman. *Geol. Soc. Lond. Spec. Publ.* **13**, 55–62.

Hager, B. H. 1991 Mantle viscosity: a comparison of models from postglacial rebound and from the geoid, plate driving forces, and advected heat flux. In *Glacial isostasy, sea-level and mantle rheology* (ed. R. Sabadini *et al.*), pp. 493–513. Dordrecht: Kluwer Academic.

Hart, S. R. 1993 Equilibrium during mantle melting: a fractal tree model. *Proc. Natn. Acad. Sci.* **90**, 11914–11918.

Hart, S. R. & Dunn, T. 1993 Experimental cpx/melt partitioning of 24 trace elements. *Contrib. Mineral. Petrol.* **113**, 1–8.

Hirschmann, M. M. & Stolper, E. M. 1996 A possible role for garnet pyroxenite in the origin of the garnet signature in MORB. *Contrib. Mineral. Petrol.* **124**, 185–208.

Hirth, G. & Kohlstedt, D. L. 1995 Experimental constraints on the dynamics of the partially molten upper mantle. 2. Deformation in the dislocation creep regime. *J. Geophys. Res.* **100**, 15441–15449.

Hirth, G. & Kohlstedt, D. L. 1995 Experimental constraints on the dynamics of the partially molten upper mantle: deformation in the diffusion creep regime. *J. Geophys. Res.* **100**, 1981–2001.

Hirth, G. & Kohlstedt, D. L. 1996 Water in the oceanic mantle: implications for rheology, melt extraction, and the evolution of the lithosphere. *Earth Planet. Sci. Lett.* **144**, 93–108.

Hoefner, M. L. & Fogler, H. S. 1988 Pore evolution and channel formation during flow and reaction in porous media. *AIChE Jl* **34**, 45–54.

Hopson, C. A., Coleman, R. G., Gregory, R. T., Pallister, J. S. & Bailey, E. H. 1981 Geologic section through the Samail Ophiolite and associated rocks along a Muscat-Ibra Transect, Southeastern Oman Mountains. *J. Geophys. Res.* **86**, 2527–2544.

Iwamori, H. 1993 A model for disequilibrium mantle melting incorporating melt transport by porous and channel flows. *Nature* **366**, 734–737.

Iwamori, H. 1994 ^{238}U/^{230}Th/^{226}Ra and ^{235}U/^{231}Pa disequilibria produced by mantle melting with porous and channel flows. *Earth Planet. Sci. Lett.* **125**, 1–16.

Jin, Z.-M., Green, H. W. & Zhou, Y. 1994 Melt topology in partially molten mantle peridotite during ductile deformation. *Nature* **372**, 164–167.

Johnson, K. T. M. & Dick, H. J. B. 1992 Open system melting and temporal and spatial variation of peridotite and basalt at the Atlantis II Fracture Zone. *J. Geophys. Res.* **97**, 9219–9241.

Johnson, K. T. M., Dick, H. J. B. & Shimizu, N. 1990 Melting in the oceanic upper mantle: an ion microprobe study of diopsides in abyssal peridotites. *J. Geophys. Res.* **95**, 2661–2678.

Jull, M. & McKenzie, D. 1996 The effect of deglaciation on mantle melting beneath Iceland. *J. Geophys. Res.* **101**, 21815–21828.

Kelemen, P. B. 1990 Reaction between ultramafic rock and fractionating basaltic magma. I. Phase relations, the origin of calc-alkaline magma series, and the formation of discordant dunite, *J. Petrol.* **31**, 51–98.

Kelemen, P. B. & Dick, H. J. B. 1995 Focused melt flow and localized deformation in the upper mantle: juxtaposition of replacive dunite and ductile shear zones in the Josephine peridotite, SW Oregon. *J. Geophys. Res.* **100**, 423–438.

Kelemen, P. B., Dick, H. J. B. & Quick, J. E. 1992 Formation of harzburgite by pervasive melt/rock reaction in the upper mantle. *Nature* **358**, 635–641.

Kelemen, P. B., Whitehead, J. A., Aharonov, E. & Jordahl, K. A. 1995*a* Experiments on flow focusing in soluble porous media, with applications to melt extraction from the mantle. *J. Geophys. Res.* **100**, 475–496.

Kelemen, P. B., Shimizu, N. & Salters, V. J. M. 1995*b* Extraction of mid-ocean-ridge basalt from the upwelling mantle by focused flow of melt in dunite channels. *Nature* **375**, 747–753.

Kelemen, P.B., Koga, K. & Shimizu, N. 1997 Geochemistry of gabbro sills in the crust/mantle transition zone of the Oman ophiolite: implications for the origin of the oceanic lower crust. *Earth Planet. Sci. Lett.* (In the press.)

Kinzler, R. J. & Grove, T. L. 1992 Primary magmas of mid-ocean ridge basalts. 2. Applications. *J. Geophys. Res.* **97**, 6907–6926.

Klein, E. & Langmuir, C. H. 1987 Global correlations of ocean ridge basalt chemistry with axial depth and crustal thickness. *J. Geophys. Res.* **92**, 8089–8115.

Kushiro, I. 1986 Viscosity of partial melts in the upper mantle. *J. Geophys. Res.* **91**, 9343–9350.

Lago, B. L., Rabinowicz, M. & Nicolas, A. 1982 Podiform chromite ore bodies: a genetic model. *J. Petrol.* **23**, 103–125.

Langmuir, C. H., Bender, J. F., Bence, A. E., Hanson, G. N. & Taylor, S. R. 1977 Petrogenesis of basalts from the Famous area, Mid-Atlantic Ridge. *Earth Planet. Sci. Lett.* **36**, 133–156.

Leblanc, M. & Ceuleneer, G. 1992 Chromite crystallization in a multicellular magma flow: evidence from a chromitite dike in the Oman ophiolite. *Lithos* **27**, 231–257.

Lippard, S. J., Shelton, A. W. & Gass, I. G. 1986 *The ophiolite of northern Oman.* (178 pages.) Oxford: Blackwell.

Lundstrom, C. C., Gill, J., Williams, Q. & Perfit, M. R. 1995 Mantle melting and basalt extraction by equilibrium porous flow. *Science* **270**, 1958–1961.

Maaloe, S. 1981 Magma accumulation in the ascending mantle. *J. Geol. Soc. Lond.* **138**, 223–236.

Maaloe, S. & Scheie, A. 1982 The permeability controlled accumulation of primary magma. *Contrib. Mineral. Petrol.* **81**, 350–357.

Magde, L. S., Detrick, R. S. & the TERA Group 1995 Crustal and upper mantle contribution to the axial gravity anomaly at the southern East Pacific Rise. *J. Geophys. Res.* **100**, 3747–3766.

McKenzie, D. 1984 The generation and compaction of partially molten rock. *J. Petrol.* **25**, 713–765.

McKenzie, D. 1985 ^{230}Th–^{238}U disequilibrium and the melting processes beneath ridge axes. *Earth Planet. Sci. Lett.* **72**, 149–157.

McKenzie, D. 1987 The compaction of igneous and sedimentary rocks. *J. Geol. Soc. Lond.* **144**, 299–307.

McKenzie, D. 1989 Some remarks on the movement of small melt fractions in the mantle. *Earth Planet. Sci. Lett.* **95**, 53–72.

McKenzie, D. & Bickle, M. J. 1988 The volume and composition of melt generated by extension of the lithosphere. *J. Petrol.* **29**, 625–679.

Menzies, M., Kempton, P. & Dungan, M. 1985 Interaction of continental lithosphere and asthenospheric melts below the Geronimo Volcanic Field, Arizona, U.S.A. *J. Petrol.* **26**, 663–693.

Mercier, J.-C. 1980 Magnitude of the continental lithospheric stresses inferred from rheomorphic petrology. *J. Geophys. Res.* **85**, 6293–6303.

Nicolas, A. 1986 A melt extraction model based on structural studies in mantle peridotites. *J. Petrol.* **27**, 999–1022.

Nicolas, A. 1989 *Structures of ophiolites and dynamics of oceanic lithosphere.* (367 pages.) Dordrecht: Kluwer Academic.

Nicolas, A. 1990 Melt extraction from mantle peridotites: hydrofracturing and porous flow, with consequences for oceanic ridge activity. In *Magma transport and storage* (ed. M. P. Ryan), pp. 159–174. New York: Wiley.

Nicolas, A. & Boudier, F. 1995 Mapping oceanic ridge segments in Oman ophiolite. *J. Geophys. Res.* **100**, 6179–6197.

Niu, Y. & Batiza, R. 1993 Chemical variation trends at fast and slow spreading mid-ocean ridges. *J. Geophys. Res.* **98**, 7887–7902.

O'Hara, M. J. 1965 Primary magmas and the origin of basalts. *Scot. J. Geol.* **1**, 19–40.

Ozawa, K. 1994 Melting and melt segregation in the mantle wedge above a subduction zone: evidence from the chromite-bearing peridotites of the Miyamori ophiolite complex, northeastern Japan. *J. Petrol.* **35**, 647–678.

Ozawa, K. & Shimizu, N. 1995 Open-system melting in the upper mantle: constraints from the Miyamori–Hayachine ophiolite, northeastern Japan. *J. Geophys. Res.* **100**, 22315–22335.

Pallister, J. S. & Hopson, C. A. 1981 Samail ophiolite plutonic suite: field relations, phase variation, cryptic variation and layering, and a model of a spreading ridge magma chamber. *J. Geophys. Res.* **86**, 2593–2644.

Paterson, M. S. 1978 *Experimental rock deformation: the brittle field.* (254 pages.) Berlin: Springer.

Phipps Morgan, J. 1987 Melt migration beneath mid-ocean spreading centers. *Geophys. Res. Lett.* **14**, 1238–1241.

Plank, T. & Langmuir, C. H. 1992 Effects of the melting regime on composition of the oceanic crust. *J. Geophys. Res.* **97**, 19749–19770.

Quick, J. E. 1981a Petrology and petrogenesis of the Trinity peridotite, an upper mantle diapir in the eastern Klamath Mountains, Northern California. *J. Geophys. Res.* **86**, 11837–11863.

Quick, J. E. 1981b The origin and significance of large, tabular dunite bodies in the Trinity peridotite, Northern California. *Contrib. Mineral. Petrol.* **78**, 413–422.

Quick, J. E. & Gregory, R. T. 1995 Significance of melt-wall rock reaction: a comparative anatomy of three ophiolites. *J. Geol.* **103**, 187–198.

Rabinowicz, M., Ceuleneer, G. & Nicolas, A. 1987 Melt segregation and flow in mantle diapirs below spreading centers: evidence from the Oman ophiolite. *J. Geophys. Res.* **92**, 3475–3486.

Richardson, C. N. & McKenzie, D. 1984 Radioactive disequilibria from 2D models of melt generation by plumes and ridges. *Earth Planet. Sci. Lett.* **128**, 425–437.

Richardson, C. N., Lister, J. R. & McKenzie, D. 1996 Melt conduits in a viscous porous matrix. *J. Geophys. Res.* **101**, 20423–20432.

Riley Jr, G. N. & Kohlstedt, D. L. 1991 Kinetics of melt migration in upper mantle-type rocks. *Earth Planet. Sci. Lett.* **105**, 500–521.

Ross, K. & Elthon, D. 1993 Cumulates from strongly depleted mid-ocean ridge basalts. *Nature* **365**, 826–829.

Ross, K. & Elthon, D. 1997a Cumulus and post-cumulus crystallization in the oceanic crust: major and trace element geochemistry of Leg 153 gabbroic rocks. *Scientific results of the Ocean Drilling Program* **153**. (In the press.)

Ross, K. & Elthon, D. 1997b Extreme incompatible trace element depletion of diopside in residual mantle from south of the Kane Fracture Zone. *Scientific Results of the Ocean Drilling Program* **153**. (In the press.)

Salters, V. J. M. & Hart, S. R. 1989 The hafnium paradox and the role of garnet in the source of mid-ocean-ridge basalts. *Nature* **342**, 420–422.

Scholz, C. H. 1990 *The mechanics of earthquakes and faulting.* (439 pages.) Cambridge University Press.

Scott, D. R. 1992 Small-scale convection and mantle melting beneath mid-ocean ridges. *Geophys. Monograph* **71**, 327–352.

Scott, D. R. & Stevenson, D. J. 1989 A self-consistent model of melting, magma migration and buoyancy-driven circulation beneath mid-ocean ridges. *J. Geophys. Res.* **94**, 2973–2988.

Shaw, D. M. 1970 Trace element fractionation during anatexis. *Geochim. Cosmochim. Acta* **34**, 237–243.

Shimizu, N. 1997 The geochemistry of olivine-hosted melt inclusions in FAMOUS basalt ALV519-4-1. *Phys. Earth Planet. Int.* (In the press.)

Sims, K. W. W., DePaolo, D. J., Murrell, M. T., Baldridge, W. S., Goldstein, S. J. & Clague, D. A. 1995 Mechanisms of magma generation beneath Hawaii and mid-ocean ridges: uranium/thorium and samarium/neodymium isotopic evidence. *Science* **267**, 508–512.

Sleep, N. H. 1975 Formation of oceanic crust: some thermal constraints. *J. Geophys. Res.* **80**, 4037–4042.

Sleep, N. H. 1984 Tapping of magmas from ubiquitous mantle heterogeneities: an alternative to mantle plumes? *J. Geophys. Res.* **89**, 10029–10041.

Sleep, N. H. 1988 Tapping of melt by veins and dikes. *J. Geophys. Res.* **93**, 10255–10272.

Sparks, D. W. & Parmentier, E. M. 1991 Melt extraction from the mantle beneath mid-ocean ridges. *Earth Planet. Sci. Lett.* **105**, 368–377.

Spera, F. J. 1980 Aspects of magma transport. In *Physics of magmatic processes* (ed. R. B. Hargraves), pp. 263–323. Princeton University Press.

Spiegelman, M. 1993*a* Flow in deformable porous media. 1. Simple analysis. *J. Fluid Mech.* **247**, 17–38.

Spiegelman, M. 1993*b* Flow in deformable porous media. 2. Numerical analysis – the relationship between shock waves and solitary waves. *J. Fluid Mech.* **247**, 39–63.

Spiegelman, M. 1993*c* Physics of melt extraction: theory, implications and applications. *Phil. Trans. R. Soc. Lond.* A **342**, 23–41.

Spiegelman, M. 1996 Geochemical consequences of melt transport in 2-D: the sensitivity of trace elements to mantle dynamics. *Earth Planet. Sci. Lett.* **139**, 115–132.

Spiegelman, M. & Elliot, T. 1993 Consequences of melt transport for uranium series disequilibrium. *Earth Planet. Sci. Lett.* **118**, 1–20.

Spiegelman, M. & Kenyon, P. 1992 The requirements for chemical disequilibrium during magma migration. *Earth Planet. Sci. Lett.* **109**, 611–620.

Spiegelman, M. & McKenzie, D. 1987 Simple 2-D models for melt extraction at mid-ocean ridges and island arcs. *Earth Planet. Sci. Lett.* **83**, 137–152.

Spiegelman, M., Aharonov, E. & Kelemen, P. B. 1996 The compaction reaction: magma channel formation by reactive flow in deformable media. *Eos* **77**, F783.

Sobolev, A. V. 1997 Melt inclusions in minerals as a source of principal petrologic information. *Petrology* **4**. (In the press.)

Sobolev, A. V. & Shimizu, N. 1993 Ultra-depleted primary melt included in an olivine from the Mid-Atlantic Ridge. *Nature* **363**, 151–154.

Sobolev, A. V. & Shimizu, N. 1994 The origin of typical N-MORB: the evidence from a melt inclusion study. *Min. Mag.* A **58**, 862–863.

Stolper, E. 1980 A phase diagram for mid-ocean ridge basalts: preliminary results and implications for petrogenesis. *Contrib. Mineral. Petrol.* **74**, 13–27.

Stevenson, D. J. 1989 Spontaneous small-scale melt segregation in partial melts undergoing deformation. *Geophys. Res. Lett.* **16**, 1067–1070.

Tait, S., Jahrling, K. & Jaupart, C. 1992 The planform of compositional convection and chimney formation in a mushy layer. *Nature* **359**, 406–408.

Takahashi, N. 1992 Evidence for melt segregation towards fractures in the Horoman mantle peridotite complex. *Nature* **359**, 52–55.

Takazawa, E. 1996 Geodynamic evolution of the Horoman peridotite, Japan: consequence of lithospheric and asthenospheric processes. (562 pp.) Ph.D. thesis, MIT.

Takazawa, E., Frey, F. A., Shimizu, N., Obata, M. & Bodinier, J. L. 1992 Geochemical evidence for melt migration and reaction in the upper mantle. *Nature* **359**, 55–58.

Toramaru, A. & Fujii, N. 1986 Connectivity of melt phase in a partially molten peridotite. *J. Geophys. Res.* **91**, 9239–9252.

Tilton, G. R., Hopson, C. A. & Wright, J. E. 1981 Uranium-lead isotopic ages of the Samail ophiolite, Oman, with applications to Tethyan ocean ridge tectonics. *J. Geophys. Res.* **86**, 2763–2775.

Turcotte, D. L. & Phipps Morgan, J. 1992 The physics of magma migration and mantle flow beneath a mid-ocean ridge. *Geophys. Monograph* **71**, 155–182.

Von Bargen, N. & Waff, H. S. 1986 Permeabilities, interfacial areas and curvatures of partially molten systems: results of numerical computations of equilibrium microstructures. *J. Geophys. Res.* **91**, 9261–9276.

Waff, H. S. & Bulau, J. R. 1979 Equilibrium fluid distribution in an ultramafic partial melt under hydrostatic stress conditions. *J. Geophys. Res.* **84**, 6109–6114.

Waff, H. S. & Faul, U. H. 1992 Effects of crystalline anisotropy on fluid distribution in ultramafic partial melts. *J. Geophys. Res.* **97**, 9003–9014.

Wang, H. F. 1993 A double medium model for diffusion in fluid-bearing rock. *Contrib. Mineral. Petrol.* **114**, 357–364.

Wang, Y. & Merino, E. 1993 Oscillatory magma crystallization by feedback between the concentrations of the reactant species and mineral growth rates. *J. Petrol.* **34**, 369–382.

Wang, Y. & Merino, E. 1995 Origin of fibrosity and banding in agates from flood basalts. *Am. Jl Sci.* **295**, 49–77.

Whitehead, J. A., Dick, H. J. B. & Schouten, H. 1984 A mechanism for magmatic accretion under spreading centres. *Nature* **312**, 146–148.

Williams, R. W. & Gill, J. B. 1989 Effects of partial melting on the uranium decay series. *Geochim. Cosmochim. Acta* **53**, 1607–1619.

Wiggins, C. & Spiegelman, M. 1995 Magma migration and magmatic solitary waves in 3-D. *Geophys. Res. Lett.* **22**, 1289–1292.

Rift-plume interaction in the North Atlantic

BY ROBERT S. WHITE

Bullard Laboratories, Madingley Road, Cambridge CB3 0EZ, UK

The style of oceanic crustal formation in the North Atlantic is controlled by interaction between the Iceland mantle plume and the lithospheric spreading. There are three main tectonic regimes comprising: (a) oceanic crust formed without fracture zones, with spreading directions varying from orthogonal up to 30° oblique to the ridge axis; (b) oceanic crust with a normal slow-spreading pattern of orthogonal spreading segments separated by fracture zones; and (c) 20–35 km thick crust generated directly above the centre of the mantle plume along the Greenland–Iceland–Færoe Ridge. I show that the main control on the tectonic style is the temperature of the mantle beneath the spreading axis. A mantle temperature increase of as little as 50 °C causes an increase of about 30% in the crustal thickness, and thereby allows the mantle beneath the crust at the ridge axis to remain sufficiently hot that it responds to axial extension in a ductile rather than a brittle fashion. This generates crust without fracture zones and with an axial high rather than a median valley at the spreading centre. Using gravity, magnetic, bathymetric and seismic refraction data I discuss the mantle plume temperatures and flow patterns beneath the North Atlantic since the time of continental breakup, and the response of the crustal generation processes to these mantle temperature variations.

---

1. Introduction

The volume of melt generated at a mid-ocean ridge spreading centre is highly sensitive to the temperature of the underlying mantle. An increase of as little as 50 °C, just a few per cent of the normal asthenospheric potential temperature of about 1320 °C, causes a 50% increase in the volume of melt that is generated by decompression of the upwelling mantle (McKenzie & Bickle 1988; White et al. 1992). Those oceanic spreading ridges that lie above a region of abnormally hot mantle caused by a mantle plume are therefore affected significantly by the increased mantle temperatures.

Since lithospheric plates move relatively fast (typically 20–150 mm yr^{-1}) with respect to the underlying mantle plumes, there are many instances where spreading centres have moved across mantle plumes and have interacted with them. Directly above the rising core of a mantle plume, the result of the interaction is usually the generation of a ridge of igneous crust, often reaching 20–30 km or more in thickness; examples include the Rio-Grande Rise and the Walvis Ridge created by the interaction between the mantle plume now beneath Tristan da Cunha and the South Atlantic spreading centre; and the Chagos–Laccadive Ridge built when the Central Indian Ridge spreading centre lay above the Réunion mantle plume. Since the thermal anomalies in the asthenospheric mantle created by plumes often extend for

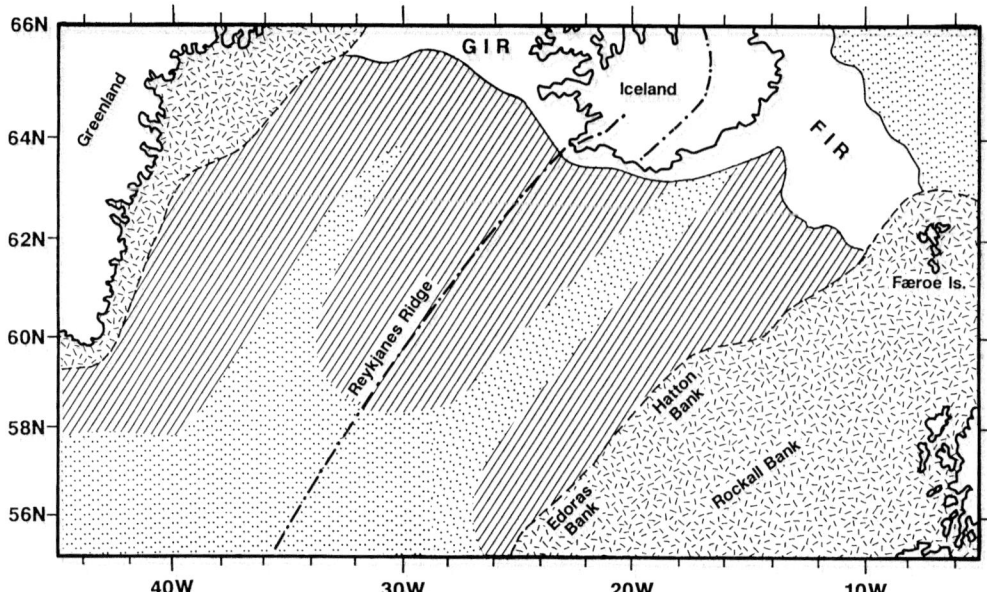

Figure 1. Outline of the main tectonomagmatic areas discussed in this paper. Shaded areas in northwest and southeast corners of the map represent stretched continental crust on the North American and Eurasian plates, respectively. Parallel shading shows oceanic crust devoid of fracture zones and dotted shading is oceanic crust with fracture zones, spreading orthogonally. Blank area is thick igneous crust above the Greenland–Iceland Ridge (GIR) and Færoe–Iceland Ridge (FIR).

distances of the order of 1000 km from the central core, the portions of the spreading axes intersecting these distal regions may also be affected. Typically, however, the temperatures in these distal regions are much lower than in the core of the plume, and the consequences are hard to detect, especially over old crust.

In this paper I discuss the interaction of the Reykjanes Ridge spreading centre in the northern North Atlantic with the Iceland mantle plume. This is a particularly good case study, because the spreading centre crosses directly above the mantle plume and is currently creating oceanic crust on the Reykjanes Ridge with relatively smooth topography and few fracture zones. This means that small fluctuations in the mantle temperature of only a few tens of °C and minor flow rate variations create marked perturbations in the gravity and topographic signatures that can be readily identified and mapped (White et al. 1995).

In the following sections I discuss the evidence from geophysical and geochemical data for the spatial and temporal variations in mantle temperature beneath the northern North Atlantic and discuss the effect of these on the crust generated at the Reykjanes Ridge spreading centre.

2. Tectonomagmatic regimes

Interaction between seafloor spreading and the Iceland mantle plume during the Tertiary has produced three distinct tectonomagmatic regimes, which I discuss below in detail. For this discussion, I restrict consideration to the region of the North Atlantic, south of, and including, Iceland. The area north of Iceland has suffered major ridge jumps in the seafloor spreading centre, leaving now extinct spreading centres, such as the Aegir Ridge, and continental fragments such as Jan Mayen. So the tectonics north of Iceland are complicated considerably by these features, obscuring the effects of ridge-plume interaction; to the south of Iceland no major ridge jumps have occurred.

The three main tectonic divisions of the oceanic crust (figure 1) are delineated by the gravity, magnetic and bathymetric fields (figures 2–4). They are distinguished by a first type exhibiting seafloor spreading magnetic anomalies largely unbroken by fracture zone offsets; a second type exhibiting normal (for slow-spreading ridges) ridge-fracture zone geometry with orthogonal spreading; and a third type with over-thickened, initially subaerial crust now found along the Greenland–Iceland–Færoe Ridge.

(a) Crust unbroken by fracture zones

This type of crust is found in two main areas: in the oldest seafloor generated in the early stages of opening of the North Atlantic; and along the northern two-thirds of the young oceanic crust on the present Reykjanes Ridge spreading axis (diagonal shading on figure 1). The distribution is not entirely symmetric in the North Atlantic. Over the crust formed immediately following continental breakup, it extends more than 1300 km from the plume centre, along the entire northern North Atlantic, whereas on the present spreading axis it extends only 1000 km from the plume centre beneath Iceland (figure 1).

A consistent characteristic of the oceanic crust generated without fracture zones is that it is thicker than normal, reaching 10–11 km thick, compared to 6–7 km for normal oceanic crust (White et al. 1992). This is indicative of mantle potential temperatures that were hotter than normal when the crust was generated. A factor which does not appear to correlate with the absence of fracture zones is the obliquity between the spreading direction and the normal to the strike of the spreading axis: over the oldest crust, the spreading direction immediately after continental breakup was normal to the ridge axis, which is the usual configuration for oceanic spreading centres. However, the youngest crust formed in this tectonomagmatic regime exhibits spreading directions up to 30° oblique to the normal from the ridge axis. So the obliquity of spreading does not appear to be a factor in controlling the formation of fracture zones.

On the present spreading axis, the young crust has little sediment cover and V-shaped ridges are prominent in the gravity field (figure 2): these cut across the isochrons marked by seafloor spreading magnetic anomalies (Vogt 1971; White et al. 1995), and are themselves split by continued spreading at the ridge axis (Keeton et al. 1997). The V-shaped ridges are thought to be caused by relatively small fluctuations in the temperature and flow rate of the mantle plume (White et al. 1995). Fine-scale structure is also present in the form of axial volcanic ridges, visible on the youngest, unsedimented crust (Laughton et al. 1979; Murton & Parson 1993; Keeton et al. 1997).

Over the oldest oceanic crust there are indications in the gravity field (figure 2) of lineations of alternating high and low gravity similar to those that mark the V-shaped ridges on the spreading axis. However, the sediment cover over the oldest crust means that no bathymetric ridges are now visible, and the magnitude of the gravity variations is greatly attenuated. They are therefore not nearly as prominent as are the V-shaped ridges near the present axis, but it is likely that they have a similar cause.

Figure 2. Free-airgravity field derived from Geosat and ERS-1 satellite geoid data (Sandwell & Smith, 1995; version 7.2). Gravity field is illuminated from the north.

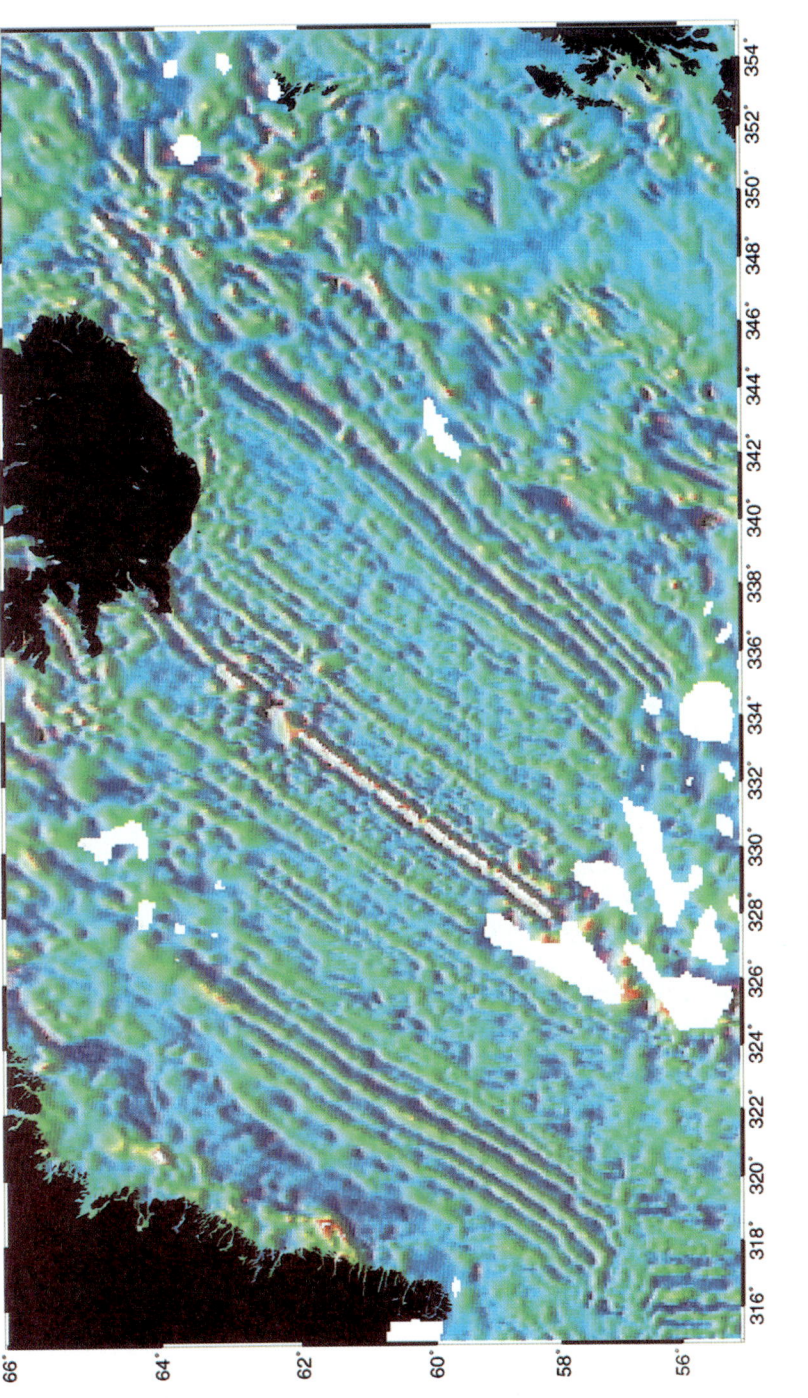

Figure 3. Magnetic anomalies derived from a compilation by Macnab *et al.* (1995). Image is illuminated from the west.

(b) Orthogonal spreading crust with fracture zones

The second tectonomagmatic type is where normal ridge segments spread in a direction close to orthogonal, with ridge segments separated by fracture zones (dotted ornament on figure 1). This is the normal pattern at slow-spreading ridges observed elsewhere in the ocean basins. In the region south of Iceland the change from spreading without fracture zones, to short orthogonal spreading segments terminated by fracture zones, occurs abruptly at magnetic anomaly 19 (42 Ma) on both sides of the ridge axis; this was recognized by Vogt (1971) from sparse ships' tracks, but is confirmed by the modern denser datasets, both of magnetic anomalies (figure 3) and of gravity anomalies (figure 2). The reversion to spreading without fracture zones is time-transgressive, and did not extend as far south as it did on the oldest oceanic crust. There was only a short interval of normal, fracture-zone dominated ridge spreading in the northern part of the area, near Iceland, but a much longer-lived period further south. Indeed, in the southernmost area we are considering here, south of 58° N, the spreading at the present day is still dominated by fracture zones (figure 2). The present day change on the ridge axis, from oblique spreading without fracture zones to orthogonal spreading with fracture zones, occurs about 1000 km from the centre of the Iceland plume at about 58° N, together with a change from a spreading axis dominated by topographic highs and axial volcanic ridges to one marked by a median valley (Keeton et al. 1997).

(c) Over-thickened oceanic crust

The third tectonomagmatic type is the crust formed above the centre of the Iceland mantle plume, which has produced abnormally thick (20–35 km) crust, whose surface was originally subaerial. This type of crust forms the present island of Iceland and the immediately surrounding shallow seafloor (figure 4), together with the bathymetric ridges known as the Greenland–Iceland Ridge and the Færoe–Iceland Ridge (marked GIR and FIR, respectively, on figure 1). Unlike the other tectonic settings, the spreading axis in this regime has suffered multiple ridge jumps: these may be caused by the spreading axis jumping to remain above the hottest upwelling region, as that region migrates with respect to the main North Atlantic spreading axis. At the present day it gives rise to a 150 km offset of the neovolcanic zone in Iceland, joined to the main North Atlantic spreading axis of the Reykjanes Ridge to the south by the South Iceland Transfer Zone, and to the Mohns Ridge in the north by a similar offset with the opposite sense of motion in the Tjörnes Fracture Zone.

3. Mantle temperatures derived from residual basement heights and crustal thickness

The three tectonomagmatic regimes reflect the effect of interaction between rifting and the underlying mantle of different temperatures and flow patterns. For the oceanic crust generated in the North Atlantic (i.e. for the first two tectonic regimes discussed above), I assume that the crust is generated by decompression melting of mantle welling up passively beneath the spreading axis. If this is the case, then the thickness of crust generated is related in a simple manner to the temperature of the mantle (McKenzie & Bickle 1988; White et al. 1992): a mantle temperature increase of 50 °C above normal increases the melt thickness and hence the crustal thickness by nearly 50% from 7.0–10.2 km (Bown & White 1994). Yet 50 °C is only a small perturbation on the normal mantle potential temperature of about 1320 °C, and is considerably less than the thermal anomalies of 200–250 °C found in the cores of mantle plumes (Watson & McKenzie 1991; White & McKenzie 1995).

The oceanic crustal thickness, and hence the temperature of the mantle at the time of crustal formation, can be measured directly by wide-angle seismic methods, assuming that the Moho marks the base of the igneous crust. There is now a considerable number of wide-angle seismic experiments in the North Atlantic (table 1), the majority of which have been interpreted using ray-tracing or synthetic seismogram methods. However, most of them are located either above the young oceanic crust on the present spreading axis, or over the oldest oceanic crust close to the continental margin. Few experiments have been done over the intermediate age crust where the tectonic regime reverted for a period to fracture-zone dominated seafloor spreading: an early experiment by Whitmarsh (1971) is the only one I report here.

However, an alternative way of inferring the oceanic crustal thickness is to measure the residual height of the basement. This is the difference between the present day water-loaded basement depth, after backstripping the sediment cover, and the depth it would be expected to have if it had followed a normal oceanic subsidence curve such as that reported by Parsons & Sclater (1977). Provided the present oceanic lithosphere is not dynamically supported by anomalously hot mantle, then the residual heights can be explained solely by the effects of isostacy operating on crust of variable thickness: normal thickness crust would produce a zero residual height anomaly, while crust that was thicker than normal would produce a positive residual height anomaly. In the next two sections I discuss the residual depth anomalies along representative isochrons and flowlines across the North Atlantic.

(a) Residual heights along isochron profiles

As is apparent from figure 5, which shows residual basement heights along a zero age isochron (solid line) and a 48 Ma isochron (magnetic anomaly 21) south of Iceland, both profiles show similar patterns: above the centre of the mantle plume (0 km on the horizontal distance scale), the crust is 4–5 km higher than normal spreading axes; there is an abrupt decrease in residual height over the next 100–200 km and then a more gentle drop over the next 1000 km; and normal seafloor depths are not reached until more than 1300 km from the plume centre. The similarity of the 0 Ma and 48 Ma curves indicates that the pattern of mantle temperature variation south of Iceland is similar today to what it was shortly after seafloor spreading started.

However, in detail we can draw out differences between the two profiles. At the Icelandic end of the profiles, the backstripped residual height of the Færoe–Iceland Ridge on the 48 Ma profile is broader, flatter-topped and more elevated than that of present day Iceland on the 0 Ma profile. For at least the first 20 Ma of its history, this portion of the Færoe–Iceland Ridge was subaerial and considerable erosion of the uppermost section has occurred. The volcanic edifices found directly above mantle plumes are generally unstable and are eroded rapidly: by analogy it is probable that the 1 km high volcanic 'relief' on the present day Icelandic profile will in due course be eroded to leave a flatter top like that of the Færoe–Iceland Ridge. So the true difference between the present day (solid line, figure 5) and the oldest oceanic (broken line, figure 5) profiles is over 1 km in height, with the present day profile being less high. This is consistent with the smaller present day Icelandic neovolcanic zone crustal thickness of 20–24 km (Bjarnason *et al.* 1993; White *et al.* 1966; Staples *et al.* 1997), compared to the Færoe–Iceland Ridge crustal thickness of about 30 km (Bott & Gunarsson 1980), or the oldest northeastern Icelandic crustal thickness of 35 km (White *et al.* 1996; Staples *et al.* 1997).

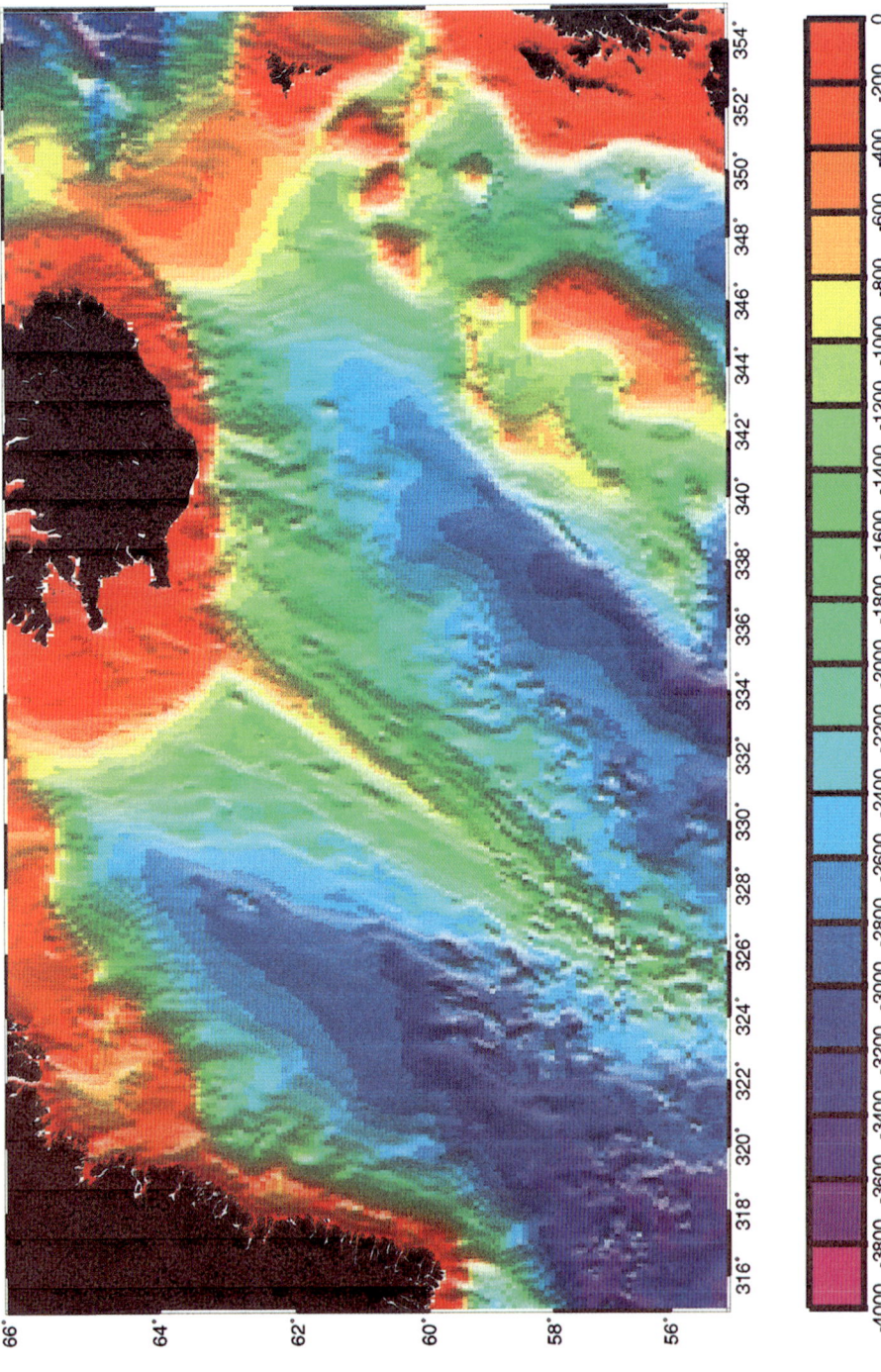

Figure 4. Bathymetry derived from five minute grid (National Geophysical Data Center 1993). Image is illuminated from the west.

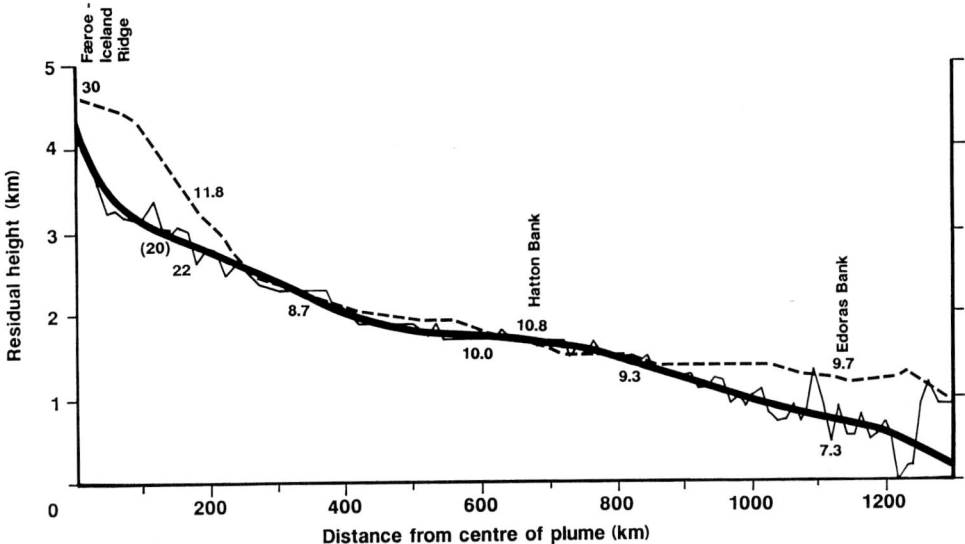

Figure 5. Residual height (i.e. height above 2.5 km below sea-level, the normal water depth at mid-ocean ridge spreading centres) along two isochrons, one along the spreading axis (zero age) and one along seafloor spreading magnetic anomaly 21 (48 Ma) on the European plate. For zero-age crust the fine line shows the actual seafloor depth and the heavier line shows the smoothed average depth. For 48 Ma crust (broken line) the average sediment thickness along the profile has been backstripped assuming Airy isostacy and LeDouran & Parson's (1982) density relationship. Parson & Sclater's (1977) curve was used to remove the effect of the increase of seafloor depth with age due to lithospheric cooling. Crustal thicknesses in km from seismic refraction experiments are shown in parentheses at appropriate distances along the curves: those above the curves refer to the oldest oceanic crust found adjacent to the rifted continental margin, and those below the curves to the zero-age crust on the spreading axis.

This difference in residual height between the youngest and oldest oceanic crust would be enhanced further if the dynamic support due to the abnormally hot mantle beneath the present axis were also to be taken into account. As much as half of the residual height along the present spreading axis can be attributed to dynamic support by hot underlying mantle rather than by crustal thickening (White et al. 1995). There is probably less dynamic support of the oldest crust, which lies further from the plume centre, so correction for the effect of dynamic support would decrease the residual height of the young crust more than that of the old crust: the enhanced difference in residual heights would indicate that mantle temperatures were hotter during the earliest phases of seafloor spreading than at the present day.

The last feature that is apparent from the isochron profiles in figure 5 is that at the greatest distances from the plume centre, the residual height of the oldest oceanic crust (broken line) remains greater than that of the spreading axis (solid line), suggesting that mantle temperatures remained abnormally high on the oldest oceanic crust even at distances in excess of 1000 km from the plume centre. Direct measurements of the crustal thickness from wide-angle seismics tell a similar story: the oldest oceanic crust off Edoras Bank is 9.7 km thick (Barton & White 1995), while that found along a flowline on the axis is only 7.3 km thick (Sinha et al., this volume). As I shall subsequently show, this is consistent with a sheet-like pattern to the early thermal anomaly during and immediately following continental breakup.

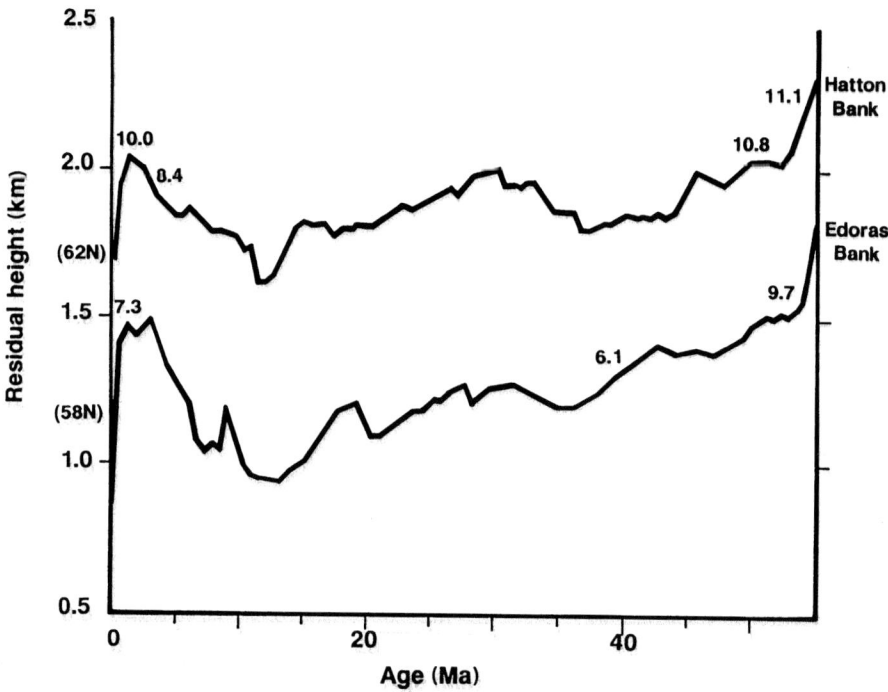

Figure 6. Residual height (i.e. height above 2.5 km below sea-level, the normal water depth at mid-ocean ridge spreading centres), of the top of oceanic basement along two flowlines: one from the spreading axis at 61.7° N to the continental margin at Hatton Bank through the locations of seismic profiles reported by Smallwood et al. (1995), Fowler et al. (1989) and Morgan et al. (1989); and the other from the axis at 57.7° N to the margin at Edoras Bank through seismic profiles reported by Sinha et al. (this volume) and Barton & White (1995, 1997). Average sediment thickness variation with age from Ruddiman (1972) was backstripped assuming Airy isostacy and Le Douran & Parson's (1983) density relationship, and increase of seafloor depth with age due to lithospheric cooling was removed using Parsons & Sclater's (1977) curve. Crustal thicknesses in km determined from seismic refraction experiments are shown at appropriate points on the curves.

(b) Residual heights along flowlines

By constructing residual height profiles of the oceanic basement along flowlines, it is possible to gain some indication of the variation of mantle temperature through time. In figure 6 I illustrate two such flowline profiles, using Srivastava & Tapscott's (1986) poles of rotation. At either end of each flowline there are good crustal thickness determinations from wide-angle seismic experiments. Although the profiles are generated from the regional bathymetry dataset ETOPO-5 (National Geophysical Data Center 1993) and from the average sediment thickness (Ruddiman 1972), and so are not reliable in detail, they do show the general features well. The residual heights along the northernmost profile, about 600 km from the plume centre, remain about 1 km more elevated than along the southern profile which is about 1000 km from the plume centre: the crustal thicknesses (small numbers on figure 6) are also consistently higher beneath the northern profile. This shows that the mantle temperature has consistently remained hotter nearer the plume centre, as would be expected.

Another significant observation is that there was a gradual decrease in residual height following continental breakup and the formation of the oldest oceanic crust at about 54 Ma, with an apparent increase over the past 5 Ma: this, too, is in agreement

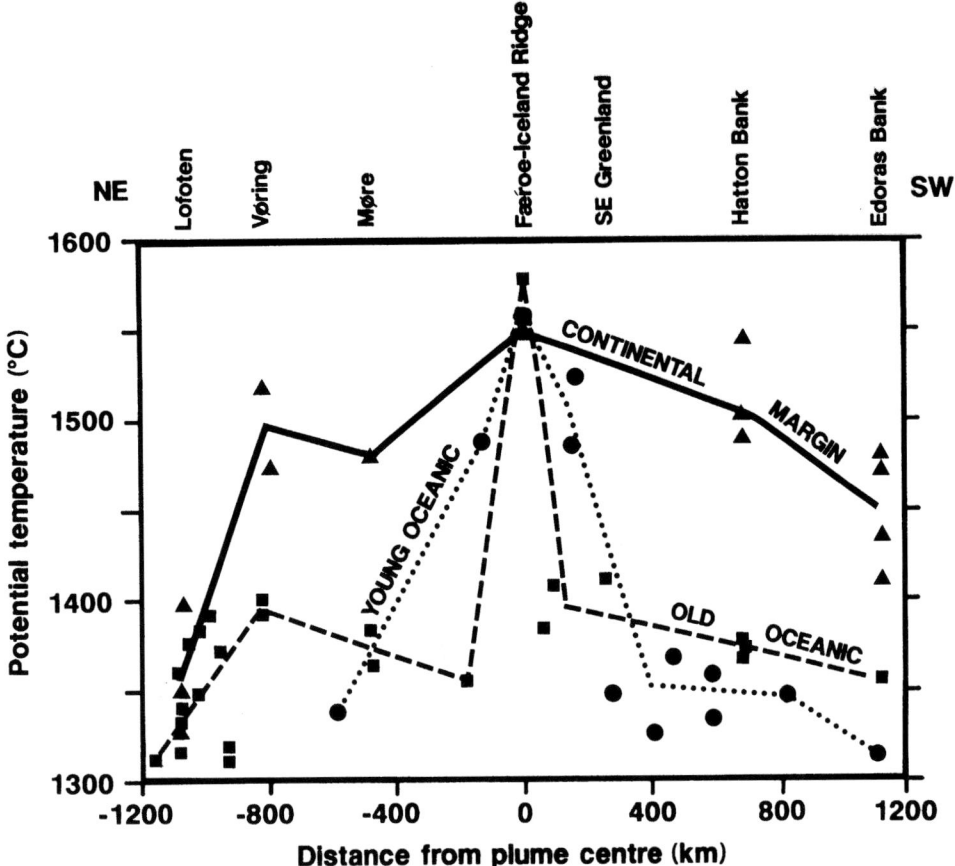

Figure 7. Inferred mantle potential temperature along three isochrons in the northern North Atlantic. 'Young oceanic' isochron is based on all crustal thickness estimates on crust aged 5 Ma or less along the North Atlantic spreading centre and the neovolcanic zones of Iceland (circles and dotted line). 'Old oceanic' isochron uses oceanic crustal thickness measurements over crust aged 50 Ma or more that exhibits clear seafloor spreading magnetic anomalies (squares and broken line). 'Continental margin' isochron, for comparison, is based on igneous crustal thickness in rifted continental margin crust generated at the time of continental breakup (triangles and solid line (Barton & White 1997)). Conversion from igneous crustal thickness to mantle temperature assumes melt generation by isentropic decompression of dry mantle using the relationships of Bown & White (1994). Sources of oceanic crustal thickness determinations are listed in table 1.

with crustal thicknesses measured from seismics. If the effect of dynamic support from underlying hot mantle were also to be removed, then the overall decrease from 54 Ma to 5 Ma would be still more marked.

(c) Mantle temperatures from crustal thickness

We can infer the mantle potential temperature from the crustal thickness determined seismically assuming dry, isentropic melting by passive mantle decompression (McKenzie & Bickle 1988; Bown & White 1994). In figure 7 I show the mantle temperatures determined in this way along three isochrons through the northern North Atlantic. The dotted line marked 'continental margin' (Barton & White 1997) is not the subject of this paper, but is shown for comparison as the temperature distribution during continental breakup immediately preceding generation of the oldest oceanic

Table 1. *North Atlantic oceanic crustal thickness*

location	profile	D^a (km)	age (Ma)	crust (km)	T_p^b (°C)	reference
Lofoten	OBS49	−1160	53	7.0	1312	Kodaira et al. (1995)
Lofoten	OBS18	−1085	53	10.1	1360	Mjelde et al. (1992)
Lofoten	OBS28	−1080	53	8.7	1340	Goldschmidt-Rokita et al. (1994)
Lofoten	OBS46	−1080	52	8.2	1332	Kodaira et al. (1995)
Lofoten	OBS27	−1080	50	7.3	1317	Goldschmidt-Rokita et al. (1994)
Lofoten	OBS19	−1050	53	11.3	1377	Mjelde et al. (1992)
Lofoten	OBS44	−1030	52	9.3	1349	Kodaira et al. (1995)
Lofoten	OBS20	−1020	53	11.8	1384	Mjelde et al. (1992)
Lofoten	OBS21	−990	53	12.3	1391	Mjelde et al. (1992)
Lofoten	OBS41	−960	53	7.4	1319	Mutter & Zehnder (1988)
NE Greenland	ESP19	−930	50	7.0	1312	Mutter & Zehnder (1988)
NE Greenland	ESP22	−820	50	12.9	1399	Mutter & Zehnder (1988)
Vøring margin	ESP13	−820	50	12.5	1394	Mutter & Zehnder (1988)
Kolbeinsey Ridge	OBS14/L3	−590	24	8.9	1342	Kodaira et al. (1997b)
Kolbeinsey Ridge	L2	−580	1	8–9	1329–1344	Kodaira et al. (1997a)
Møre margin	E45	−480	55	11.7	1383	Olafsson et al. (1992)
Møre margin	E46	−480	54	10.4	1365	Olafsson et al. (1992)
NE Greenland	82–27	−260	19	9.6	1353	Larsen & Jakobsdóttir (1988)
Færoe–Iceland Ridge	OBS43	−180	43	9.8	1356	Makris et al. (1995)

Table 1. Cont.

location	profile	D^a (km)	age (Ma)	crust (km)	T_p^b (°C)	reference
Iceland	FIRE	−120	0	20	1483	Staples et al. (1997)
Færoe–Iceland Ridge	NASP	0	50	30	1580	Bott & Gunnarsson (1980)
Færoe–Iceland Ridge	OBS10	55	52	13.7	1409	Makris et al. (1995)[c]
SE Greenland	81–20	70	50	11.9	1385	Larsen & Jakobsdóttir (1988)[d]
Færoe–Iceland Ridge	OBS01	90	52	11.8	1384	Makris et al. (1995)[c]
Iceland	SIST	150	0–3	20–24	1483–1525	Bjarnason et al. (1993)
SE Greenland	82–01	240	52	13.9	1412	Larsen & Jakobsdóttir (1988)[d]
Reykjanes Ridge	BI01	270	12	9.3	1348	Ritzert & Jacoby (1985)
Reykjanes Ridge	BI02	400	15	7.9	1327	Ritzert & Jacoby (1985)
Reykjanes Ridge	1	440	0	11.2	1376	Mochizuki (1995)
Reykjanes Ridge	CAM71	580	0	10.0	1359	Smallwood et al. (1995)
Reykjanes Ridge	CAM73	580	4	8.4	1335	Smallwood et al. (1995)
Hatton Bank	ESPG	680	53	11.2	1376	Fowler et al. (1989)
Hatton Bank	ESPH	680	53	11.1	1374	Spence et al. (1989)
Hatton Bank	OBS4	680	50	10.8	1370	Morgan et al. (1989)
Iceland Basin	PUBS	740	38	6.1	1297	Whitmarsh (1971)
Reykjanes Ridge	Z	810	9	9.3	1348	Bunch & Kennett (1980)
Reykjanes Ridge	Line 2	1100	0	7.3	1317	Sinha et al. (this volume)
Edoras Bank	OBH14	1120	53	9.7	1354	Barton & White (1995)

[a] Distance from the inferred centre of the Iceland mantle plume, taken as the highest point on Iceland for zero-age crust, and the centre of the Greenland–Iceland–Færoe Ridge for older crust. Negative values are to the north of the plume centre, positive values to the south.
[b] Potential temperature of mantle at the time of crustal generation assuming isentropic decompression of dry mantle, using values from Bown & White (1994).
[c] I assume the crust here is oceanic, though Makris et al. (1995) assume it is continental.
[d] Moho depth is only poorly constrained from weak signals close to, or at noise level at offsets of 15–30 km.

crust. It makes two bold assumptions, neither of which is likely to be completely true: these are, first, that decompression melting beneath the rifting margin was passive, without melt enhancement by active convection in the mantle; and, second, that all the melt was retained on the margin and none flowed out of the developing rift. Departure from the first assumption would lead to lower inferred mantle temperatures, while departure from the second would lead to somewhat higher inferred temperatures. For our purposes the important observation is that along a 2000 km-long portion of the North Atlantic continental margin, the melt thickness, and hence the inferred mantle temperature, did not vary significantly, and was considerably hotter than the mantle temperature when the oldest oceanic crust was formed.

The relative uniformity of the high temperatures along the entire margin suggests that the pattern of the thermal anomaly was in the shape of a rising sheet of abnormally hot mantle rather than an axisymmetric plume with a hot rising core only at its centre. Numerical and laboratory convection experiments indicate that boundary layer instabilities that develop into plumes may often originate at depth as radial sheets: these often develop subsequently into an axisymmetric system centred on the junction between the sheets (White & McKenzie 1995). It appears likely that some such system of rising sheets occurred beneath the North Atlantic prior to breakup: one sheet would have extended southward from the region of the present Færoe Islands beneath the western margin of Rockall Bank and southeast Greenland; another northward beneath the Norwegian-northeast Greenland boundary; and a third at a high angle from the Færoe Islands region to the area of Disko Island off western Greenland. This can explain the distribution of high-temperature contemporaneous igneous rocks on the eastern and western Greenland margins and along the North Atlantic margins.

As the mantle plume developed and North Atlantic opening started, the mantle temperatures dropped everywhere along the margins except beneath the Færoe–Iceland Ridge. The oldest oceanic crust (broken line, figure 7) was formed from lower temperature mantle than that which produced the outburst of magmatism on the margins. This may have been due to the mantle upwelling developing into a more axisymmetric system centred on the junction between the rising sheets beneath the Færoe–Iceland Ridge. Nevertheless, the interaction between rifting and the upwelling mantle plume may have kept the upwelling in a broadly sheet-like pattern beneath the early ocean basin. This is evidenced by the broadly constant temperatures inferred beneath the ocean basin southward away from the Færoe–Iceland Ridge and by the vestiges of gravity lineations visible in the free air gravity field over the old oceanic crust and interpreted as minor fluctuations in the temperature of the plume.

The present day temperatures under the spreading axis (solid line, figure 7) are somewhat lower than those inferred for the oldest oceanic crust. The V-shaped ridges, which propagate south from Iceland at apparent rates of 75–150 mm yr^{-1} (compared to the full spreading rate of $ca.$ 20 mm yr^{-1}) represent temperature fluctuations of about 30 °C superimposed on this general pattern on a timescale of 3–5 Ma (White et al. 1995).

4. The evidence from geochemistry

The geochemistry of the basaltic rocks in the North Atlantic provides strong constraints on the nature and temperature of the mantle from which the melts were generated. I here highlight two features that bear on the mantle plume temperature and circulation. First, the neodymium isotopic content of Icelandic basalts indicate that they consist on average of only 10–15% primitive mantle, with a majority of depleted MORB-source mantle (Condomines et al. 1983; Elliot et al. 1991; Hemond et al. 1993; Meyer et al. 1985; O'Nions et al. 1977; Zindler et al. 1979). This is indicative that there is considerable mixing between a relatively small amount of primitive plume mantle and the surrounding depleted upper mantle.

Second, rare earth element inversions on analyses of basalts along the Greenland and European continental margins suggest that they were formed from mantle with a potential temperature of 1450–1500 °C (Brodie 1995; White & McKenzie 1995), the same as is found from analyses of neovolcanic basalts from the present day core of the plume beneath Iceland (Nicholson & Latin 1992; White & McKenzie 1995). By contrast, basalts from the DSDP holes on the Reykjanes Ridge south of Iceland indicate considerably lower parent mantle temperatures (White et al. 1995). All this is consistent with the pattern of mantle temperatures discussed in the previous section during continental breakup and at the present day.

Combination of the melt thicknesses inferred from rare earth element inversions with the crustal thicknesses measured by seismic techniques provides insight into whether the mantle upwelling that caused decompression melting was a passive response to the lithospheric extension, or whether it was a result of forced convection that cycled mantle through the melt region. So, for example, a global review of normal mid-ocean ridge basalts (White et al. 1992) showed that estimates of the melt thicknesses from rare earth element inversions are, within error, the same as measurements of melt thickness from seismic crustal thickness determinations. This indicates that mantle upwelling beneath oceanic spreading centres is predominantly a passive response to lithospheric separation. By contrast, rare earth element inversions of basalts from the Hawaiian islands (Watson & McKenzie 1991) show that considerable volumes of melt are produced by mantle being forced through the melting region beneath Hawaii by active convection in the core of the underlying mantle plume, despite there being no lithospheric extension in this mid-plate location.

In the case of Iceland, the rare earth element inversions indicate melt thicknesses of 15–20 km (White et al. 1995; White & McKenzie 1995), while the seismic measurements indicate crustal thicknesses of about 20 km spread across the 300 km long neovolcanic zone of Iceland. If the central rising core of the mantle plume beneath Iceland has a diameter of about 100–150 km, as is indicated by the extent of the most highly elevated region, then I conclude that the mantle convection that causes melt generation beneath Iceland is somewhat more active than a purely passive response to plate separation (i.e. there is forced convection analogous to that under, for example, Hawaii), but that the forced convection is not particularly vigorous. It is probable that melt flows laterally at crustal levels from the main locus of melt generation above the core of the plume, so that it becomes distributed along the neovolcanic zones across the width of Iceland. A simple mass calculation suggests that the bulk of the mantle that is convected to shallow levels in the plume beneath Iceland and which becomes partially depleted by melting, eventually becomes incorporated into the lithospheric plates that absorb the cooling mantle and which spread away from the neovolcanic zones in Iceland.

The V-shaped ridges mapped on the Reykjanes Ridge suggest that there is some lateral flow of asthenospheric mantle away from Iceland, but the geochemistry of the basalts on the ridge axis suggest that they have come from mantle that has not been through the melting column beneath Iceland. The relatively small temperature anomalies and high percentage of depleted MORB-source mantle found beneath the Reykjanes Ridge are consistent with the asthenospheric mantle in this region representing a sheath of only slightly hotter than normal mantle that surrounded the central plume core beneath Iceland (White et al. 1995).

5. Influence of mantle temperature on oceanic crustal formation

The characteristics of the three main tectonomagmatic regimes in which oceanic crust is generated in the northern North Atlantic are governed by the temperature and flow patterns of the mantle beneath the spreading centres at the time of crustal formation.

(a) Spreading axis unbroken by fracture zone offsets

Oceanic crust unbroken by fracture zones was formed within the entire ocean basin immediately following continental breakup, and again, after a short reversion to a phase of fracture zone dominated crustal formation, is being formed at the present day. However, the present day crust without fracture zones is found only along the northern portion of the Reykjanes Ridge closest to the centre of the Iceland plume. The oldest crust of this type was generated when the spreading direction was orthogonal to the strike of the ridge axis, as is normal for oceanic spreading centres. The present day Reykjanes Ridge, however, is spreading at a high obliquity, of about 30°. So it is clear that the obliquity of the spreading is not the controlling factor as to whether fracture zones are formed.

The consistent characteristic of the crust formed without fracture zones is its thickness: everywhere this type of crust is found in the North Atlantic, the crust is thicker than 8 km, and is on average 10–11 km thick. Reversion to the fracture-zone dominated type of crust that is normal at low spreading rates occurs south of 57° N, where the crustal thickness decreases to a normal value of about 7 km. Away from mantle plumes oceanic crust exhibiting similar characteristics, namely an absence of fracture zones and an axial high rather than a median valley, is found predominantly on fast-spreading ridges such as the East Pacific Rise. However, on the fast-spreading ridges the oceanic crust exhibits normal thicknesses of 6–7 km (White et al. 1992). So it cannot be simply the crustal thickness that is the controlling factor in this tectonomagmatic regime, but rather a combination of crustal thickness and spreading rate.

The formation of median valleys (Phipps Morgan et al. 1987) and of fracture zones both represent the brittle response of the lithosphere at the spreading axis. If neither a median valley nor fracture zones are present, it suggests that conditions are such that the lithosphere is weak and can respond ductilely to the forces at the spreading centre. The uppermost portion of the oceanic crust, down to a depth of about 2 km below the seafloor, is quenched by hydrothermal circulation. Below the limit of hydrothermal penetration the crust remains hot from igneous intrusions and loses heat mainly by conduction (e.g. Henstock et al. 1993). At any given temperature the mantle beneath the crust is considerably stronger than the crust because it has a lower homologous temperature, but the cooling of the mantle is controlled by conductive cooling through the overlying crust.

The strength of the lithosphere beneath the axial region is therefore controlled by the rate of cooling of the lower crust and of the underlying mantle. At normal slow-spreading ridges, with intrusion events occurring, typically, at intervals of

10 000–50 000 years, the crust and underlying mantle near the spreading axis cools sufficiently between igneous injections to allow brittle behaviour, and the formation of a median valley, of fracture zones, and of earthquakes extending to depths of up to 8 km (Toomey et al. 1985; Kong et al. 1992). On fast-spreading ridges with injections of melt occurring at intervals an order of magnitude smaller, the crust remains hotter and weaker and earthquakes are not only rarer, but extend down to only about 3 km at the axis (Riedesel et al. 1982; Orcutt et al. 1984).

There is apparently a delicate balance between the spreading rate and the crustal cooling rate. On normal slow-spreading ridges the rate of cooling exceeds the rate of heat input from igneous injections such that faulting can extent down to the stronger mantle near the axis. On fast-spreading ridges the higher rate of igneous injection means that conductive cooling cannot cool the mantle sufficiently for it to behave brittly before the new lithosphere has moved away from the axis.

On the Reykjanes Ridge the increased crustal thickness means that even at slow spreading rates, the mantle beneath the spreading axis remains sufficiently hot to behave ductilely. There are three factors which all tend to keep the mantle hotter. First, the increased crustal thickness means that there is an increase in the frequency of melt injection, assuming that the melt volume per injection event remains the same. Second, the mantle temperature itself is somewhat hotter than it is in areas away from mantle plumes: this is only a small increase, of perhaps 50 °C, but nevertheless it means that the mantle starts off hotter so has further to cool before it becomes brittle. Third, and probably most importantly, the increased crustal thickness provides a thicker insulating layer which means that it takes longer for the underlying mantle to cool. Bell & Buck (1992), using similar arguments, suggest that the lower crust beneath the Reykjanes Ridge remains sufficiently hot to flow ductilely. Chen & Morgan (1990) similarly show that an increase in crustal thickness and in mantle temperature can account for the absence of a median valley on the Reykjanes Ridge. Direct evidence for a relatively thin brittle layer restricted to the upper crust comes from earthquake hypocentral determinations on the ridge axis immediately south of Reykjanes Peninsula which show earthquakes extending down to only 8 km beneath the seafloor, well within the crust (Mochizuki 1995).

(b) Oceanic crust broken by fracture zones

The normal slow-spreading oceanic crust with abundant fracture zones and orthogonal spreading is found in the North Atlantic in areas where the crustal thickness, and hence the underlying mantle temperature, were normal. This type of crust is found at the present day in the distal regions more than 1000 km from the plume centre, and also throughout the area in the mid-Tertiary. A rather poorly constrained crustal thickness determination over crust formed during the mid-Tertiary phase yields a crustal thickness of 6.3 km (Whitmarsh 1971), which is consistent with the idea discussed above that it is the thermal state of the lithosphere at the spreading axis which controls the tectonic style of the oceanic crust.

(c) Crust created directly above the mantle plume

The crust of the Greenland–Iceland–Færoe Ridge created directly above the core of the mantle plume is considerably thicker than normal due to the enhanced mantle temperatures and to the forced convection in the plume, which exceeds the mantle upwelling caused solely by plate separation. Its thickness varies from 20–35 km through its history, suggesting variations in the temperature or the mantle flow-rate, or both.

A particular feature of this crust formed directly above the plume centre is that it exhibits multiple jumps of the spreading axis, in a way not seen on the Reykjanes Ridge to the south. These probably occur because the centre of the plume is currently migrating eastward with respect to the centre of the North Atlantic. So the neovolcanic zone on Iceland jumps to keep above the axis of the plume: the most recent eastward jump occurred about 4 Ma ago.

6. History of mantle plume-ridge interaction

There is a close link between the history of the Iceland mantle plume and oceanic crustal formation in the North Atlantic. The continental breakup phase was apparently associated with abnormally hot mantle, probably welling up in a pattern of sheets meeting beneath the central region where Iceland now lies. This sheet-like pattern may have marked the onset of a new plume instability. The new ocean basin broke open along the weak line marking the western edge of the string of Mesozoic basins along the northwest European margin.

The first-formed oceanic crust was 10–11 km thick, reflecting high underlying mantle temperatures beneath the entire oceanic basin. It formed without fracture zones. However, at about the time that the Labrador Sea stopped opening, the mantle temperatures beneath most of the oceanic basin dropped, and oceanic crust formed for a period with a normal slow-spreading pattern of fracture zones and orthogonal spreading. At about the same time that the neovolcanic zone on the Greenland–Færoe Ridge jumped westward toward Greenland there was also renewed uplift in east Greenland. It is possible that an original upwelling sheet of hot mantle beneath the Greenland–Færoe axis developed into a more localized upwelling region under the east Greenland coastal area: this would explain both the westward jump in the neovolcanic zone and the temporary drop in mantle temperatures beneath the North Atlantic, which allowed the formation of normal slow-spreading oceanic crust in this area which lay distant from the new focus of upwelling.

Over the past 25 Ma the centre of the mantle plume has been migrating eastward and the neovolcanic zone has been jumping eastward along the Greenland–Færoe Ridge to keep up with it. Temperatures under the North Atlantic have again increased, producing thick oceanic crust without fracture zones, but mantle temperatures are somewhat cooler now than in the early phase of seafloor spreading following continental breakup. Small fluctuations in the mantle temperature and flow rate are recorded by the V-shaped ridges on the present Reykjanes Ridge, on timescales of 3–5 Ma. Such fluctuations are probably present in all mantle plumes, but can be readily detected here because the spreading axis cuts directly above the plume, and so the crustal thickness directly reflects the mantle temperature.

There is forced convection in the present mantle plume, with the upwelling rate being somewhat faster than the rate that would result from passive upwelling beneath the spreading plates. However, a simple mass balance suggests that the bulk of the mantle plume is absorbed into the thickening lithosphere beneath Iceland. It is likely that the interaction between the North Atlantic lithospheric spreading and the upwelling plume created a sheet-like pattern to the upwelling under the Reykjanes Ridge, which generates the characteristic V-shaped ridges as the asthenospheric mantle moves along the ridge axis and up toward the surface.

It is clear from the history of seafloor spreading in the North Atlantic that on slow-spreading ridges the normal pattern of crustal generation with axial valleys, fracture zones and orthogonal spreading is easily perturbed to a pattern without an axial valley and without fracture zones. There is a delicate thermal balance which controls the temperature of the lithosphere at the spreading axis, and which determines whether there is a ductile or brittle response to the lithospheric extension at the axis. A mantle temperature increase of as little as 50 °C beneath the spreading axis causes a change from a brittle, fracture-zone dominated regime to a ductile spreading regime capable of supporting highly oblique spreading and without fracture zones or an axial valley.

I thank A. J. Barton, J. Bown, D. McKenzie, K. R. Richardson, M. C. Sinha, J. R. Smallwood and R. K. Staples for discussions on the results of their researches on various aspects of this work and C. Enright for help with the figures. Department of Earth Sciences, Cambridge, contribution number 4736.

References

Barton, A. J. & White, R. S. 1995 The Edoras Bank margin: continental break-up in the presence of a mantle plume. *J. Geol. Soc. Lond.* **152**, 971–974.

Barton, A. J. & White, R. S. 1997 Crustal structure of the Edoras Bank continental margin and mantle thermal anomalies in the North Atlantic. *J. Geophys. Res.* **102** (In the press.)

Bell, R. E. & Buck, W. R. 1992 Crustal control of ridge segmentation inferred from observations of the Reykjanes Ridge. *Nature* **357**, 583–586.

Bjarnason, I. Th., Menke, W., Flóvenz, Ó. G. & Caress, D. 1993 Tomographic image of the Mid-Atlantic plate boundary in southwestern Iceland. *J. Geophys. Res.* **98**, 6607–6622.

Bott, M. H. P. & Gunnarsson, K. 1980 Crustal structure of the Iceland–Færoe Ridge. *J. Geophys.* **47**, 221–227.

Bown, J. W. & White, R. S. 1994 Variation with spreading rate of oceanic crustal thickness and geochemistry. *Earth Planet Sci. Lett.* **121**, 435–449.

Brodie, J. A. 1995 Early Tertiary volcanism in the North Atlantic. Ph.D. thesis, University of Cambridge.

Bunch, A. W. H. & Kennett, B. L. N. 1980 The crustal structure of the Reykjanes Ridge at 59° 30′ N. *Geophys. J. R. Astron. Soc.* **61**, 141–166.

Chen, Y. & Morgan, W. J. 1990 A nonlinear rheology model for mid-ocean ridge axis topography. *J. Geophys. Res.* **95**, 17 583–17 604.

Condomines, M., Grönvold, K., Hooker, P. J., Muehlenbachs, K., O'Nions, R. K., Oskarsson, N. & Oxburgh, E. R. 1983 Helium, oxygen, strontium and nedymium relationships in Icelandic volcanics. *Earth Planet. Sci. Lett.* **66**, 125–136.

Elliot, T. R., Hawkesworth, C. J. & Grönvold, K. 1991 Dynamic melting of the Iceland plume. *Nature* **351**, 201–206.

Fowler, S. R., White, R. S., Spence, G. D. & Westbrook, G. K. 1989 The Hatton Bank continental margin. II. Deep structure from two-ship expanding spread seismic profiles. *Geophys. J.* **96**, 295–309.

Goldschmidt-Rokita, A., Hansch, K. J. F., Hirschleber, H. B., Iwasaki, T., Kanazawa, T., Shimanmur, H. & Sellevol, M. A. 1994 The ocean/continent transition along a profile through the Lofoten Basin, Northern Norway. *Mar. Geophys. Res.* **16**, 201–224.

Hemond, C., Arndt, N. T., Lichtenstein, U., Hofmann, A. W., Oskarsson, N. & Steinthorsson, S. 1993 The heterogeneous Iceland plume: Nd-Sr-O isotopes and trace element constraints. *J. Geophys. Res.* **98**, 15 833–15 850.

Henstock, T. J., Woods, A. W. & White, R. S. 1993 The accretion of oceanic crust by episodic sill intrusion. *J. Geophys. Res.* **98**, 4143–4161.

Keeton, J. A., Searle, R. C., Parsons, B., White, R. S., Murton, B. J., Parson, L. M., Pierce, C. & Sinha, M. C. 1997 Bathymetry of the Reykjanes Ridge. *Mar. Geophys. Res.* (In the press.)

Kodaira, S., Goldschmidt-Rokita, A., Hartmann, J. M., Hirschleber, H. B., Iwasaki, T., Kanazawa, T., Krahn, H., Tomita, S. & Shimamura, H. 1995 Crustal structure of the Lofoten continental margin, off northern Norway, from ocean-bottom seismographic studies. *Geophys. J. Int.* **121**, 907–924.

Kodaira, S., Mjelde, R., Gunnarsson, K., Shiobara, H. & Shimamura, H. 1997a Crustal structure of the Kolbeinsey Ridge, N. Atlantic, obtained by use of ocean-bottom seismographs. *J. Geophys. Res.* (In the press.)

Kodaira, S., Mjelde, R., Gunnarsson, K., Shiobara, H. & Shimamura, H. 1997b Structure of the Jan Mayen microcontinent and implications for its evolution. *Geophys. J. Int.* (In the press.)

Kong, I. A., Solomon, S. C. & Purdy, G. M. 1992 Microearthquake characteristics of a mid-ocean ridge along-axis high. *J. Geophys. Res.* **97**, 1659–1685.

Larsen, H. C. & Jakobsdóttir, S. 1988 Distribution, crustal properties and significance of seaward dipping subbasement reflectors off east Greenland. In *Early Tertiary volcanism and the opening of the NE Atlantic* (ed. A. C. Morton & L. M. Parson), pp. 95–114. Geol. Soc. Lond. Spec. Publ., no. 39.

Laughton, A. S., Searle, R. C. & Roberts, D. G. 1979 The Reykjanes Ridge crest and the transition between its rifted and non-rifted regions. *Tectonophys.* **55**, 173–177.

Le Douran, S. & Parsons, B. 1982 A note on the correction of ocean floor depths for sediment loading. *J. Geophys. Res.* **87**, 4715–4722.

Macnab, R., Verhoef, J., Roest, W. & Arkani-Hamed, J. 1995 New database documents the magnetic character of the Arctic and North Atlantic. *Eos* **76**, 449, 458.

McKenzie, D. & Bickle, M. J. 1988 The volume and composition of melt generated by extension of the lithosphere. *J. Petrol.* **29**, 625–679.

Makris, J., Lange, K., Savostin, L. & Sedov, V. 1995 A wide-angle reflection profile across the Iceland–Færoe Ridge. In *The petroleum geology of Ireland's offshore basins* (ed. P. F. Croker & P. M. Shannon), pp. 459–466. Geol. Soc. Lond. Spec. Pub., no. 93.

Meyer, P. S., Sigurdsson, H. & Schilling, J. G. 1985 Petrological and geochemical variations along Iceland's neovolcanic zones. *J. Geophys. Res.* **90**, 10027–10042.

Mjelde, R., Sellevol, M. A., Shimamura, H., Iwasaki, T. & Kanazawa, T. 1992 A crustal study off Lofoten, W. Norway, by use of 3-component ocean bottom seismographs. *Tectonophys.* **212**, 269–288.

Mochizuki, M. 1995 Crustal structure and micro-seismicity of the Mid-Atlantic Ridge, near Iceland, derived from ocean bottom seismographic observations. Ph.D. dissertation, Hokkaido University, Japan, p. 162.

Morgan, J., Barton, P. J. & White, R. S. 1989 The Hatton Bank continental margin. III. Structure from wide-angle OBS and multichannel seismic refraction profiles. *Geophys. J. Int.* **89**, 367–384.

Murton, B. J. & Parson, L. M. 1993 Segmentation, volcanism and deformation of oblique spreading centres: a quantitative study of the Reykjanes Ridge. *Tectonophys.* **222**, 237–257.

Mutter, J. C. & Zehnder, C. M. 1988 Deep crustal structure and magmatic processes: The inception of seafloor spreading in the Norwegian–Greenland Sea. In *Early Tertiary volcanism and the opening of the NE Atlantic* (ed. A. C. Morton & L. M. Parson), pp. 35–48. Geol. Soc. Lond. Spec. Publ., no. 39.

National Geophysical Data Center. 1993 *GEODAS CD-ROM worldwide marine geophysical data*, 2nd edn. Data Announcement 93-MGG-04, National Oceanic and Atmospheric Administration, U.S. Department of Commerce, Boulder, CO.

Nicholson, H. & Latin, D. 1992 Olivine tholeiites from Krafla, Iceland: evidence for variations in melt fraction within a plume. *J. Petrol.* **33**, 1105–1124.

Olafsson, I., Sundvor, E., Eldholm, O. & Grue, K. 1992 Møre margin: crustal structure from analysis of ESPs. *Mar. Geophys. Res.* **14**, 137–162.

O'Nions, R. K., Hamilton, P. J. & Evensen, N. M. 1977 Variations in ^{143}Nd/^{144}Nd and ^{87}Sr/^{86}Sr ratios in oceanic basalts. *Earth Planet. Sci. Lett.* **34**, 13–22.

Orcutt, J. A., McClain, J. S. & Burnett, M. 1984 Evolution of the oceanic crust, results from seismic experiments. In *Ophiolites and Oceanic Lithosphere* (ed. I. G. Gass, S.J. Lippard & A. W. Shelton), pp. 7–16. Geol. Soc. Lond. Spec. Pub., no. 13.

Parsons, B. & Sclater, J. G. 1977 An analysis of the variation of ocean floor bathymetry and heat flow with age. *J. Geophys. Res.* **82**, 803–827.

Phipps Morgan, J., Parmentier, E. M. & Lin, J. 1987 Mechanisms for the origin of mid-ocean ridge axial topography: implications for the thermal and mechanical structure of accreting plate boundaries. *J. Geophys. Res.* **92**, 12 823–12 836.

Riedesel, M., Orcutt, J. A., Macdonald, K. C. & McClain, J. S. 1982 Microearthquakes in the black smoker hydrothermal field, East Pacific Rise at 21° N. *J. Geophys. Res.* **87**, 10 613–10 624.

Ritzert, M. & Jacoby, W. R. 1985 On the lithospheric seismic structure of Reykjanes Ridge at 62.5° N. *J. Geophys. Res.* **90**, 10 117–10 128.

Ruddiman, W. F. 1972 Sediment redistribution on the Reykjanes Ridge: seismic evidence. *Geol. Soc. Am. Bull.* **83**, 2039–2062.

Sandwell, D. T. & Smith, W. H. F. 1995 Marine gravity anomaly from satellite altimetry. Geological Data Center, Scripps Institution of Oceanography, La Jolla, CA.

Smallwood, J. R., White, R. S. & Minshull, T. A. 1995 Seafloor spreading in the presence of the Iceland mantle plume: the structure of the Reykjanes Ridge at 61° 40' N. *J. Geol. Soc. Lond.* **152**, 1023–1029.

Spence, G. D., White, R. S., Westbrook, G. K. & Fowler, S. R. 1989 The Hatton Bank continental margin. I. Shallow structure from two-ship expanding spread profiles. *Geophys. J.* **96**, 273–294.

Srivastava, S. P. & Tapscott, C. R. 1986 Plate kinematics of the North Atlantic. In *The geology of North America*, The western North Atlantic region (ed. P. R. Vogt & B. E. Tucholke), vol. M., pp. 379–404. Geological Society of America.

Staples, R. K., White, R. S., Brandsdottir, B., Menke, W. H., Maguire, P. K. H., Smallwood, J. R. & McBride, J. 1997 Færoe–Iceland Ridge experiment. I. The crustal structure of northeastern Iceland. *J. Geophys. Res.* **102**. (In the press.)

Toomey, D. R., Solomon, S. C., Purdy, G. M. & Murray, M. H. 1985 Micro-earthquakes beneath the median valley of the Mid-Atlantic Ridge near 23° N: hypocenters and focal mechanisms. *J. Geophys. Res.* **90**, 5443–5485.

Vogt, P. R. 1971 Asthenosphere motion recorded by the ocean floor south of Iceland. *Earth Planet Sci. Lett.* **13**, 153–160.

Watson, S. & McKenzie, D. 1991 Melt generation by plumes: a study of Hawaiian volcanism. *J. Petrol.* **32**, 501–537.

White, R. S., McKenzie, D. & O'Nions, R. K. 1992 Oceanic crustal thickness from seismic measurements and rare earth element inversions. *J. Geophys. Res.* **97**, 19 683–19 715.

White, R. S. & McKenzie, D. 1995 Mantle plumes and flood basalts. *J. Geophys. Res.* **100**, 17 543–17 585.

White, R. S., Bown, J. W. & Smallwood, J. R. 1995 The temperature of the Iceland plume and origin of outward propagating V-shaped ridges. *J. Geol. Soc. Lond.* **152**, 1039–1045.

White, R. S., McBride, J. H., Maguire, P. K. H., Brandsdóttir, B., Menke, W., Minshull, T. A., Richardson, K. R., Smallwood, J. R., Staples, R. K. and the FIRE Working Group 1996 Seismic images of crust beneath Iceland contribute to long-standing debate. *Eos* **77**, 197 199–197 200.

Whitmarsh, R. B. 1971 Seismic anisotropy of the uppermost mantle absent beneath the east flank of the Reykjanes Ridge. *Bull. Seism. Soc. Am.* **61**, 1351–1368.

Zindler, A., Hart, S. R., Frey, F. A. & Jakobsson, S. P. 1979 Nd and Sr isotope ratios and REE abundances in Reykjanes Peninsula basalts: evidence for mantle heterogeneity beneath Iceland. *Earth Planet. Sci. Lett.* **5**, 249–262.

The ultrafast East Pacific Rise: instability of the plate boundary and implications for accretionary processes

By Marie-Helène Cormier

Lamont-Doherty Earth Observatory, Columbia University, NY 10964-8000, USA

The Pacific–Nazca plate boundary evolves continuously through the frequent, rapid propagation of ridge segments and through the growth or abandonment of microplates. Propagation events can initiate at overlapping spreading centres only a few kilometres wide as well as within large transform faults. This instability of the ultrafast East Pacific Rise (EPR) probably results from the presence of a hot, thin lithosphere in the axial region, coupled with a melt supply that may be temporally or spatially variable. It indicates that along-axis magma transport can be efficient at rates corresponding to propagation rates, up to 1000 mm yr^{-1}. To a first-order, the tectonic segmentation of the ridge correlates with along-axis variations of the axial morphology and other physical parameters suggesting a diminished magmatic budget near offsets larger than a few kilometres. A similar correlation between axial segmentation and variations in physical characteristics at the Mid-Atlantic Ridge (MAR) is commonly interpreted to indicate that mantle upwelling is focused near mid-segment at slow-spreading ridges (three dimensional). Accordingly, mantle upwelling may be focused at discrete intervals along the ultrafast EPR. However, fluctuations of the along-axis characteristics are considerably more subdued at the EPR than at the MAR. This has been interpreted to reflect smoothing of the structural variations by efficient transport within the shallow crust and upper mantle of the material brought up through focused upwelling. Alternatively, it has been argued that mantle flow is essentially uniform along-axis (two dimensional) at the faster spreading centres. It is proposed here that the actual pattern of mantle flow along the EPR combines aspects of both models. Fast spreading centres may be supplied by vertical mantle flow nearly continuously along-axis, but the intensity of this upwelling can fluctuate both temporally and spatially, hence favouring along-axis transport away from the magmatically most robust areas.

1. Introduction

The Pacific–Nazca plate boundary spreads apart at 125–152 mm yr^{-1}, the fastest rate of the entire mid-ocean ridge system (DeMets *et al.* 1994). Over the approximately 4000 km spanned by this plate boundary, the morphology of the axial region varies remarkably little (Macdonald *et al.* 1988a; Lonsdale 1989; Sinton *et al.* 1991). It is defined by a linear high, 5–15 km wide and rising 300–500 m above the surrounding seafloor. This axial high continues uninterrupted for distances of 25–250 km (figure 1). Except in the vicinity of Easter hot spot or along short intratransform spread-

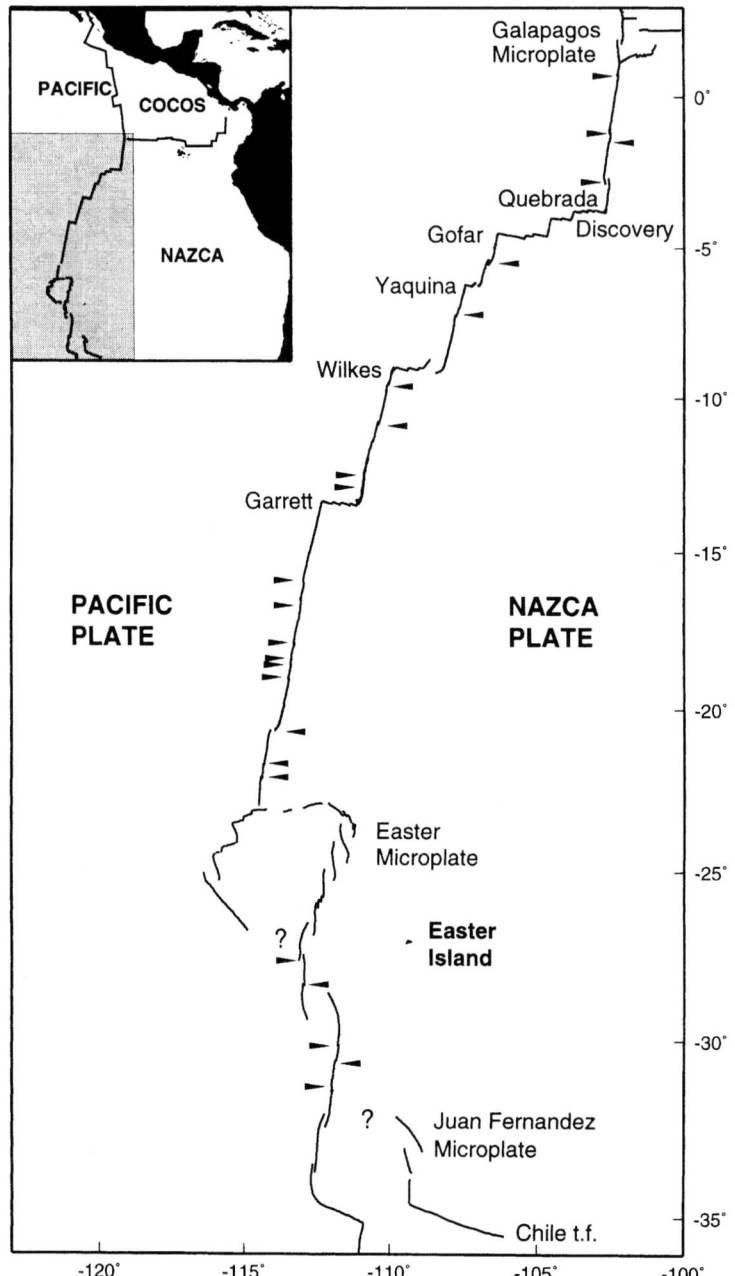

Figure 1. Map view of the southern EPR. Inset shows location relative to South America. The plate boundaries are compiled from Tighe *et al.* (1988), Macdonald *et al.* (1988*b*), Lonsdale (1989), Naar & Hey (1991), Larson *et al.* (1992) and Hey *et al.* (1995). Arrow heads indicate the location of OSCs with offset greater than 3 km (microplates excluded). Arrows point to the west for right-stepping OSCs, and to the east for left-stepping OSCs.

ing centres, axial depths undulate gently between 2500–3000 m (figure 2). Direct observations from submersible and towed-camera surveys indicate that the narrow axial summit (0.2–4 km wide) is the locus of the most recent volcanism (Macdon-

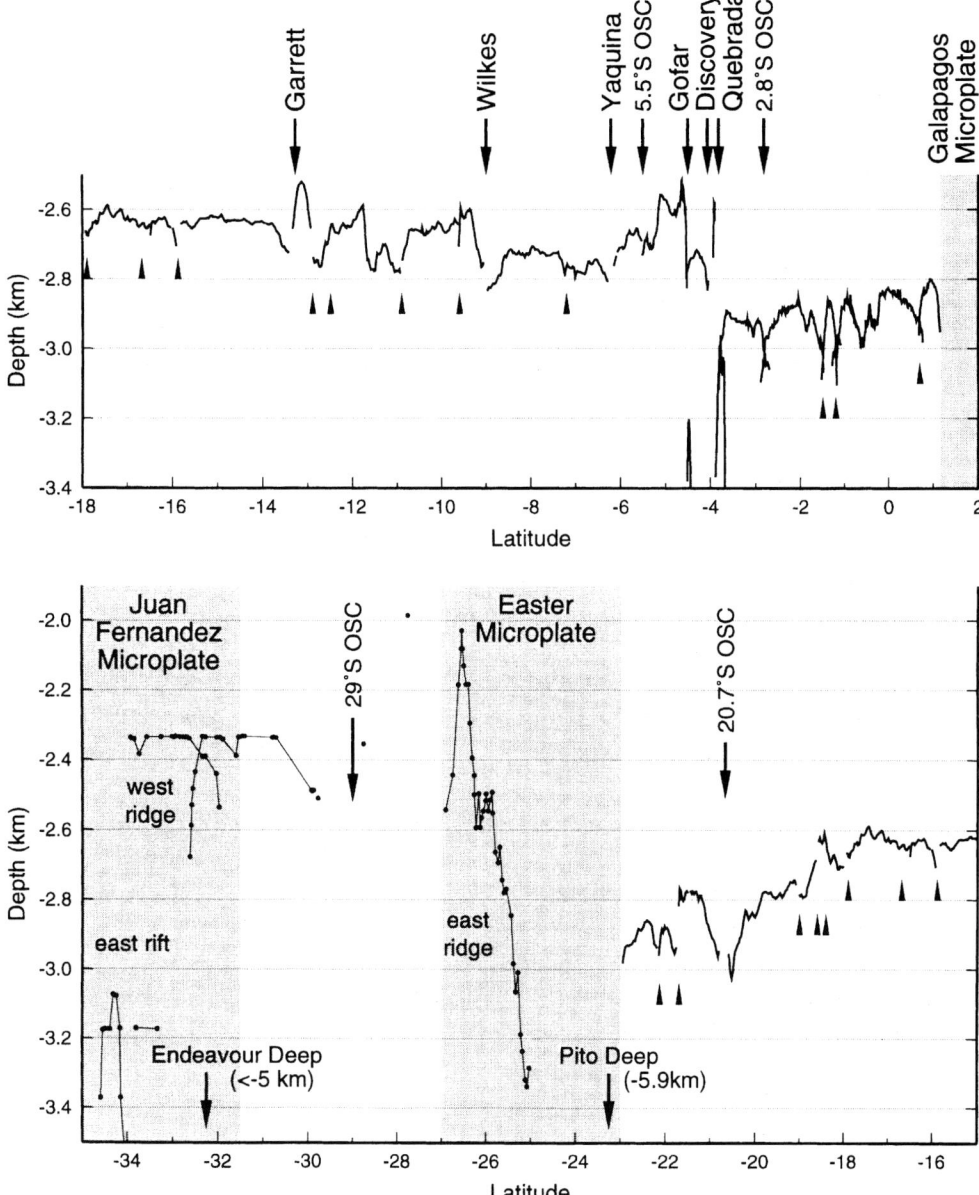

Figure 2. Zero-age depth variations as a function of latitude. North of 23° S, depths are compiled from continuous SeaBeam swath bathymetry data (Macdonald et al. 1988; Lonsdale 1989; Scheirer et al. 1993). South of 23° S, published bathymetry data are incomplete, and axial depths are derived form discrete Sea Beam crossings of the ridge axis (Naar & Hey 1986; Francheteau et al. 1987). Small arrow heads indicate discontinuities with 3–8 km offset. Axial depths near discontinuities are not systematically displayed; they typically increase by a 100–300 m at large OSCs, and vary by 100–1500 m at intratransform spreading centres (Lonsdale 1989; Fox & Gallo 1989).

ald et al. 1982; Francheteau & Ballard 1983; Renard et al. 1985; Morton & Ballard 1986; Bicknell et al. 1987; Holler et al. 1990; Macdonald et al. 1988b; Urabe et al. 1995; Auzende et al. 1996). A zone of crustal accretion at most 1–2 km wide is also indicated by the narrowness of the axial magma chamber (less than 1 km) detected by seismic methods (Kent et al. 1994), the sharpness of the magnetic reversal boundaries on the rise flanks (Sempéré et al. 1987), and the rapid transition from extrusives to sheeted dikes in the upper crust (Hooft et al. 1996).

This uniformity and orderliness of the axial morphology at the ultrafast EPR is similar to that of other sections of the EPR spreading at slightly slower rates (80–125 mm yr^{-1}), along the Pacific–Cocos plate boundary to the north (Macdonald et al. 1984; Macdonald et al. 1992) and along 2400 km of the Pacific–Antarctic plate boundary to the south (Lonsdale 1994). It contrasts with the dramatic variations in morphology of the rift valley along slow spreading ridges. At the Mid-Atlantic Ridge, seismic, gravity and geological studies reveal that the systematic 500–2000 m deepening of the rift valley toward the extremities of ridge segments correlates with the accretion of a progressively thinner crust (White et al. 1984; Lin et al. 1990; Tolstoy et al. 1993; Detrick et al. 1995; Cannat et al. 1995). These crustal thickness variations reach a few kilometres in amplitude, and are thought to reflect the three-dimensional geometry of mantle and magma circulation that feed slowly spreading ridge segments. In comparison, the subdued morphological variations along the fast spreading EPR suggest that magma supply is relatively steady-state and uniform. Nonetheless, although subdued, these morphological variations correlate with the tectonic segmentation of the EPR. This systematic correlation has been used to argue that, on the contrary, mantle upwelling is focused at intervals along the ridge, and that melt is subsequently transported along-strike within the crust or upper mantle. This paper reviews the arguments for and against uniform accretionary processes at the EPR. In particular, the detailed kinematic evolution of the southern EPR reveals that axial segmentation is unstable at time scales as small as 0.1 Ma, and that ridge segment propagation across ridge offsets of any size is frequent. This suggest that along-strike magma transport of material can be efficient at rates at least as high as propagation rates, up to ca. 1000 mm yr^{-1}.

2. The present plate boundary

At intermediate and fast spreading centres, 2–30 km ridge offsets are accommodated by overlapping spreading centres (OSCs) rather than by the classic ridge-transform-ridge geometry (Macdonald et al. 1988a). Along the ultrafast EPR, OSCs are irregularly spaced at intervals of 25–300 km (Macdonald et al. 1988; Lonsdale 1989; Sinton et al. 1991; Scheirer et al. 1996a). They generally have offset smaller than 8 km, and their presence does not significantly affect the regional linearity of the plate boundary (figure 1). North of the Easter microplate, only three OSCs have offsets larger than 10 km: the 2° 45′ S OSC (27 km offset), the 5° 30′ S OSC (15 km offset), and the 20° 40′ S OSC (15–20 km offset) (Rea 1981; Lonsdale 1983; Macdonald et al. 1988b). Offsets larger than 30 km are clustered along two sections of the EPR. From 3° 50′ S to 13° S, six right-stepping transform faults gradually offset the EPR by about 750 km (Searle 1983; Fox & Gallo 1989; Lonsdale 1989). These are the only transform faults along the ultrafast EPR. Each one is a 'multiple' transform fault, consisting of a few parallel transform strands linked by short (less than 15 km) intratransform spreading centres. South of 23° S, where full spreading rates peak at 147–152 mm yr^{-1}, the plate boundary displays a more complex geometry. It is straddled by two microplates a few hundred kilometres across, the Easter microplate and the Juan Fernandez microplate (Searle et al. 1989; Larson et al. 1992). Between these two microplates, the EPR is offset by a 120 km wide OSC, the largest non-transform offset mapped along the mid-ocean ridge system (Hey *et al.* 1995). Each of these three features imparts a left-stepping jog to the ridge axis, and together they correspond to a cumulative offset of *ca.* 350 km. From 23–35° S, the full spreading rate is accommodated at one simple ridge segment only along the 30–32° S section. Outside of this latitudinal range, crustal accretion is partitioned between two spreading centres, either between the east and west ridges of the Easter and Juan Fernandez microplates, or between the overlapping ridges of the large 29° S OSC.

3. Kinematic evolution of the southern EPR since 7 Ma

Recent swath surveys of the southern EPR which extend off-axis on to seafloor at least 1 my old reveal that although the axis of accretion is narrowly defined everywhere, it can rapidly shift location. On a time scale of a few hundred thousand years, some ridge segments propagated hundreds of kilometres, while entire sections of ridge became abandoned. These propagation events initiated at discontinuities ranging from OSCs only a few kilometres wide to transforms with a few hundred kilometre offsets. This section and table 1 summarize those events that have been recognized, following the ultrafast EPR from south to north.

Magnetic and swath sonar data indicate that both the Juan Fernandez and Easter microplates evolved from propagating ridges at offsets of the EPR. The ridges which bound these two microplates to the east propagated northward, and progressively overlapped a few hundred km of the EPR (Francheteau *et al.* 1987; Searle *et al.* 1989; Naar & Hey 1991; Larson *et al.* 1992; Rusby & Searle 1995). Recent, extensive side scan surveys indicate that both east ridges originated at about 6 Ma from within transform faults rather than at OSCs (figure 3*d*; Bird & Naar 1994). The east ridge of the Juan Fernandez microplate propagated from within the Chile transform, and the east ridge of the Easter microplate propagated from within the now abandoned SOEST fracture zone, the fossil trace of which is located about 60 km south of Easter Island (Hey *et al.* 1995). Crust initially formed at both east ridges is truncated by these fracture zones, indicating that propagation did not originate at a classic ridge-transform intersection. Rather, intratransform spreading centres must have served as the focal points for these propagation events (Bird & Naar 1994).

The large left-stepping 29° S OSC located between these two microplates has been migrating southward at *ca.* 120 mm yr^{-1} (figure 3*c*; Hey *et al.* 1995). In the course of this migration, the overlap region is regularly rafted onto the Nazca plate; therefore, despite its large width (120 km), the overlap region does not define a rigidly rotating microplate. A detailed kinematic analysis shows that propagation originated at the SOEST transform fault (Korenaga & Hey 1996). The SOEST offset evolved from a classic transform fault to an OSC configuration at *ca.* 1.95 Ma, and initiated its rapid southward migration at *ca.* 1.5 Ma (figure 3*c*). The propagating ridge segment is roughly aligned with the east rift of the Easter microplate, which propagates in the opposite direction. Both propagation events are directed away from the Easter hot spot.

The right-stepping OSC at 20° 40′ S is a 'duelling propagator', the overlapping ridge tips of which have alternatively propagated and retreated over distances greater than 50 km (Macdonald *et al.* 1988*b*; Perram *et al.* 1993). The net migration rate since 2 Ma is only *ca.* 20 mm yr^{-1}, southward. Based on sparser magnetic coverage

Table 1. *Summary of off-axis swath sonar surveys and their kinematic interpretations*

latitudes	geometry of plate boundary	swath sonar data collected off-axis	kinematic interpretation	references
1° 10′ N–3° S Galapagos triple junction	Galapagos microplate	Some GLORIA and SeaBeam coverage	Microplate initiated at ca. 1 Ma	Searle & Francheteau (1986) Lonsdale (1988)
3–5° S Quebrada, Discovery and Gofar transform faults	Right-stepping multiple transform faults	GLORIA side-scan coverage, out to 1–2 Ma	ca. 1 Ma old Nazca seafloor appears rotated, which is suggestive of large scale propagation event: present transforms' geometry is less than 2 Ma old	Searle (1983, 1984), Lonsdale (1989)
7–9° 30′ S Wilkes transform fault and 'nanoplate'	Right-stepping multiple transform fault and nanoplate	Hydrosweep out to ca. 1.6 Ma	Propagation of northern segment into transform domain since 3 Ma, and formation of nanoplate	Cochran et al. (1993) Goff et al. (1993)
15° 30–19° S MELT experiment area	Staircases of left-stepping OSCs	SeaMARC II, HMR-1 and Sea Beam 2000, out to 6 Ma	Several propagation events at rates greater than 1000 mm yr^{-1}. Propagating segments often initiated within OSCs, bisecting them into smaller OSCs	Cormier & Macdonald (1994) Cormier et al. (1996) Scheirer et al. (1996a)

Table 1. *Cont.*

latitudes	geometry of plate boundary	swath sonar data collected off-axis	kinematic interpretation	references
20–21° S dueling propagator	Right-stepping 20° 40′ S OSC (15–20 km offset)	SeaBeam and SeaMARC II, out to 2 Ma	OSC alternatively migrates northward and southward, with an overall 20 mm yr^{-1} southward migration rate	Macdonald et al. (1988b) Perram et al. (1993)
23–27° S Easter microplate	Microplate ca. 450 km across	GLORIA coverage of entire microplate; some Sea Beam and SeaMARC II coverage	Northward propagation of east rift initiated at ca. 6 Ma within SOEST transform	Francheteau et al. (1988) Naar & Hey (1991) Searle (1989) Rusby & Searle (1995)
27–32° S	Left-stepping 29° S OSC (120 km offset)	GLORI-B and SeaBeam 2000 out to 2.5 Ma	Ridge axis propagated south across SOEST transform at 1.9 Ma, and formed large OSC. Since then, OSC has net southward migration rate of 120 mm yr^{-1}	Klaus et al. (1991) Hey et al. (1995) Korenaga & Hey (1996)
31–35° S Juan Fernandez microplate	Microplate ca. 350 km across, at triple junction	GLORIA coverage of entire microplate; some SeaBeam and Hydrosweep coverage	Northward propagation of east rift initiated at ca. 6 Ma within Chile transform	Francheteau et al. (1987) Larson et al. (1992) Bird & Naar (1994) Bird et al. (1996)

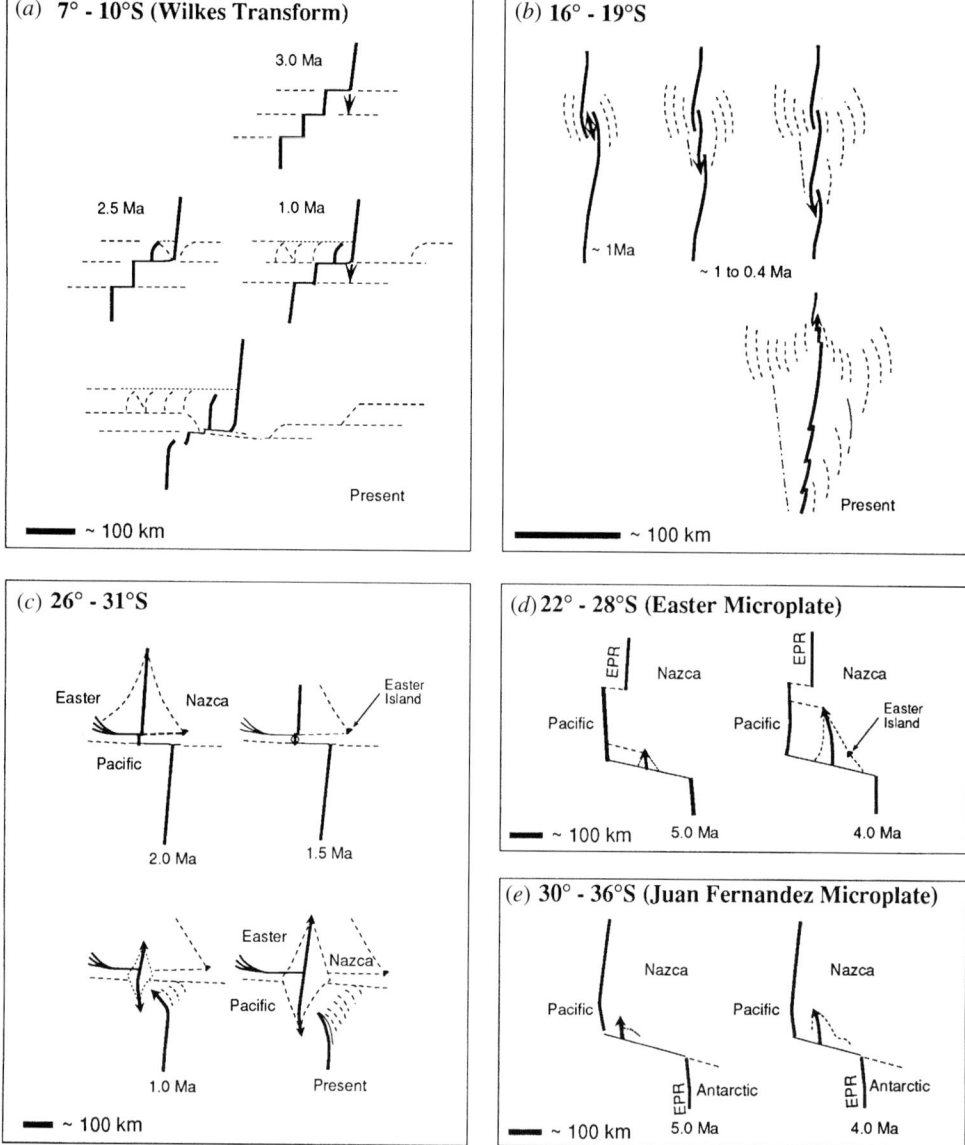

Figure 3. Schematic time frames illustrating the recent kinematic evolution of several sections of the ultrafast EPR. Thick lines indicate actively spreading EPR, and dash lines abandoned ridges or fracture zone traces. Approximate scale bars are indicated for each area. (a) Evolution of Wilkes transform domain, after Goff et al. (1993). (b) Evolution of EPR at 16–19° S, after Cormier et al. (1996). (c) Evolution of 29° S propagator, after Hey et al. (1995). (d) and (e) Origin of Easter and Juan Fernandez microplates, after Bird & Naar (1994).

beyond seafloor 2 Ma old, this slow overall migration rate appears to have been sustained since at least 6 Ma (D. S. Wilson, personal communication).

Although only small offset (less than 8 km) left-stepping OSCs presently dot the EPR between 15 and 19° S, combined analysis of side-scan and magnetic data shows that up to 1 Ma this section of the ridge was marked instead by a few left-stepping OSCs with 15–20 km offsets (figure 3b; Cormier & Macdonald 1994; Cormier et al.

1996). These OSCs were bisected into smaller OSCs by new spreading segments forming within their overlap basins. The smaller OSCs proceeded to migrate rapidly (greater than 500 mm yr^{-1}) and were further bisected by newly spawned ridge segments until the present staircase of small, left-stepping OSCs was achieved.

The Wilkes transform geometry has been continuously evolving since at least 3 Ma, as revealed by a large scale Hydrosweep survey (figure 3a; Goff et al. 1993). Stepwise southward propagation of the EPR into the Wilkes transform domain initiated at ca. 3 Ma, as indicated by the successive abandonment of two transform strands. Secondary rifting west of the EPR has formed a 'nanoplate' about 50–60 km across which is rotating counterclockwise and may eventually evolve into a microplate.

GLORIA side scan images collected between 3 and 5° S over the Quebrada, Gofar and Discovery transforms system show that the seafloor fabric of ca. 1 Ma old Nazca crust is rotated counterclockwise with respect to the ambient orientation (Searle 1983). Accordingly, poorly resolved magnetic lineations suggest that the spreading system north of Yaquina Transform had an overall strike highly oblique to Pacific–Nazca relative motion as recently as 0.7 Ma (Lonsdale 1983). Hence, the EPR between 3 and 7° S probably reached its present configuration since 1 Ma.

The origin of all six transform faults between 3 and 13° S may be related to the creation and abandonment of Bauer paleo-microplate (Lonsdale 1989). Large scale northward propagation of the EPR beyond the right-stepping Bauer transform between 10 and 7 Ma created the Bauer microplate. Spreading ceased on the east ridge of the microplate at about 6–5 Ma, and a staircase of new transforms developed on the remaining west ridge as the relative motion rotated counterclockwise from Pacific–Bauer to Pacific–Nazca (Lonsdale 1989). Although not imaged with swath sonar system, the boundaries of the Bauer paleo-microplate on the Nazca plate are plainly visible on recent maps of free-air anomaly derived from satellite altimetry data (Sandwell et al. 1994). These boundaries delimit an area about 1100 km across, three to four times wider than Easter or Juan Fernandez microplates.

Hence, based on detailed kinematic evolution of some of its sections, the ultrafast EPR appears to be unstable at time scale of 1 Ma or less. Rapid ridge propagation is common, and occurs at axial discontinuities with offsets ranging anywhere from a few kilometres to a few hundred kilometres. In several instances, propagation apparently initiated at short spreading segments located within axial discontinuities. Averaged over several 100 000 yr, propagation rates range from 10–20 mm yr^{-1} (near-stationary) to greater than 500 mm yr^{-1} (ultra rapid).

4. Instability of the tectonic segmentation

Although large scale propagation events are also recognized at slow and intermediate spreading centres (Phipps Morgan & Sandwell 1994; Gente et al. 1995; Auzende et al. 1995), the stability of first order segmentation apparently decreases with increasing spreading rates. This is indicated primarily by the fact that off-axis traces of transform and non-transform discontinuities at slow spreading rates remain subparallel to the spreading direction for many million years, and give the seafloor a distinct 'crenulated' texture (Phipps Morgan & Parmentier 1995). Ridge propagation along the Mid-Atlantic Ridge has not succeeded in disrupting the overall plate boundary geometry inherited from the initial continental break-up, even though the spreading direction has at times rotated by ca. 30° (Tucholke & Schouten 1989). In contrast, the paucity of transform faults along the ultrafast EPR may reflect the tendency for ridge segments to episodically propagate across transform domains (Naar & Hey 1989a). In this way, transform faults, or sections of transform fault defined by intratransform spreading centres, may evolve into large OSCs (see, for example, the Yaquina transform fault), nanoplates (see, for example, the Wilkes transform fault), or microplates (see, for example, the SOEST and Chile transform faults).

Ridge propagation at faster spreading rates is favoured by the rheology of the lithosphere. For equivalent spatial offsets of the ridge axis, the lithosphere abutting the ridge tips is about an order of magnitude younger (thinner) at the EPR than at the MAR. A thinner lithosphere will decrease the viscous resistant forces to propagation and thus promote propagation of a ridge tip (Phipps Morgan & Parmentier 1995). However, no general model has yet emerged which can consistently predict the direction of propagation. Each proposed model can account for some regional propagation patterns, but fails to satisfactorily predict other propagation patterns, suggesting that a diversity of factors must govern propagation, or that the principal control on propagation has yet to be understood.

Changes in spreading direction are often associated with large scale propagation events (Wilson et al. 1984; Atwater et al. 1989). A proposed clockwise change in the Pacific–Nazca relative motion of a few to several degrees in the past several my could explain why all the right-stepping transforms are multiple, 'leaky' transforms (Searle 1983), could account for the existence of a staircase of small left-stepping OSCs between the Garrett transform fault and 20° 40′ S (Lonsdale 1989), and could have initiated the growth of the Easter and Juan Fernandez microplates through the propagation of intratransform spreading centres (Bird & Naar 1994). Indeed, variations in the orientation of fault scarps and abyssal hills with seafloor age indicate that the spreading direction along the southern EPR has rotated clockwise by a few to several degrees since 5–6 Ma (Goff et al. 1993; Cormier et al. 1996). However, although ridge propagation represents an effective mechanism for adjusting the orientation of the ridge axis to a new spreading-normal direction (Wilson et al. 1984), the direction of propagation cannot be predicted from the change in spreading direction. The associated change in the far field stress is expected to affect two offset ridge segments equally rather than favouring one of them (Phipps Morgan & Parmentier 1985).

Hey et al. (1980), Phipps Morgan & Parmentier (1985) and Phipps Morgan & Sandwell (1994) suggest that ridge propagation is driven by the excess relief of a ridge segment, such that the most prominent segment will prevail over the adjacent ones. This model applies well to the pattern of propagation at 16–19° S, where all OSCs are migrating away from the shallow magmatically robust 17–18° S area, and to the east rift of the Easter microplate and the 29° S OSC, which are both propagating away from the shallow region surrounding the Easter hot spot. However, this mechanism cannot explain why new propagating segments sometime initiate within or close to OSCs (Cormier et al. 1996) and transform faults (Bird & Naar 1994), where the ridge axis was presumably deeper.

Based on fracture mechanics theory, Macdonald et al. (1991) propose that the relative lengths of the segments on either side of an OSC may govern its migration direction, so that the longer ridge segment will lengthen at the expense of the shorter one. Although this suggestion applies relatively well to several first- and second-order segments of the EPR, it, again, cannot explain the spawning of propagating segments from within OSCs or transform faults.

Lonsdale (1994) suggests that the migration direction of an OSC is governed by its offset direction (left- or right-stepping) and the migration direction of the ridge axis relative to the asthenosphere. According to this model, an OSC will migrate in a direction which transfers the lithosphere from the 'leading flank' to the 'trailing flank' of a laterally migrating spreading centre. Over time, this would lead to systematic spreading asymmetry of the ridge axis. Indeed, the southern EPR is generally spreading faster to the east, while it is slowly migrating westward (Gripp & Gordon 1990). The above model predicts that left-stepping OSCs should migrate south, and right-stepping should migrate north. This prediction is not supported by the slow southward migration of the right-stepping 20° 40′ S OSC since 2 Ma (Macdonald et al. 1988b; Perram et al. 1993), the northward propagations of the east rifts of the Easter and Juan Fernandez microplates (Bird & Naar 1994), or the northward propagation of the left-stepping 16° 55′ S OSC (Cormier et al. 1996).

Ridge propagation may also be triggered and sustained by local increase in the melt supplied to the EPR. Between 16 and 19° S, propagation events since 1 Ma have been directed away from 17–17° 30′ S, where the ridge presently is in a robust magmatic stage (Cormier et al. 1996). Prior to 1 Ma, a large offset (15–20 km) OSC was located at that latitude, and was, in all likelihood, associated with a relatively starved magmatic supply. Waxing and waning of the melt supply may be cyclical and related to the dynamics of mantle upwelling, as has been suggested for the MAR (Tucholke & Lin 1994; Jha et al. 1995; Gente et al. 1995). Alternatively, the westward migrating EPR may be approaching a mantle thermal anomaly or an 'easily melted' mantle heterogeneity located beneath the Pacific plate, which would be recently diverted toward the ridge axis near 16–19° S. There are several lines of evidence for the presence of such a mantle anomaly. Between Garrett transform fault and Easter microplate, the Pacific plate has an anomalously low subsidence rate, consistent with the presence of a hot thermal anomaly in the mantle beneath it (Rea 1978), or with a lateral temperature gradient in the mantle with temperatures increasing to the west (Cochran 1986). A lateral temperature variation in the mantle could also explain the westward decrease in residual gravity anomalies (Cormier et al. 1995; Magde et al. 1995). Seamounts are anomalously numerous on the Pacific plate within that region, which may indicate the presence of off-axis mantle heterogeneities (Shen et al. 1993, 1995; Scheirer et al. 1996). Finally, based on isotopic and trace element characteristics of zero-age basalts, Mahoney et al. (1994) have suggested that a discrete mantle heterogeneity may be entering into the axial melt zone between 16 and 19° S, although from which direction is uncertain.

5. Structure and rheology of the axial region

Seismic studies along several sections of the EPR have documented the existence of a seismic low velocity zone (LVZ) beneath the ridge crest, which extends from 1–2 km below the seafloor down to the base of the crust and is a few to several kilometres wide (Detrick et al. 1993; Harding et al. 1989; Vera et al. 1990; Toomey et al. 1990; Caress et al. 1992; Mutter et al. 1995). At 9° 30′ N, the largest velocity anomaly is confined to a zone less than 2 km wide and less than 1.5 km thick in the mid crust (Toomey et al. 1990). The overall small velocity anomaly (less than 1 km s^{-1}) associated with the LVZ precludes the existence of a large melt fraction, and the LVZ is generally interpreted as a mostly solidified plutonic section (Toomey et al. 1990; Vera et al. 1990; Harding et al. 1989). In about 60% of the surveyed sections of the EPR, a seismic reflector is tied to the top of the LVZ which can be traced nearly continuously for several tens of kilometres along the ridge axis (Detrick *et al.* 1993). The characteristics of this reflector are consistent with those expected from an interface between magma and overlying crustal rocks. Further analysis of this reflector constrains the axial magma chamber (AMC) to be a narrow sill-like body probably less than a few hundred metres thick (Kent *et al.* 1990). Between the Garrett transform fault and 20° S, its width is only 400–1050 m and it is as shallow as 0.7–1.5 km below seafloor (Detrick *et al.* 1993; Kent *et al.* 1994; Mutter *et al.* 1995; Tolstoy *et al.* 1996).

Diverse magma types reflecting distinct mantle source compositions are often sampled at close intervals along the EPR, which constrains chemically coherent magma bodies in the crust to be a few kilometres long only, sometimes less (Langmuir *et al.* 1986; Sinton *et al.* 1991). This length scale is significantly shorter than the distance over which a continuous AMC reflector is observed, and probably indicates that along-strike mixing of melts is inhibited by the small thickness of the reservoir (Macdonald *et al.* 1988*a*; Sinton & Detrick 1992). Furthermore, based on detailed analysis of the seismic data, Hussenoeder *et al.* (1996) argue that this melt lens is actually partly crystallized, and that in some cases it may contain as much as 60% crystal. High crystal content would also inhibit along-strike mixing within the magma reservoir.

A few models have been proposed for the thermal structure of the EPR which takes into account the seismic results (Wilson *et al.* 1988; Phipps Morgan & Chen 1993; Henstock *et al.* 1993). Although they differ slightly in their details, all three models predict that the crustal volume which underlies the AMC down to the base of the crust is at temperatures close to that of the melt lens, *ca.* 1100–1200 °C. The rheology of gabbros at these high temperatures is not well constrained. The melt fraction is generally inferred to increase up section, but the estimated range varies from 5–10% (Nicolas 1993), to 0–100% (Sinton & Detrick 1992). The rigidus is the temperature below which the magma is a crystal bounded aggregate which behaves rheologically like a solid. Marsh (1989) proposed that the rigidus for MORBs corresponds to 50–60% crystal fraction, implying that the volume underlying the melt lens will behave mostly rigidly. Nicolas *et al.* (1993) suggest on the contrary that the gabbroic crystal mush (90% crystal volume) is close to its rigidus temperature and could dynamically convect. This inference is based on the foliation patterns in ophiolitic gabbros and on laboratory experiment.

It is largely assumed that crustal thicknesses vary little along the EPR, averaging 6–7 km (White *et al.* 1992; Chen 1992). Yet, actual direct crustal thickness measurements are rare. Admittedly, seismic refraction experiments in the eastern Pacific have only occasionally detected P_n arrivals from the moho. Crustal thicknesses are usually inferred (rather than constrained) from vertical two-way travel times to the moho and from the pattern of precritical mantle reflection arrivals (Harding *et al.* 1989; Vera *et al.* 1989). However, rather than marking a petrologic boundary between crust and mantle, the seismic reflection moho may indicate the top of the crust–mantle transition zone; that is, a boundary between the purely mafic portion of the crust and the region with ultramafic presence (Collins *et al.* 1986; Barth & Mutter 1996). Vertical two-way travel times to the moho are quite variable along portions of the northern EPR, and short-wavelength crustal thickness variations of up to *ca.* 2.6 km has been recently proposed to exist near 9 and 13° N (Barth & Mutter 1996). If the seismic moho marked the top of the crust–mantle transition zone rather than the base of the magmatic crust, these short wavelength variations could also reflect the variable thickness of the transition zone rather than the total thickness of crust plus transition zone.

6. Ridge segmentation and variability of axial characteristics

Ridge segments probably represent the primary units of crustal accretion, and their spacing, morphology and migration along axis provide clues to the architecture of magmatic plumbing that feed the ridge axis. The best surveyed section of the southern EPR stretches from the Garrett transform fault to the Easter microplate, and the following focuses mostly on the variability of the geological and geophysical parameters along that section of the ridge and compares it to its axial segmentation (figure 4).

Zero-age depth variations at wavelengths of a few to several hundred km most likely reflect regional anomalies in the temperature and composition of the mantle (LeDouaran & Francheteau 1980; Klein & Langmuir 1987). Temperature variations in the mantle may account for the greater depths (2800–3000 m) of the equatorial EPR (Bonatti et al. 1995) and the shallower depths (2000–2600 m) south of 26° S near the Easter hotspot (Schilling et al. 1985) (figure 2). Superimposed over these regional variations, smaller fluctuations in zero-age depths generally correlate to the tectonic segmentation of the ridge axis (Macdonald & Fox 1988; Macdonald et al. 1988a). Axial depths tend to plunge by a few to several hundred metres toward OSCs and transform faults, and the magnitude of this deepening increases with the width of the offset (Macdonald et al. 1991). However, other factors than proximity to ridge offsets must also control the short wavelength fluctuations in zero-age depths along the superfast EPR. Depth variations may be greater from one segment to the next than along any individual segment. In particular, intratransform spreading centres often stands a few to several hundred metres deeper or shallower that the neighbouring ridge segments (Lonsdale 1989). Some segments also maintain a constant depth (± 20 m) almost right up to their extremities. The series of ridge segments between Garrett transform and 18° S have a nearly flat summit, at ca. 2630 m. Seemingly minor discontinuities can also represent significant boundaries in axial depth characteristics. Hence, the small 8° 38' S OSC (ca. 1.2 km offset) marks a boundary between a 150 km long northern segment which has a constant axial depth (ca. 2725 m) and a southern segment that steadily deepens by a few hundred metres toward the Wilkes transform (Cochran et al. 1993).

The axial segmentation of the EPR is in fact better correlated to the width of the axial high than to the axial depth. For most segments, the axial high narrows significantly toward segment ends, as indicated by the pinching of bathymetry contours. In particular, even though the zero-age depth is nearly constant between 7° 12'–8° 38' S or 14–18° S, depths measured just 2 km from the ridge axis (along the 0.02–0.03 Ma isochrons) increase by a few 100 m toward segment ends (Cochran et al. 1993; Scheirer & Macdonald 1993). This decoupling between the on-axis depth variations and the near-axis depth variations suggest that they are responding to somewhat different processes.

The measure of the cross-sectional area of the ridge axis integrates both axial depth and axial width information, and has proven to be a sensitive indicator of magma supply along the EPR, as confirmed from several other parameters (Scheirer & Macdonald 1993). However, large cross-sectional areas do not correlate with thicker accumulation of extrusives. Although data are few, along- and across-axis

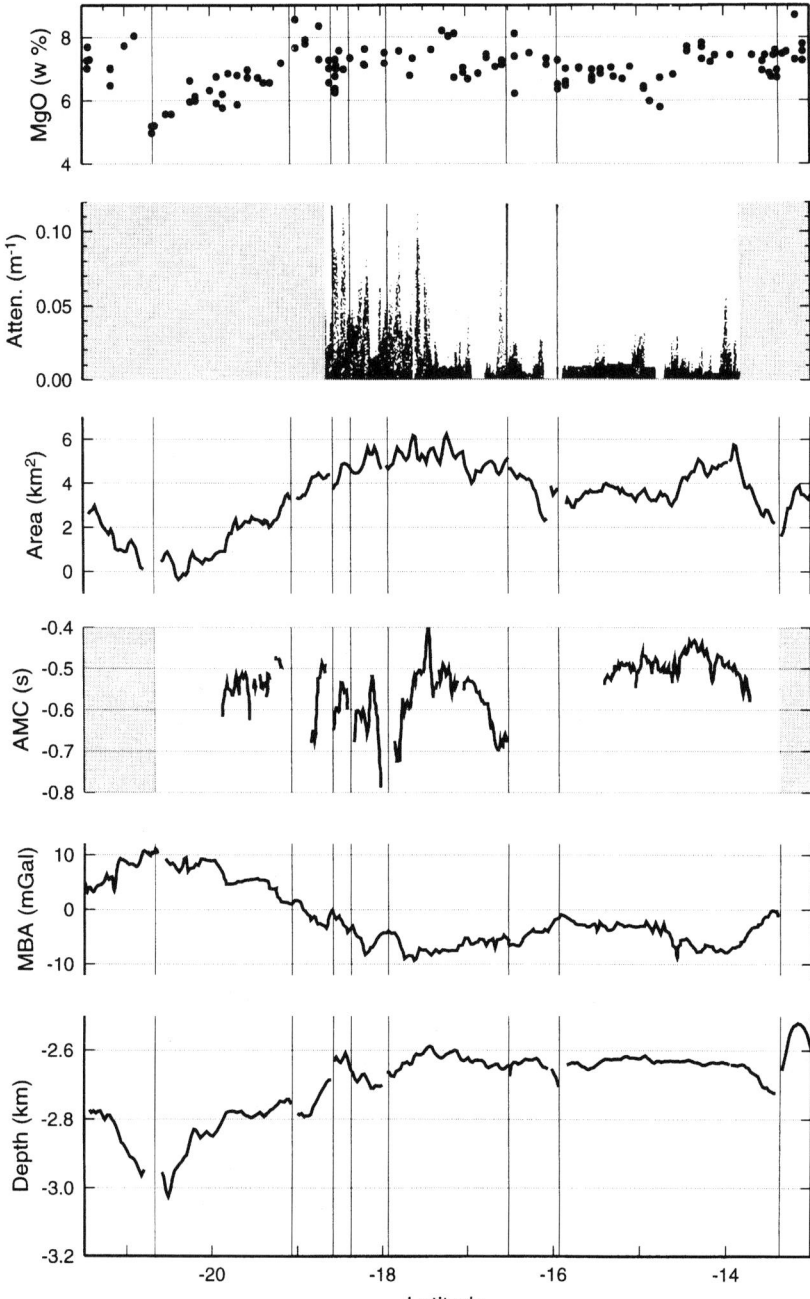

Figure 4. Along-axis variations of several parameters between Garrett transform and Easter microplate. From bottom to top, parameters are zero-age depths (Scheirer & Macdonald 1993), mantle Bouguer gravity anomalies (Cormier et al. 1995; Magde et al. 1995), subseafloor two-way travel times to the axial magma chamber seismic reflector (Detrick et al. 1993), cross-sectional area of the ridge axis (Scheirer & Macdonald 1993), light attenuation anomalies caused by suspended hydrothermal precipitates (Baker 1996), and MgO number for normal basalts (Sinton et al. 1991). Vertical lines mark the location of axial discontinuities with offset greater than 2 km. Shaded intervals indicate data gaps.

seismic lines actually show a tendency for the extrusive layer to be thinner where the axial high is most inflated (Mutter et al. 1995; Carbotte et al. 1996; Tolstoy et al. 1996). This counter-intuitive relation can be explained if erupted lavas were more fluid (hotter) where the ridge is more inflated and ponded several kilometres off-axis rather than within the axial summit caldera (Hooft et al. 1996). The axial high seems to inflate in response to the size of the subcrustal mass anomaly rather than shallow crustal processes (Scheirer & Macdonald 1993; Wang & Cochran 1993): larger cross-sectional areas would reflect the presence of a subcrustal column of partial melt which is taller, wider or with a higher melt content. As a general rule, the cross-sectional area decreases toward segment ends (figure 4). The 18° 35′–19° 04′ S and the 19° 04′–20° 40′ S segments are notable exceptions to this rule. Both are shallowest and widest at their northern ends and plunge toward their southern ends. The former segment has initiated within the 18° 35′ S OSC less than 40 000 yr ago and has since propagated southward at rates exceeding 1000 mm yr^{-1} (Cormier & Macdonald 1994); Conversely, the latter segment is retreating at the same rate from its northern end, and terminates into the large 20° 40′ S OSC at its southern end. This recent, transient behaviour of the two segments probably accounts for their unusual characteristics.

Based on isotopic data and minor element ratios of basaltic glasses, Sinton et al. (1991) and Mahoney et al. (1994) propose that the three sections of the ridge bounded by the Garrett transform, the 16° S OSC, the 20° 40′ S OSC and the northern boundary of the Easter microplate have distinct mantle source compositions. This primary magmatic segmentation happens to correspond to the long-wavelength variations in cross-sectional area (figure 4). Higher MgO content of basaltic glasses, which is thought to reflect higher temperatures of erupted basalts, also correlates well with the long-wavelength variations in cross-sectional area (Scheirer & Macdonald 1993). A similar relationship between mean MgO content and axial width was noted earlier for the northern EPR (Langmuir et al. 1986). Isotopic variations between 16–20° 40′ S define a broad peak which may reflect the recent entry of a mantle plume or mantle heterogeneity beneath the ridge at 17–17° 30′ S (Mahoney et al. 1994; Bach et al. 1994). Accordingly, ridge segments have propagated away from 17–17° 30′ S since 1 Ma, and 16° 30′–20° 40′ S corresponds to the outer limit of this propagation pattern (Cormier et al. 1996). While there exists a good correlation between first-order tectonic and magmatic segmentations, the existence of such a correlation for the finer scale segmentation is open to discussion. Based on glass compositional variations, Sinton et al. (1991) propose that OSCs and some other minor discontinuities define segments with distinct parental magma composition. In contrast, Bach et al. (1994) argue that melting conditions are basically uniform between 13 and 18° S.

Hydrothermal vents have been detected in most submersible dives (Renard et al. 1984; Auzende et al. 1996), towed camera surveys (Bäcker et al. 1985; Morton & Ballard 1986; Macdonald et al. 1988b; Sinton et al. 1991) and hydrographic tow-yos (Urabe et al. 1995; Baker & Urabe 1996) carried out along the southern EPR, attesting to the high level of hydrothermal activity at ultrafast spreading rates. Hydrothermal plume incidence is generally higher along portions of the ridge crest with inflated cross-sectional areas (figure 4), suggesting that hydrothermal circulation is more vigorous where the volume of interstitial melt in the mantle is high (Baker & Urabe 1996).

The two-way travel time to the AMC reflector is highly variable (figure 4). Near OSCs, the increasing two-way travel time to the AMC is probably an artifact of a wandering ship track and of the high lateral velocity variations in the shallow crust near axis (Harding et al. 1993; Kent et al. 1993). Nonetheless, where additional control is provided by cross-axis profiles or away from OSCs, the depth to the AMC varies along-axis by up to 0.6–0.8 km (assuming an average velocity for the upper crust of 4.5–5 km s^{-1}), an order of magnitude more than the corresponding seafloor depths. A strong AMC reflector is systematically detected along sections of the ridge with large cross-sectional areas (Scheirer & Macdonald 1993). Between Garrett transform and 16° S, the depth to the AMC correlates well with the cross-sectional area of the ridge. However, this simple relationship does not really apply elsewhere. For instance, a shallow AMC reflector is present between 19 and 20° S, where the ridge crest is relatively deep and narrow (figure 4).

The mantle Bouguer anomaly (MBA), which is the gravity anomaly corrected for the effect of bathymetry, of a constant crustal thickness and density, is inversely correlated to the axial volume (Scheirer & Macdonald 1993, and figure 4). Lower values of the MBA are indicative of a relative mass deficiency, and are generally interpreted to reflect the presence of a thicker crustal section, or excess heat and melt in the subaxial crust and mantle. MBA values increase toward the Wilkes transform fault, the Garrett transform fault, and the large 20° 40′ S OSC (Wang & Cochran 1993; Magde et al. 1995; Cormier et al. 1995), suggesting that the magma supply is relatively starved in their vicinity. Small OSCs are generally not associated with any detectable MBA anomalies. Overall, the amplitude of the MBA variations is significantly smaller along the EPR (10–20 mgal) than those present along the slow spreading Mid-Atlantic Ridge (20–80 mgal) (Lin & Phipps Morgan 1992).

The fluctuations of the above parameters are good indicators of the magmatic budget of the EPR (Langmuir et al. 1986; Macdonald & Fox 1988; Scheirer & Macdonald 1993; Baker 1996). Shallow axial depths, large cross-sectional areas, high hydrothermal plume incidence, strong AMC reflectors, high MgO content of the glass and low residual gravity anomalies all correspond to a robust magmatic budget, such as occurs between 16° 30′ and 19° S. The opposite trends, such as are present between 19° 30′ and 21° 30′ S, are indicators of a starved magmatic budget. Present magmatic regimes of the EPR seem to be long-lived, as revealed by the constancy of several parameters along flow lines for up to a few million years. Narrow axial highs are systematically flanked by blocky lineated abyssal hills, and broad axial highs by smoother abyssal hill terrain (Cochran et al. 1993; Goff et al. 1993; Cormier et al. 1994). Smoother seafloors are thought to result from the combined effects of blanketing by voluminous lava flows issued from the ridge axis (Macdonald et al. 1996) and lower fault throws developing in areas where the crust is hotter, leading to a thinner brittle layer (Goff et al. 1993; Carbotte & Macdonald 1994). Careful rock sampling across the northern EPR indicates that where the ridge is magmatically robust, the temperatures of eruption remained nearly constant since 0.8 Ma, and that where the ridge is magmatically starved, the temperature of erupted magma changes significantly with time (Batiza et al. 1996). This is interpreted to indicate that magmatically robust ridges have steady state magma chambers, whereas magmatically starved ridges have smaller, transient magma chambers. Finally, the broad, high residual gravity anomalies associated with the slowly migrating 20° 40′ S OSC persist off-axis for at least 2 my, indicating that a starved regime has persisted at this OSC during that entire time interval (Cormier et al. 1995). Ridge propagation events are expected to disrupt flow-line trends, and their associated oblique discordant zones on the ridge flank may sometimes correspond to boundaries between seafloors with opposing characteristics.

7. Accretionary processes along the ultrafast EPR: two or three-dimensional?

Two extreme models of mantle upwelling pattern could explain the relatively uniform characteristics of the EPR (Lin & Phipps Morgan 1992; Bell & Buck 1992). Mantle could be upwelling somewhat evenly beneath the ridge axis, in a two-dimensional or 'sheet-like' pattern. Alternatively, mantle could be upwelling at discrete intervals along the ridge axis and the magma being transported efficiently along-axis within the crust and upper mantle, forming a three-dimensional or 'plume-like' pattern. Mantle upwelling could also assume any pattern intermediate to these two or three-dimensional models. Which style of mantle upwelling actually feeds fast spreading ridges is still debatable, the following sections review the arguments for and against each model.

(a) Arguments in favour of uniform accretion

The concept of two-dimensional accretion at the fast spreading EPR initiated from comparing the variations of its axial characteristics (figure 4) to those of the slow-spreading Mid-Atlantic Ridge (Lin & Phipps Morgan 1992). Although many of the characteristics of the MAR display the same trends as those of the EPR with respect to tectonic segmentation, their variations are not only more systematic, they also have much larger magnitudes. Along the MAR, the ridge axis deepens by 700–1700 m and the residual gravity anomalies increase by 20–80 mGal toward transform faults, while the crustal thickness is often a few kilometres thinner beneath the transforms faults (White *et al.* 1984; Lin *et al.* 1992; Detrick *et al.* 1995). The strong correlation between axial characteristics and axial segmentation of the MAR is commonly thought to reflect the three-dimensional geometry of mantle and magma circulation beneath the ridge. Magma supply is inferred to be focused beneath the shallower part of the ridge axis near mid-segment, and transform faults to mark the distal extremities of the magmatic segments. This model predicts that the magmatic budget will change from robust near mid-segment to relatively starved near ridge-transform intersection, consistent with the observations along the MAR (White *et al.* 1984; Cannat *et al.* 1995). By contrast, the smoothly varying EPR characteristics suggest that magma supply is more homogenous.

Satellite altimetry data reveal that the seafloor created at the EPR singularly lacks the 'crenulated' appearance of the seafloor created at slower spreading ridges (Phipps Morgan & Parmentier 1995). 'Crenulations' are regularly spaced gravity lineations in the spreading direction, which reflect the stationary segmentation of the ridge axis over long periods of time. The absence of crenulations along the EPR could result from an essentially two-dimensional mantle flow structure (Phipps Morgan & Parmentier 1995). Accordingly, numerical experiments indicate that with increasing spreading rates, passive upwelling will eventually prevail over buoyant upwelling (Parmentier & Phipps Morgan 1990; Lin & Phipps Morgan 1992; Sparks & Parmentier 1993; Jha *et al.* 1994; Kincaid *et al.* 1996). Passive upwelling is that induced by the separation of the plates, and is predicted to be mostly uniform along a ridge segment (Phipps Morgan & Forsyth 1988). Buoyant active upwelling initiates as along-axis instabilities in the mantle, and is enhanced by the lowered densities of the thermally expanded, depleted mantle and of the retained melt within the mantle matrix. Whether buoyant upwelling is a cause or a consequence of the segmentation still remains to be determined.

The axial high stands 200–300 m higher than predicted by simple conductive cooling of the lithosphere with age (Madsen et al. 1984; Wilson 1992). Detailed, two-dimensional analysis of both gravity and bathymetric data indicate that the magma chamber contributes negligibly to these anomalous bathymetry signals; on the other hand, the gravity signal of the axial region is well accounted for by the thermal structure of the crust and shallow mantle (Madsen et al. 1990; Wilson 1992; Wang & Cochran 1993; Magde et al. 1995; Wang et al. 1996). These studies suggest that the axial high is isostatically compensated by a low density body located below the crust. By taking into account some realistic thermal models for the lithosphere and asthenosphere, this compensating body is estimated to be narrower than 10 km and taller than 20 km (Wilson 1992; Wang & Cochran 1993; Magde et al. 1995). If this low density body represented a column of partial melt feeding the ridge axis, a melt fraction of 1–3% would imply a column at least 30 km tall. Because an axial high is a ubiquitous feature along the EPR, the above gravity studies imply that a tall column of partial melt underlies the entire length of the ridge axis, consistent with a two-dimensional model for mantle upwelling. Nonetheless, the width and height of that column and its melt content would need to vary along-axis in accordance with the varying depth and cross-sectional area of the axial high, and with its varying MBA (Wang & Cochran 1993).

If all axial discontinuities represented the distal ends of a three-dimensional magmatic plumbing system, intuition would predict a decrease in extrusive thickness in their vicinities. Yet, two-way travel times to the base of seismic layer 2A, interpreted to correspond to the base of the extrusive layer, do not vary in any systematic way near a *ca.* 1 km offset of the EPR at 14° 27′ S (Kent et al. 1994). This suggests that minor offsets of the ridge axis do not correspond to long-term reduction in magma supply, and are probably 'transparent' features in terms of mantle upwelling. Accordingly, smaller offsets are generally not associated with any recognizable 'scars' on the rise flanks, implying that most of them are transient, surficial features of the plate boundary (Macdonald et al. 1988a) or that they migrate too rapidly to generate easily detected discordant zones (Cormier & Macdonald 1994). Furthermore, the small OSCs between 18° 15′ and 20° S are not associated with any significant residual MBA (figure 4), as would be expected if they were associated with a long-lived segmentation in mantle upwelling (Cormier et al. 1995).

(b) Arguments in favour of large scale along-axis magma transport

The common occurrence of ridge propagation events along the southern EPR, as summarized above, attests to one form of along-strike magma transport. Based on detailed studies at propagator tips, spreading centres seem to lengthen initially by tectonic rupture of the older lithosphere, extrusive volcanism following at a later stage only (Kleinrock et al. 1989; Naar et al. 1991). This progression toward full seafloor spreading suggests that, at least initially, magma is being transported along-strike toward the propagator tip rather than being fed vertically from the mantle. Large scale propagation events often occur with average sustained rates comparable to the half spreading rates, or higher. For instance, the west ridge of the Bauer microplate propagated northward *ca.* 1000 km sometime between 10 and 5 Ma (Lonsdale 1989), yielding a conservative estimate for its mean propagation rate of 200 mm yr^{-1}. The large 29° S OSC averages a net propagation rate of 120 mm yr^{-1} since 2 Ma (Korenaga & Hey 1996). Since 1 Ma, a large OSC has propagated southward between 17–19° S at 200–1000 mm yr^{-1} (Cormier et al. 1996). Through propagation, magmatic segments lengthen or shorten through time, implying that underlying mantle upwelling patterns are unstable on the same time scale. Hence, the absence of spreading-parallel crenulations in the altimetry data on the EPR flanks may reflect the frequency of ridge propagation events rather than the two dimensionality of mantle upwelling (Phipps Morgan & Parmentier 1995).

While recent gravity analyses argue for the existence of a continuous column of partial melt beneath the ridge axis, they also imply that the melt content, width or height of that column must vary significantly along axis. Although axial depths are constant between 7° 12′ and 8° 38′ S, the changes in cross-sectional area and MBA requires that the mass deficiency beneath the central section is 20–60% more than near segment ends (Wang & Cochran 1993). Similarly, axial depths vary little between 13° 30′ and 17° 56′ S, yet MBA variations imply the presence of a hotter crust or greater amount of interstitial melt at 17° 20′ S than at 14° 15′ S (Magde et al. 1995). The constant axial depth along sections of the EPR requires the presence of an efficient shallow magmatic plumbing system that extends the length of these ridge segments and redistributes the magma evenly (Wang & Cochran 1993).

Magma supply seems slightly lower in the vicinity of large axial discontinuities, in agreement with them marking the distal ends of an efficient magmatic plumbing system. Hence, while axial depths remain constant as the Garrett transform fault is approached from the south, the axial magma chamber progressively deepens by about 240 m over a 60 km distance (Tolstoy et al. 1996). Interpretation of gravity anomalies over the ridge axis and its flanks indicates that a crust about 500 m thinner than typical has been consistently accreted near the large 20° 40′ S OSC since 2 Ma (Cormier et al. 1995).

Topography and gravity signals along the ultra-slow spreading Reykjanes Ridge display the same subdued fluctuations as the EPR (Bell & Buck 1992). Because the crust of the Reykjanes Ridge is much thicker than typical, its lower section may be reaching temperatures as high as those beneath the EPR. This suggests that mantle upwelling may be diapiric at all spreading rates, but that wherever the lower crust is hot enough, it undergoes rapid ductile flow which evens out any significant crustal thickness variations (Bell & Buck 1992).

Two-way travel time variations to the seismic moho are interpreted as crustal thickness variations of up to 2.6 km at 9–10° N along the EPR (Barth & Mutter 1996). Counter to conventional wisdom, the thinnest crust seems to coincide with the lowest MBA values and minimum axial depth (Wang et al. 1996). A self-consistent interpretation of both the seismic and gravity data near 9–10° N is that mantle upwelling is focused beneath 9° 50′ N and that through efficient along-axis transport, magma accumulates near the ends of the segment. However, moho reflections may arise from the top of the moho transition zone rather than from the crust–mantle boundary (Collins et al. 1986), and the variability in two-way travel time may reflect that of thickness of the magmatic crust only, rather than thickness of the crust plus transition zone. Accordingly, detailed mapping in the Oman ophiolite reveals that the thickest moho transition zones occur on top of mantle diapirs (Nicolas et al. 1996).

A seismic tomography experiment at 9° 30′ N along the EPR reveals a local segmentation of the low velocity zone which is strikingly similar in extent and dimension to the fine scale tectonic segmentation of the ridge (Toomey et al. 1990). This apparent thermal segmentation of the ridge is consistent with injection of mantle-derived melt midway along the 12 km long ridge segment. The small wavelength of this segmentation is also consistent with observations in ophiolites. Detailed mapping of the flow structures preserved in the peridotites of the Oman ophiolites, which are thought to have been accreted at a fast spreading ridge, reveals three-dimensional patterns with 10–20 km wavelength (Nicolas et al. 1994). These are interpreted as frozen-in mantle diapirs that fed the ridge axis before cessation of spreading and obduction.

8. Discussion and conclusions

Although models for two- or three-dimensional mantle upwelling beneath the EPR seem incompatible, evidence presented in support of each indicates that, to some extent, both can be active along the EPR. Which model best applies for a given area may depend on the observation scale, or on local fluctuations in melt supply and thermal structure of the ridge.

The variability in axial morphology and the ridge propagation patterns suggest that along-strike magma transport can occur and is directed away from the most inflated, most robust sections of the southern EPR (Wang & Cochran 1993; Cormier et al. 1996). However, how widespread lateral transport is as an accretionary mechanism remains to be constrained. If lateral melt flow were entirely confined within the crustal layer, simple-minded computations indicate that flow rates necessary to feed a ridge segment would be reasonable. For example, in the extreme case where the entire melt supply to a ridge segment would upwell in its centre, the maximum rate at which magma would need to flow along-axis to accommodate the accretion of a crust 6.5 km thick is directly proportional to the half segment length and the spreading rate, and inversely proportional to the cross-sectional area of the conduit through which magma travels. Hence, if flow were confined to a 500 m wide and 10 m thick melt lens, the maximum flow rate necessary to feed a 200 km long segment would be about 2 m h^{-1} above the locus of upwelling, and would decrease to zero toward segment ends. These rates seem feasible for mid-ocean ridge basalts, which are inferred to flow significantly faster along dikes (Dziak et al. 1995) and during eruptions (Griffiths & Fink 1992). If the conduit were extended to include the entire thickness of the crystal mush zone below the magma lens (about 2 km thick and 2 km wide), the maximum flow rate required for a 200 km long segment would drop to ca. 25 m yr^{-1} above the centre of upwelling, becoming stagnant near segment ends. This maximum rate is somewhat higher than documented ridge propagation rates along the southern EPR (up to 1 m yr^{-1}), but is not unlikely for partially crystallized gabbros deforming ductilely near their rigidus temperature (Bell & Buck 1992).

Axial depth characteristics may reflect whether along-axis magma transport is vigorous or stagnant. Over the two long sections of ridge at 7° 12′–8° 38′ S and 14–18° S, axial depths undulate by less than 50 m. This remarkable flatness suggests either efficient along-axis flow (Lonsdale 1989; Cochran et al. 1993), magmastatic equilibrium (Sinton & Detrick 1992) or isostatic compensation over a uniform partial melt conduit 50–70 km tall (Magde et al. 1995). In contrast, only 2 km from the ridge crest in the same areas, seafloor depths fluctuate by a few hundred metres along-strike. The very narrowness of the zone of flat topography suggests that the controlling process is a shallow one, probably occurring within the crust. Because the depth to the top of the AMC varies by 240–600 m between 14–18° S (Detrick et al. 1993; Mutter et al. 1995; Tolstoy et al. 1996), an order of magnitude higher than the axial depths, constant axial depths cannot result from efficient horizontal transport within the thin melt lens. Rather, magmastatic equilibrium or lateral flow must occur within the lower crust, in the crystal mush zone that underlies the melt lens. Temperatures within this crystal mush zone are inferred to be near the rigidus, and slow magmatic flow could occur (Bell & Buck 1992; Sinton & Detrick 1992; Nicolas et al. 1993).

Although minor discontinuities are not associated with any large signal in the MBA or axial volume, they can represent significant barrier to along-axis flow. The 1.2 km offset at 8° 38′ S marks the boundary between a ridge segment with constant axial depths to the north, and one with steadily decreasing axial depths to the south. Provided that constant axial depths reflect active ductile flow within the partially crystallized gabbros, the small 8° 38′ S discontinuity would mark the southern extent of that process for this area. Accordingly, there is a jump in MgO content across the minor 14° 27′ S discontinuities although the AMC appears continuous, suggesting it represents a mixing boundary within a continuous magma chamber (Sinton et al. 1991).

The existence of a tall column of partial melt feeding the entire EPR is not required by the gravity anomalies (Wilson 1992), and has been inferred because it could isostatically support the axial high (Madsen et al. 1984; Wilson 1992; Wang & Cochran 1993; Magde et al. 1995). Alternatively, the axial high may owe its existence to the dynamic moments acting in the axial region (Forsyth et al. 1994). The shallow magma chamber and frequent intrusion of dikes above the chamber may relieve the regional stresses, creating a weak zone at the top of the extending lithosphere. This would induce dynamic moments which tend to uplift the plate boundary. If this model for the origin of the axial high were correct rather than that of isostatic equilibrium, there would not be any requirements for the presence of a tall column of partial melt. Further evidence for or against the presence of a partial melt column beneath the EPR should come from analysis of the data acquired during the recent MELT experiment at 16–18° S, a large scale experiment designed precisely to address this problem (Forsyth & Chave 1994).

The following model is proposed, which could reconcile most observations. It is envisioned that mantle upwells vertically everywhere along the ridge axis, but that the intensity of this upwelling varies significantly both spatially and temporally. Larger volume of melt could be produced where 'easily melted' mantle heterogeneities are embedded within the asthenosphere (Wilson 1992). As the EPR migrates westward relative to the asthenosphere, such heterogeneities may be fed into the ridge axis somewhat randomly. Because the gabbros beneath the ridge crest are around their rigidus temperatures, minor temperature fluctuations might be sufficient to favour or inhibit their ductile flow. Where melt supply is locally abundant, gabbro temperatures would become higher than their rigidus, favouring along-axis flow toward regions with lower melt supply. On occasion, gabbros may be hot enough and contain sufficient amount of interstitial melt to allow magmastatic equilibrium or efficient along-axis flow, resulting in a subhorizontal ridge crest. Where melt supply becomes particularly robust, lateral flow may be sufficiently vigorous to allow propagation across large ridge offsets. In contrast, locally cooler mantle or low melt supply will result in gabbro temperatures remaining below their rigidus, implying a sluggish or non-existent along-axis flow.

I thank Joe Cann and Harry Eldefield for having motivated this paper. I am indebted to Ken

Macdonald for introducing me to the southern EPR and for many interesting discussions. This paper benefited from insightful remarks by John Mutter and Bill Menke. Lamont–Doherty Earth Observatory contribution # 5546.

References

Atwater, T. & Severinghaus, J. 1989 Tectonic maps of the northeast Pacific. In *The geology of North America, the East Pacific Ocean and Hawaii* (ed. E. L. Winterer, D. M. Hussong & R. W. Decker), vol. N, pp. 15–20. Geological Society of America.

Auzende, J.-M., Hey, R. N., Pelletier, B., Rouland, D., Lafoy, Y., Gracia, E. & Huchon, P. 1995 Propagating rift west of the Fiji Archipelago (North Fiji Basin, SW Pacific). *J. Geophys. Res.* **100**, 17 823–17 835.

Auzende, J.-M., Ballu, V., Batiza, R., Bideau, D., Charlou, J.-L., Cormier, M.-H., Fouquet, Y., Geistdoerfer, P., Lagabrielle, Y., Sinton, J. & Spadea, P. 1996 Recent tectonic, magmatic and hydrothermal activity on the East Pacific Rise between 17° and 19° S: submersible observations. *J. Geophys. Res.* **101**, 17 995–18 010.

Bach, W., Hegner, E., Erzinger, J. & Satir, M. 1994 Chemical and isotopic variations along the superfast spreading East Pacific Rise from 6° to 30° S. *Contr. Miner. Petr.* **116**, 365–380.

Bäcker, H., Lange, J. & Marchig, V. 1985 Hydrothermal activity and sulfide formation in axial valleys of the East Pacific Rise crest between 18° and 22° S. *Earth Planet. Sci. Lett.* **72**, 9–22.

Baker, E. T. & Urabe, T. 1996 Extensive distribution of hydrothermal plumes along the superfast spreading East Pacific Rise, 13° 30′–18° 40′ S. *J. Geophys. Res.* **101**, 8685–8695.

Baker, E. T. 1996 Geological indexes of hydrothermal venting. *J. Geophys. Res.* **101**, 13 741–13 753.

Barth, G. A. & Mutter, J. C. 1996 Variability in oceanic crustal thickness and structure: Multichannel seismic reflection results from the East Pacific Rise, 8° 50′ N to 9° 50′ N and 12° 30′ N to 13° 30′ N. *J. Geophys. Res.* **101**, 17 951–17 975.

Batiza, R., Niu, Y., Karsten, J. L., Potts, E., Norby, L. & Buttler, R. 1996 Steady and non-steady magma chambers below the East Pacific Rise. *Geophys. Res. Lett.* **23**, 221–224.

Bell, R. E. & Buck, W. R. 1992 Crustal control of ridge segmentation inferred from observations of the Reykjanes Ridge. *Nature* **357**, 583–586.

Bicknell, J. D., Sempèrè, J.-C., Macdonald, K. C. & Fox, P. J. 1987 Tectonics of a fast spreading centre: a Deep-Tow and Sea Beam survey on the East Pacific Rise at 19° 30′ S. *Mar. Geophys. Res.* **9**, 25–46.

Bird, R. T. & Naar, D. F. 1994 Intratransform origins of mid-ocean ridge microplates. *Geology* **22**, 987–990.

Bird, R. T., Naar, D. F., Larson, R. L., Searle, R. C. & Scotese, C. R. 1997 Plate tectonic reconstructions of the Juan Fernandez microplate: transformation from internal shear to rigid rotation. *J. Geophys. Res.* (Submitted.)

Cannat, M., Mèvel, C., Maia, M., Deplus, C., Durand, C., Gente, P., Agrinier, P., Belarouchi, A., Dubuisson, G., Humler, E. & Reynolds, J. 1995 Thin crust, ultramafic exposures, and rugged faulting patterns at the Mid-Atlantic Ridge (22°–24° N). *Geology* **23**, 49–52.

Carbotte, S. M. & Macdonald, K. C. 1994 Comparison of sea floor tectonic fabric created at intermediate, fast, and superfast spreading ridges: influence of spreading rate, plate motions and ridge segmentation on fault patterns. *J. Geophys. Res.* **99**, 13 609–13 631.

Carbotte S. M., Mutter, J. C. & Xu, L. 1997 Contribution of tectonism and volcanism to axial and flank morphology of the southern EPR, 17° 10′–17° 40′, from a study of layer 2A geometry. *J. Geophys Res.* (In the press.)

Caress, D. W., Burnett, M. S. & Orcutt, J. A. 1992 Tomographic image of the axial low velocity zone at 12° 50′ N on the East Pacific Rise. *J. Geophys. Res.* **97**, 9243–9263.

Charlou, J. L., Fouquet, Y., Donval, J.-P., Auzende, J.-M., Jean-Baptiste, P., Stievenard, M. & Michel, S. 1996 Mineral and gas chemistry of hydrothermal fluids on an ultrafast spreading ridge: East Pacific Rise, 17°–19° S (Naudur cruise 1993) phase separation processes controlled by volcanic and tectonic activity. *J. Geophys. Res.* **101**, 15 899–15 919.

Chen, Y. J. 1992 Oceanic crustal thickness versus spreading rate. *Geophys. Res. Lett.* **19**, 753–756.

Cochran, J. R. 1986 Variations in subsidence rates along intermediate and fast spreading mid-ocean ridges. *Geophys. Jl R. Astr. Soc.* **87**, 421–454.

Cochran, J. R., Goff, J. A., Malinverno, A., Fornari, D. J., Keeley, C. & Wang, X. 1993 Morphology of a 'superfast' mid-ocean ridge crest and flanks: the East Pacific Rise 7°–9° S. *Mar. Geophys. Res.* **15**, 65–75.

Collins, J. A., Brocher, T. M. & Karson, J. A. 1986 Two-dimensional seismic reflection modeling of the inferred fossil oceanic crust/mantle transition in the Bay of Islands ophiolite. *J. Geophys. Res.* **91**, 12 520–12 538.

Cormier, M.-H. & Macdonald, K. C. 1994 East Pacific Rise 18°–19° S: asymmetric spreading and ridge reorientation by ultra-fast migration of axial discontinuities. *J. Geophys. Res.* **99**, 543–564.

Cormier, M.-H., Macdonald, K. C. & Wilson, D. S. 1995 A three-dimensional gravity analysis of the East Pacific Rise, 18°–21° 30′ S. *J. Geophys. Res.* **100**, 8063–8082.

Cormier, M.-H., Scheirer, D. S. & Macdonald, K. C. 1996 Evolution of the East Pacific Rise at 16°–19° S since 5 Ma: bisection of overlapping spreading centres by new, rapidly propagating ridge segments. *Mar. Geophys. Res.* **18**, 53–84.

DeMets, C., Gordon, R. G, Argus, D. F. & Stein, S. 1994 Effect of recent revisions to the geomagnetic reversal time scale on estimates of current plate motions. *Geophys. Res. Lett.* **21**, 2191–2194.

Detrick, R. S., Harding, A. J., Kent, G. M., Orcutt, J. A., Mutter, J. C. & Buhl, P. 1993 Seismic structure of the southern East Pacific Rise. *Science* **259**, 499–503.

Detrick, R. S., Needham, H. D. & Renard, V. 1995 Gravity anomalies and crustal thickness variations along the Mid-Atlantic Ridge between 33° N and 40° N. *J. Geophys. Res.* **100**, 3767–3787.

Dziak, R. P., Fox, C. G. & Schreiner, A. E. 1995 The June–July 1993 seismo-acoustic event at CoAxial segment, Juan de Fuca Ridge: evidence for a lateral dike injection. *Geophys. Res. Lett.* **22**, 135–138.

Forsyth, D. W. & Chave, A. D. 1991 Experiment investigates magma in the mantle beneath mid-ocean ridge. *Eos* **75**, 537–540.

Forsyth, D. W., Eberle, M. A. & Parmentier, E.M. 1994 An alternative explanation for the origin of the axial high on fast spreading ridges. *Eos* **75**, 640 (abstract).

Fox, P. J. & Gallo, D. J. 1989 Transforms of the eastern central Pacific. In *The Geology of North America, the Eastern Pacific Ocean and Hawaii*, vol. N, pp. 111–124. Geological Society of America.

Francheteau, J. & Ballard, R. D. 1983 The East Pacific Rise near 21° N, 13° N and 20° S: inferences for along-strike variability of axial processes of the mid-ocean ridge. *Earth Planet. Sci. Lett.* **64**, 93–116.

Francheteau, J., Yelles-Chaouche, A. & Craig, H. 1987 The Juan Fernandez microplate north of the Pacific-Nazca-Antarctic plate unction at 35° S. *Earth Planet. Sci. Lett.* **86**, 253–268.

Francheteau, J., Patriat, P., Segoufin, J., Armijo, R., Doucoure, M., Yelles-Chaouche, A., Zukin, J., Calmant, S., Naar, D. F. & Searle, R. 1988 Pito and Orongo fracture zones: the northern and southern boundaries of the Easter microplate (southeast Pacific). *Earth Planet. Sci. Lett.* **89**, 363–374.

Gente, P., Pockalny, R. A., Durand, C., Deplus, C., Maia, M., Ceuleneer, G., Mèvel, C., Cannat, M. & Laverne, C. 1995 Characteristics and evolution of the segmentation of the Mid-Atlantic Ridge between 20° N and 24° N during the last 10 million years. *Earth Planet. Sci. Lett.* **129**, 55–71.

Goff, J. A., Malinverno, A., Fornari, D. J. & Cochran, J. R. 1993a Abyssal hills segmentation—quantitative analysis of the East Pacific Rise flanks 7° S–9° S. *J. Geophys. Res.* **98**, 13 851–13 862.

Goff, J. A., Fornari, D. J., Cochran, J. R., Keeley, C. & Malinverno, A. 1993b Wilkes transform system and 'nannoplate'. *Geology* **21**, 623–626.

Griffiths, R. W. & Fink, J. H. 1992 Solidification and morphology of submarine lavas: a dependence on extrusion rate. *J. Geophys. Res.* **97**, 19 729–17 737.

Gripp, A. E. & Gordon, R. G. 1990 Current plate velocities relative to the hot spots incorporating the NUVEL-1 global plate motion model. *Geophys. Res. Lett.* **17**, 1109–1112.

Harding, A. J., Orcutt, J. A., Kappus, M. E., Vera, E. E., Mutter, J. C., Buhl, P., Detrick, R. S. & Brocher, T. M. 1989 Structure of young oceanic crust at 13° N on the East Pacific Rise from expanding spread profiles. *J. Geophys. Res.* **94**, 12 163–12 196.

Henstock, T. J., Woods, A. W. & White, R. S. 1993 The accretion of oceanic crust by episodic sill injection. *J. Geophys. Res.* **98**, 4143–4161.

Hey, R., Duennebier, F. K. & Morgan, W. J. 1980 Propagating rifts on mid-ocean ridges. *J. Geophys. Res.* **85**, 3647–3658.

Hey, R. N., Johnson, P. D., Martinez, F., Korenaga, J., Somers, M. L., Huggett, Q. J., LeBas, T. P., Rusby, R. I. & Naar, D. F. 1995 Plate boundary reorganization at a large-offset, rapidly propagating rift. *Nature* **378**, 167–170.

Holler, G., Marchig, V. & the shipboard scientific party 1990 Hydrothermal activity on the East Pacific Rise: stages of development. *Geol. Jl B* **75**, 3–22.

Hooft, E. E., Schouten, H. & Detrick, R. S. 1996 Constraining crustal emplacement processes from the variation in seismic layer 2A thickness at the East Pacific Rise. *Earth Planet. Sci. Lett.* **142**, 289–309.

Hussenoeder, S. A., Collins, J. A., Kent, G. M., Detrick, R. S. & the TERA Group 1996 Seismic analysis of the axial magma chamber reflector along the southern East Pacific Rise from conventional reflection profiling. *J. Geophys. Res.* **101**, 22087–22105.

Jha, K., Parmentier, E. M. & Phipps Morgan, J. 1994 The role of mantle-depletion and melt-retention buoyancy in spreading-centre segmentation. *Earth Planet. Sci. Lett.* **125**, 221–234.

Kent, G. M., Harding, A. J. & Orcutt, J. A. 1990 Evidence for a smaller magma chamber beneath the east Pacific Rise at 9° 30' N. *Nature* **344**, 650–663.

Kent, G. M., Harding, A. J. & Orcutt, J. A. 1993 Distribution of magma beneath the east Pacific Rise between the Clipperton transform and the 9° 17' N deval from forward modeling of common depth point data. *J. Geophys. Res.* **98**, 13 945–13 969.

Kent, G. M., Harding, A. J., Orcutt, J. A., Detrick, R. S., Mutter, J. C. & Buhl, P. 1994 Uniform accretion of oceanic crust south of the Garrett transform at 14° 15' S on the East Pacific Rise. *J. Geophys. Res.* **99**, 9097–9116.

Kincaid, C., Sparks, D. W. & Detrick, R. S. 1996 The relative importance of plate-driven and buoyancy-driven flow at mid-ocean ridges. *J. Geophys. Res.* **101**, 16 177–16 193.

Klaus, A., Icay, W., Naar, D. & Hey, R. N. 1991 SeaMARC II survey of a propagating limb of a nontransform offset near 29° S along the fastest spreading East pacific Rise segment. *J. Geophys. Res.* **96**, 9985–9998.

Kleinrock, M. C. & Hey, R. N. 1989 Detailed tectonics near the tip of the Galapagos 95.5° W propagator: how the lithosphere tears and a spreading axis develops. *J. Geophys. Res.* **94**, 13 801–13 838.

Korenaga, J. & Hey, R. N. 1996 Recent dueling propagation history at the fastest spreading centre, the East Pacific Rise, 26°–32° S. *J. Geophys. Res.* **101**, 18 023–18 041.

Langmuir, C. H., Bender, J. F. & Batiza, R. 1986 Petrological and tectonic segmentation of the East Pacific Rise, 5° 30'–14° 30' N. *Nature* **322**, 422–429.

Larson, R. L., Searle, R. C., Kleinrock, M. C., Schouten, H., Bird, R. T., Naar, D. F., Rusby, R. I., Hooft, E. E. & Lasthiotakis, H. 1992 Roller-bearing tectonic evolution of the Juan Fernandez microplate. *Nature* **356**, 571–576.

Lin, J., Purdy, G. M., Schouten, H., Sempèrè, J.-C. & Zervas, C. 1990 Evidence from gravity data for focused magmatic accretion along the Mid-Atlantic ridge. *Nature* **344**, 627–632.

Lin, J. & Phipps Morgan, J. 1992 The spreading rate dependence of three-dimensional mid-ocean ridge gravity structure. *Geophys. Res. Lett.* **19**, 13–16.

Lonsdale, P. 1983 Overlapping rift zones at the 5.5° S offset of the East Pacific Rise. *J. Geophys. Res.* **88**, 9393–9406.

Lonsdale, P. 1988 Structural pattern of the Galapagos microplate and evolution of the Galapagos triple junctions. *J. Geophys. Res.* **93**, 13 551–13 574.

Lonsdale, P. 1989 Segmentation of the Pacific-Nazca spreading centre, 1° N–20° S. *J. Geophys. Res.* **94**, 12 197–12 226.

Lonsdale, P. 1994 Geomorphology and structural segmentation of the crest of the southern (Pacific–Antarctic) East Pacific Rise. *J. Geophys. Res.* **99**, 4683–4702.

Macdonald, K. C. 1982 Mid-ocean ridges: fine scale tectonic, volcanic and hydrothermal processes within the plate boundary zone. *Ann. Rev. Earth Planet. Sci.* **10**, 155–190.

Macdonald, K. C., Sempèrè, J.-C. & Fox, P. J. 1984 East Pacific Rise from Siqueiros to Orozco fracture zones: along-strike continuity of axial neovolcanic zone and structure and evolution of overlapping spreading centres. *J. Geophys. Res.* **89**, 6049–6069.

Macdonald, K. C., Fox, P. J., Perram, L. J., Eisen, M. F., Haymon, R. M., Miller, S. P., Carbotte, S. M., Cormier, M.-H. & Shor, A. N. 1988*a* A new view of the mid-ocean ridge from the behaviour of ridge axis discontinuities. *Nature* **335**, 217–225.

Macdonald, K. C., Haymon, R. M., Miller, S. P. & Sempèrè, J.-C. 1988*b* Deep-tow and sea beam studies of dueling propagating ridges on the East Pacific Rise near 20° 40′ S. *J. Geophys. Res.* **93**, 2875–2898.

Macdonald, K. C. & Fox, P. J. 1988 The axial summit graben and cross-sectional shape of the East Pacific Rise as indicators of axial magma chambers and recent volcanic eruptions. *Earth Planet. Sci. Lett.* **88**, 119–131.

Macdonald, K. C., Scheirer, D. S. & Carbotte, S. M. 1991 Mid-ocean ridges: discontinuities, segments and giant cracks. *Science* **253**, 986–994.

Madsen, J. A., Forsyth, D. W. & Detrick, R. S. 1984 A new isostatic model for the East Pacific Rise crest. *J. Geophys. Res.* **89**, 9997–10015.

Madsen, J. A., Detrick, R. S., Mutter, J.-C., Buhl, P. & Orcutt, J. A. 1990 A two and three-dimensional analysis of gravity anomalies associated with the East Pacific Rise at 9° N and 13° N. *J. Geophys. Res.* **95**, 4967–4987.

Magde, L. S., Detrick, R. S. & the TERA group 1995 Crustal and upper mantle contribution to the axial gravity anomaly at the southern East Pacific Rise. *J. Geophys. Res.* **100**, 3747–3766.

Mahoney, J. J., Sinton, J. M., Macdougall, J. D., Spencer, K. J. & Lugmair, G. W. 1994 Isotope and trace element characteristics of a superfast spreading ridge: East Pacific Rise, 13–23° S. *Earth Planet. Sci. Lett.* **121**, 173–193.

Marsh, B. D. 1989 Magma chambers. *A. Rev. Earth Planet. Sci.* **17**, 439–474.

Morton, J. L. & Ballard, R. D. 1986 East Pacific Rise at latitude 19° S: evidence for a recent ridge jump. *Geology* **14**, 111–114.

Mutter, J. C., Carbotte, S. M., Su, W., Xu, L., Buhl, P., Detrick, R. S., Kent, G. M., Orcutt, J. A. & Harding, A. J. 1995 Seismic images of active magma systems beneath the East Pacific Rise between 17° 05′ and 17° 35′ S. *Science* **268**, 391–395.

Naar, D. F. & Hey, R. N. 1989*a* Speed limit for oceanic transform faults. *Geology* **17**, 534–537.

Naar, D. F. & Hey, R. N. 1989*b* Recent Pacific–Easter–Nazca plate motions. *Evolution of mid-ocean ridges* (ed. J. M. Sinton). Geophysical monograph 57, pp. 9–30.

Naar, D. F. & Hey, R. N. 1991 Tectonic evolution of the Easter microplate. *J. Geophys. Res.* **96**, 7961–7993.

Naar, D. F., Martinez, F., Hey, R. N., Reed IV, T. B. & Stein, S. 1991 Pito rift: how a large offset rift propagates. *Mar. Geophys. Res.* **13**, 287–309.

Nicolas, A., Freydier, C., Godard, M. & Vauchez, A. 1993 Magma chambers at oceanic ridges: how large? *Geology* **21**, 53–56.

Nicolas, A., Boudier, F. & Ildefonse, B. 1994 Evidence from the Oman ophiolite for active mantle upwelling beneath a fast spreading ridge. *Nature* **370**, 51–53.

Nicolas, A., Boudier, F. & Ildefonse, B. 1996 Variable crustal thickness in the Oman ophiolite: implication for oceanic crust. *J. Geophys. Res.* **101**, 17 941–17 950.

Parmentier, E. M. & Phipps Morgan, J. 1990 Spreading rate dependence of three-dimensional structure in oceanic spreading centres. *Nature* **348**, 325–328.

Perram, L. J., Cormier, M.-H. & Macdonald, K. C. 1993 Magnetic and tectonic studies of the dueling propagating spreading centres at 20° 40′ S on the East Pacific Rise: evidence for crustal rotations. *J. Geophys. Res.* **98**, 13 835–13 850.

Phipps Morgan, J. & Parmentier, E. M. 1985 Causes and rate-limiting mechanisms of ridge propagation: a fracture mechanics model. *J. Geophys. Res.* **90**, 8603–8612.

Phipps Morgan, J. & Forsyth, D. W. 1988 Three dimensional flow and temperature pertubations due to a transform offset: effect on oceanic crustal and upper mantle structure. *J. Geophys. Res.* **93**, 2955–2966.

Phipps Morgan, J. & Chen, Y. J. 1993 Dependence of ridge axis morphology on magma supply and spreading rate. *Nature* **364**, 706–708.

Phipps Morgan, J. & Sandwell, D. T. 1994 Systematics of ridge propagation south of 30° S. *Earth Planet. Sci. Lett.* **121**, 245–258.

Phipps Morgan, J. & Parmentier, E. M. 1995 Crenulated seafloor: evidence for spreading-rate dependent structure of mantle upwelling and melting beneath a mid-ocean spreading centre. *Earth Planet. Sci. Lett.* **129**, 73–84.

Rea, D. K. 1978 Asymmetric sea-floor spreading and a nontransform axis offset: the East Pacific Rise 20° S survey area. *Geol. Soc. Am. Bull.* **89**, 836–844.

Rea, D. K. 1981 Tectonics of the Nazca–Pacific divergent plate boundary. *Nazca plate: crustal formation and Andean convergence* (ed. L. D. Kulm et al.), pp. 27–62. Geol. Soc. Am. Memoir no. 154.

Renard, V., Hèkinian, R., Francheteau, J., Ballard, R. D. & Bäcker, H. 1985 Submersible observations at the axis of the ultra-fast-spreading East Pacific Rise (17° 30' to 21° 30' S). *Earth Planet. Sci. Lett.* **75**, 339–353.

Rusby, R. I. & Searle, R.C. 1995 A history of the Easter microplate, 5.25 Ma to present. *J. Geophys. Res.* **100**, 12617–12640.

Sandwell, D. T., Yale, M. M. & Smith, W. H. F. 1994 ERS-1 geodetic mission reveals detailed tectonic structures. *Eos* **75**, 155.

Scheirer, D. S. & Macdonald, K. C. 1993 The variation in cross-sectional area of the axial ridge along the East Pacific Rise: evidence for the magmatic budget of a fast-spreading centre. *J. Geophys. Res.* **98**, 7871–7885.

Scheirer, D. S., Macdonald, K. C., Forsyth, D. W., Miller, S. P., Wright, D. J., Cormier, M.-H. & Weiland, C. M. 1996a A map series of the southern East Pacific Rise and its flanks, 15° S to 19° S. *Mar. Geophys. Res.* **18**, 1–12.

Scheirer, D. S., Macdonald, K. C., Forsyth, D. W. & Shen, Y. 1996b Abundant seamounts of the Rano Rahi seamount field near the southern East Pacific Rise, 15°–19° S. *Mar. Geophys. Res.* **18**, 13–52.

Searle, R. C. 1983 Multiple, closely space transform faults in fast-slipping fracture zones. *Geology* **11**, 607–610.

Searle, R. C. 1984 Gloria survey of the East Pacific Rise near 3.5° S: tectonic and volcanic characteristics of a fast spreading mid-ocean rise. *Tectonophys.* **101**, 319–344.

Searle, R. C. & Francheteau, J. 1986 Morphology and tectonics of the Galapagos triple junction. *Mar. Geophys. Res.* **8**, 98–129.

Searle, R. C., Rusby, R. I., Engeln, J., Hey, R. N., Zukin, J., Hunter, P. M., LeBas, T. P., Hoffman, H. J. & Livermore, R. 1989 Comprehensive sonar imaging of the Easter microplate. *Nature* **341**, 701–705.

Sempèrè, J.-C., Macdonald, K. C., Miller, S. P. & Shure, L. 1987 Detailed study of the Brunhes/Matuyama reversal boundary on the East Pacific Rise at 19° 30' S: implication for crustal emplacement processes at an ultrafast spreading centre. *Mar. Geophys. Res.* **9**, 1–23.

Shen, Y., Forsyth, D. W., Scheirer, D. S. & Macdonald, K. C. 1993 Two forms of volcanism: implications for the volume of off-axis crustal production on the west flank of the East Pacific Rise. *J. Geophys. Res.* **98**, 17875–17889.

Shen, Y., Scheirer, D. S., Forsyth, D. W. & Macdonald, K. C. 1995 Trade-off in production between adjacent seamount chains near the East Pacific Rise. *Nature* **373**, 14–142.

Sinton, J. M., Smaglik, S. M., Mahoney, J. J. & Macdonald, K. C. 1991 Magmatic processes at superfast mid-ocean ridges: glass compositional variations along the East-Pacific Rise 13°–23° S. *J. Geophys. Res.* **96**, 6133–6155.

Sinton, J. M. & Detrick, R. S. 1992 Mid-ocean ridge magma chambers. *J. Geophys. Res.* **97**, 197–216.

Sparks, D. W. & Parmentier, E. M. 1993 The structure of three-dimensional convection beneath oceanic spreading centres. *Geophys. J. Int.* **112**, 81–92.

Tolstoy, M., Harding, A. J. & Orcutt, J. A. 1993 Crustal thickness on the Mid-Atlantic Ridge: bull's eye gravity anomalies and focused accretion. *Science* **262**, 726–728.

Tolstoy, M., Harding, A. J., Orcutt, J. A. & the TERA Group 1997 Deepening of the axial magma chamber on the southern East Pacific Rise toward the Garrett fracture zone. *J. Geophys. Res.* (In the press.)

Toomey, D. R., Purdy, G. M., Solomon, S. C. & Wilcock, W. S. D. 1990 The three-dimensional seismic velocity structure of the East Pacific Rise near latitude 9° 30′ N. *Nature* **347**, 639–645.

Tucholke, B. E. & Lin, J. 1994 A geological model for the structure of ridge segments in slow spreading ocean crust. *J. Geophys. Res.* **99**, 11 937–11 958.

Tucholke, B. E. & Schouten, H. 1989 Kane fracture zone. *Mar. Geophys. Res.* **10**, 1–40.

Urabe, T., Baker, E. T. & others 1995 The effect of magmatic activity on hydrothermal venting along the superfast-spreading East Pacific Rise. *Science* **269**, 1092–1095.

Vera, E. E., Mutter, J. C., Buhl, P., Orcutt, J. A., Harding, A. J., Kappus, M. E., Detrick, R. S. & Brocher, T. M. 1990 The structure of 0 to 0.2 my old oceanic crust at 9° N on the East Pacific Rise from expanded spread profiles. *J. Geophys. Res.* **95**, 15 529–15 556.

Wang, X. & Cochran, J. R. 1993 Gravity anomalies, isostasy, and mantle flow at the East Pacific Rise crest. *J. Geophys. Res.* **98**, 19 505–19 531.

Wang, X., Cochran, J. R. & Barth, G. A. 1995 Gravity anomalies, crustal thicknesses and the pattern of mantle flow at the fast spreading East Pacific Rise, 9°–10° N: evidence for three-dimensional upwelling. *J. Geophys. Res.* **101**, 17 927–17 940.

White, R. S., Detrick, R. S., Sinha, M. C. & Cormier, M.-H. 1984 Anomalous seismic crustal structure of oceanic fracture zones. *Geophys. Jl R. Astr. Soc.* **79**, 779–798.

White, R. S., McKenzie, D. & O'Nions, R. K. 1992 Oceanic crustal thickness from seismic measurements and rare earth element inversions. *J. Geophys. Res.* **97**, 19 683–19 715.

Wilson, D. S. 1992 Focused mantle upwelling beneath mid-ocean ridges: evidence from seamount formation and isostatic compensation of topography. *Earth Planet. Sci. Lett.* **113**, 41–55.

Wilson, D. S., Hey, R. N. & Nishimura, C. 1984 Propagation as a mechanism of reorientation of the Juan de Fuca plate. *J. Geophys. Res.* **91**, 9215–9225.

Wilson, D. S., Clague, D. A., Sleep, N. H. & Morton, J. L. 1988 Implications of magma convection for the size and temperature of magma chambers at fast spreading ridges. *J. Geophys. Res.* **93**, 11 974–11 984.

Seafloor eruptions and evolution of hydrothermal fluid chemistry

By D. A. Butterfield[1], I. R. Jonasson[2], G. J. Massoth[3], R. A. Feely[3], K. K. Roe[1], R. E. Embley[4], J. F. Holden[5], R. E. McDuff[5], M. D. Lilley[5] and J. R. Delaney[5]

[1] Joint Institute for the Study of Atmosphere and Ocean, University of Washington, Seattle, WA 98195, USA
[2] Geological Survey of Canada, Ottawa, Ontario, Canada
[3] Pacific Marine Environmental Laboratory, National Oceanic and Atmospheric Administration, Seattle, WA 98115, USA
[4] Pacific Marine Environmental Laboratory, National Oceanic and Atmospheric Administration, Newport, OR 97365, USA
[5] School of Oceanography, University of Washington, Seattle, WA 98195, USA

A major challenge confronting geochemists is to relate the chemistry of vented hydrothermal fluids to the local or regional tectonic and volcanic state of mid-ocean ridges. After more than 15 years of sampling submarine hydrothermal fluids, a complex picture of spatial and temporal variability in temperature and composition is emerging. Recent time-series observations and sampling of ridge segments with confirmed recent volcanic eruptions (CoAxial and North Cleft on the Juan de Fuca ridge and 9–10° N on the East Pacific Rise) have created a first-order understanding of how hydrothermal systems respond to volcanic events on the seafloor. Phase separation and enhanced volatile fluxes are associated with volcanic eruptions, with vapour-dominated fluids predominating in the initial post-eruption period, followed in time by brine-dominated fluids, consistent with temporary storage of brine below the seafloor. Chemical data for CoAxial vents presented here are consistent with this evolution. Rapid changes in output and composition of hydrothermal fluids following volcanic events may have a profound effect on microbiological production, macrofaunal colonization, and hydrothermal heat and mass fluxes. Size and location of the heat source are critical in determining how fast heat is removed and whether subseafloor microbial production will flourish. CoAxial event plumes may be a direct result of dyking and eruption of lavas on the seafloor.

---

1. Introduction

Seawater evolves into hydrothermal fluid through a series of reactions as it is heated below the seafloor (see review by Seyfried & Mottl 1995). Magnesium and sulphate are removed from solution and many other elements are extracted from the rock through dissolution and exchange reactions. Hydrothermal fluids commonly pass through two-phase conditions in their subseafloor reaction path (Massoth *et al.* 1989; Butterfield *et al.* 1990, 1994; Von Damm *et al.* 1995) and separate into a low-chlorinity vapour phase and a high-chlorinity liquid or brine phase. The differing physical properties of vapour- and brine-like fluids (density, viscosity, surface tension) provide a mechanism to segregate them and produce vents on the seafloor with a wide range of chlorinities (Goldfarb & Delaney 1988; Butterfield et al. 1990; Fox 1990). To a first approximation, the ratios of most major elements to chloride are changed very little by the phase separation process (Berndt & Seyfried 1990), while gases are enriched in the low-chlorinity phase. Most of the variation in major element composition between different mid-ocean ridge (MOR) hydrothermal fluids can be explained by phase separation and segregation of brine and vapour. Different source rock chemistry, reaction zone conditions, kinetic factors and continued reactions in the segregated fluids are required to explain the remaining variation.

It has been recognized since seafloor hydrothermal vents were first sampled that they represent an important part of the geochemical cycles of many elements (Edmond et al. 1979; Von Damm et al. 1985; Palmer & Edmond 1989). Attempts to extrapolate to global chemical fluxes (using independent estimates of MOR fluxes of heat or helium and their correlations to elemental concentrations) have been tempered by an ever-increasing range of endmember fluid concentrations and by the lack of information on how fluid concentrations vary with time (see recent discussion by Von Damm 1995).

Large changes over time in chemical composition of vent fluids were not seen until 1990, when hydrothermal systems affected by volcanic activity were first sampled and it became apparent that rapid and significant changes in the style of venting and fluid compositions occurred immediately following an eruption (Haymon et al. 1993; Butterfield & Massoth 1994; Von Damm et al. 1995). Seafloor volcanic eruptions clearly have dramatic and interconnected consequences including formation of megaplumes (event plumes), microbiological blooms and rapid evolution of the temperature and composition of vent fluids (Haymon et al. 1993; Embley & Chadwick 1994; Baker 1995; Baker et al. 1995; Embley et al. 1995; Lupton et al. 1995; Von Damm et al. 1995; Holden 1996; Charlou et al. 1996). With the evidence accumulating from seafloor eruption events at N. Cleft, EPR at 9° 50′ N, CoAxial segment, and the EPR near 17° 30′ S (Auzende et al. 1996), it is possible to describe how hydrothermal systems change immediately following a volcanic event and to estimate how hydrothermal fluxes are affected. In this paper, we present the first detailed chemical data on hydrothermal fluids from the CoAxial segment of the Juan de Fuca Ridge, and relate it to a general model of hydrothermal response and chemical evolution of fluids following a volcanic event.

2. Description of CoAxial site

A seafloor volcanic event was detected by the US Navy SOSUS array beginning near 46° 15′ N, 129° 51′ W on 26 June 1993 and ending near 46° 32′ N, 129° 35′ W on about 16 July 1993 (Dziak et al. 1995; Fox et al. 1995). In July and August 1993, the Canadian remotely operated vehicle ROPOS was deployed from the NOAA ship *Discoverer* to explore and sample the area affected by the seafloor eruption (Embley et al. 1995) (figure 1). Three event plumes were found in July 1993 (Baker et al. 1995). Given the locations and approximate ages of the plumes (Baker et al. 1995; Massoth et al. 1995) and the prevailing current speed and direction (Cannon et al. 1995), it is likely that all three event plumes originated near the Flow site during the seismically active period. We were able to obtain one pair of titanium major samples

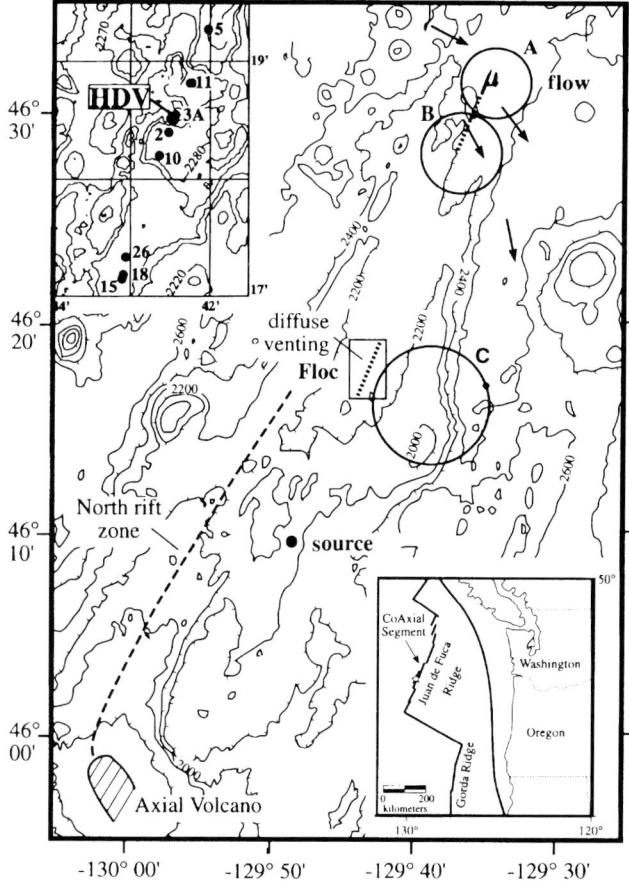

Figure 1. Map of CoAxial area, showing regional setting (inset bottom), location of lava Flow site, Floc site, and Source site. Circles are event plumes labelled A, B and C in order of both discovery and increasing age at the time of sampling (Baker et al. 1995; Massoth et al. 1995). Arrows show current direction at 1800 m depth (Cannon et al. 1995). A detailed map of the Floc site (inset top) shows the location of markers referred to in the text. Bathymetric contour interval is 200 m.

during ROPOS dive 234 on 1 August (approximately 2 weeks after the end of the volcanic event) from a 22 °C vent at the lava Flow site (marker P1, 46° 31.42′ N, 129° 34.85′ W; a 51 °C vent was found later on the same dive). Two more venting areas (Floc site centred near 46° 18′ N, 129° 42.5′ W, and Source site centred near 46° 9.3′ N, 129° 48.6′ W, which is ca. 10 km to the SSE of the initial event swarms) were found and many additional seafloor vent fluid samples were obtained with Alvin in October 1993, July 1994 and July 1995 to provide time series data on temperature and fluid chemistry. A multidisciplinary collection of papers describes some of the initial results of research on the CoAxial segment (Fox 1995).

3. Results

(a) General comments

Table 1 contains the chemical data as analysed in individual samples. Selected elements for Flow and Floc sites are shown in figure 2. Note that the measured Mg

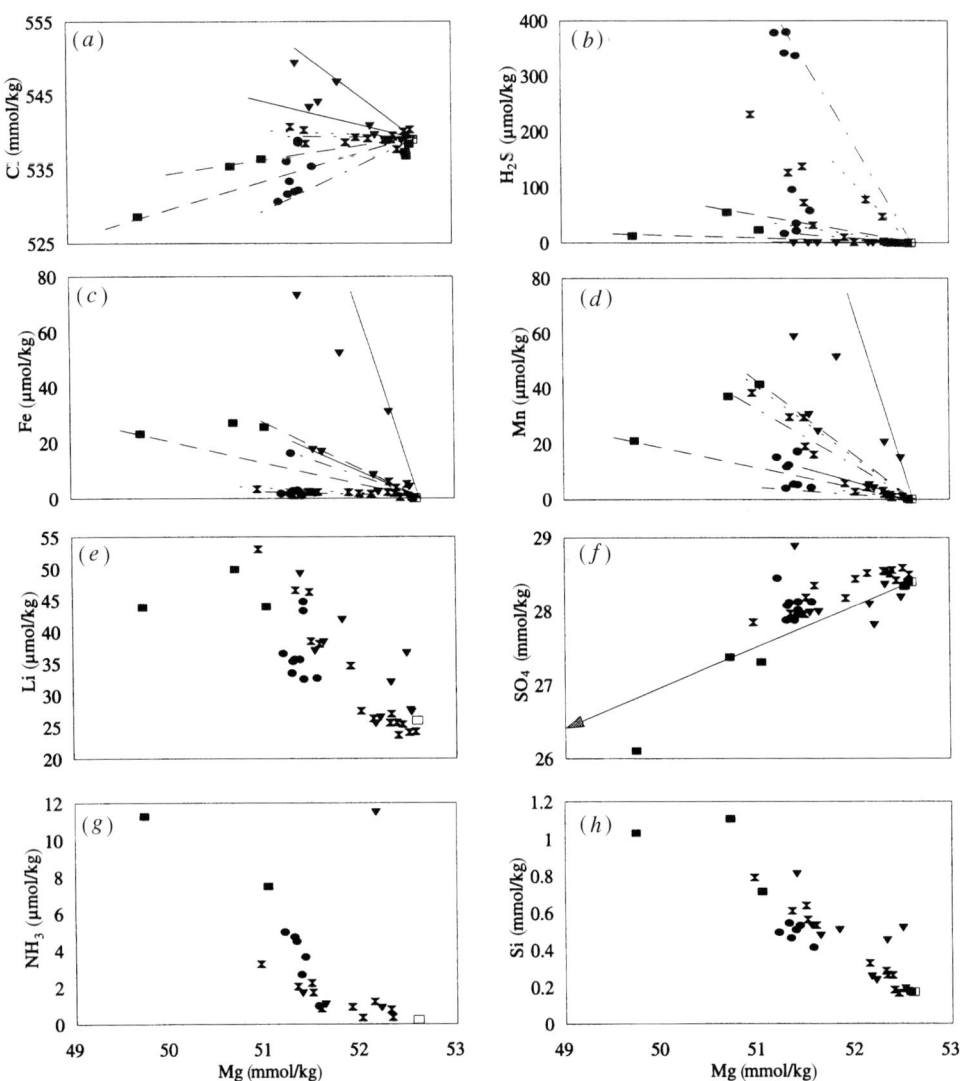

Figure 2. Plots of (a) Cl, (b) H_2S, (c) Fe, (d) Mn, (e) Li, (f) SO_4, (g) NH3 and (h) Si versus Mg for Flow and Floc diffuse fluids. Symbols: triangle, Flow site 1993; rectangle, Floc site 1993; circles, Floc site 1994; hourglass, Floc site 1995; open square, ambient seawater. Lines drawn to show range of element–magnesium trends: thin line, Flow site; dashed line, Floc 1993; dot-dashed line, Floc 1994; dotted line, Floc 1995. Arrow in (f) represents conservative mixing of seawater with a zero-sulphate, zero-magnesium endmember.

concentration in diffuse fluids is affected by subseafloor dilution of a Mg-depleted source fluid and by entrainment of ambient seawater at the vent orifice during sampling (i.e. sample quality).

No high-temperature fluids were found in the Flow or Floc areas. Observed post-eruption venting was strictly diffuse and low temperature (maximum 51 °C on 1 August 1993). Evidence for high-temperature (greater than 200 °C) reactions is present in the diffuse fluids in the depleted Mg and elevated Li, Fe, Mn, Ca and Si concentrations. Flow, Floc, and Source sites each have distinctive chemical characteristics reflecting differences in hydrothermal conditions in the three locations.

Table 1. *Vent fluid chemical composition* (continued overleaf)

H$_2$S, pH, alkalinity and ammonia in hydrothermal fluids collected in 755 ml titanium major samplers were determined within 12 h of submersible recovery, and dissolved silica was analysed within 48 h on diluted aliquots. In 1994 and 1995, anions were analysed shipboard using a Dionex DX500 ion chromatograph. Other methods are the same as in Butterfield & Massoth (1994) except that Li, Na, K, Mg and Ca were also measured by ion chromatography. For diffuse fluid samples, high-precision (less than 0.3% rsd) titration methods were used for Mg, Ca and Cl, with IAPSO seawater as standard. Alk is in units of meq l^{-1}; H$_2$S, NH$_3$, Li, Fe and Mn are in units of µmol kg^{-1}; Cl, SO$_4$, Mg, Ca and Si are in units of mmol kg^{-1}.

sample#	vent name or marker	date	temp °C	pH at 25°C	Alk	H$_2$S	NH$_3$	Cl	SO$_4$	Mg	Ca	Li	Fe	Mn	Si
source area															
2680-14	diffuse	10/93	6	7.41	2.73	0	3.1	539.5	28.27	51.65	10.58		84		
2681-10	Beard	10/93	276	5.79	2.10	404	2.9	594.1	18.94	35.08	30.10	334	43	72	
2681-12	Church	10/93						630.7		21.03	43.14	481	52	108	
2681-4	Beard	10/93						545.0		49.99	12.27	49	3	8	
2681-6	Church	10/93	284	6.89	2.69	34	0.4	545.6	26.86	50.85	12.57	59	5	8	
2787d14	Mongo, top	7/94	280	5.93	1.80	402		588.6	20.37	33.71	28.39	297	37	63	
2787d4	Church, base	7/94	270	5.16	0.71	1326		654.8	7.52	12.66	52.88	686	56	184	
2787m10	Church, base	7/94	294	4.87	0.45	1450		674.8	3.71	5.74	60.34	811	116	194	
2787m15	Chuch, base	7/94	293	4.84	0.40	1554		677.0	3.35	4.88	61.31	831	113	199	
2790d14	Twin Spires	7/94		5.50	1.27	704		625.3	13.05	23.19	42.51	516	29	121	
2790d15	Beard	7/94	255	4.81	0.44	1435		670.2	3.95	7.17	58.85	776	72	209	
2790d9	Beard	7/94	255	6.38	2.24	145		559.1	24.62	45.32	17.61	133	15	34	
2790m10	Twin Spires	7/94	223	4.77	0.37	1193		687.3	1.71	2.60	65.20	859	47	190	
2790m13	Mongo	7/94	284	4.72	0.33	1427		687.4	1.46	1.98	64.92	874	80	206	
2790m5	Mongo	7/94	284	4.75	0.40	1387		684.5	1.98	2.89	63.91	871	79	198	
2945d10	Mongo	7/95	292.5	4.58	0.36	1140	13.7	688.5	1.65	1.78	65.59	866	75	186	
2945d12	Mongo	7/95	292.5			1050	14.2	688.3	1.35	2.27	65.06	868	84	183	
2945d4	Twin Spires	7/95	253	4.69	0.45	885	12.2	690.7	1.42	2.07	65.94	864	44	188	
2945d5	Mongo	7/95	292.5	4.70	0.44	1080	12.1	683.0	1.91	3.36	63.73	844	75	188	
2945d6	Twin Spires	7/95	253	4.70	0.50	905	12.0	690.6	1.39	2.40	65.88	853	40	184	
2945m11	Church	7/95	281.5	4.73	0.46	1060	12.9	674.7	2.55	6.15	59.61	791	84	187	
2945m14	Church	7/95	281.5	4.94	0.63	90	12.2	660.7	4.68	10.49	54.48	731	67	162	

Table 1. Cont.

sample#	vent name or marker	date	temp °C	pH at 25°C	Alk	H$_2$S	NH$_3$	Cl	SO$_4$	Mg	Ca	Li	Fe	Mn	Si
Flow area															
hys234l-Ti	P1	8/93	22	6.80	2.66	0		538.7	28.36	52.33	10.53	32	31	21	0.452
hys234r-Ti	P1	8/93	22	6.65	2.68	0	0.9	538.9	28.19	52.49	10.63	37	108	15	0.520
2671-12	5A	10/93	4.2	7.46		0		539.6	27.81	52.22	10.45	27	2	4	0.239
2677-10	6A	10/93	32	6.86	3.06	0		546.7	29.10	51.83	12.93	42	52	51	0.508
2670-15	6A	10/93	36	6.74	2.93	0	1.7	549.3	28.88	51.40	13.28	49	73	59	0.812
2670-14	7A	10/93	15	6.91		0		543.3	27.98	51.55	11.46	37	18	31	0.534
2670-10	7A	10/93	6.5	6.81	2.75	0	1.1	544.0	27.99	51.64	11.37	38	17	25	0.479
2671-4	7A	10/93	6.5	7.33		0	11.5	540.8	28.09	52.17	10.83	26	8	5	0.259
2788m10	0	7/94	9.2	7.74	2.58	0	tr	536.8	28.35	52.54	10.25	27	4	0	0.180
2788m15	0	7/94	9.2	7.82	2.63	0	tr	537.3	28.34	52.53	10.19	28	1	0	0.181
Floc area															
2676-4	11	10/93	22	6.11	2.96	13	11.3	528.6	26.11	49.74	10.47	44	23	21	−.033
2673-10	3A	10/93	18.2	6.30	3.00	55		535.4	27.38	50.72	11.03	50	27	37	−.109
2676-10	3A	10/93	16	6.41	2.92	24	7.5	536.4	27.31	51.04	10.92	44	26	42	0.715
2793m10	11	7/94	7.2	7.24	2.54	17		536.1	27.88	51.31	10.28	33	2	4	
2791m5	2	7/94	8.2	6.82	2.46	379	5.0	530.6	28.45	51.22	10.47	37	2	15	0.496
2793m5	10	7/94	6.2	7.10	2.46	22		538.9	28.02	51.43	10.77	43	2	17	
2793m13	10	7/94	6.2	7.00	2.54	36		538.6	27.96	51.43	10.81	45	2	18	
2791m13	hdv	7/94	5.7	7.01	2.51	337	3.7	532.2	28.13	51.43	10.30	33	1	5	0.533
2791d9	hdv	7/94	4.6	7.12	2.81	342	4.8	531.7	28.08	51.32	10.39	35	16	12	0.545
2791d14	2	7/94	8.4	6.85	2.51	380	4.5	533.4	28.12	51.34	10.44	36	3	13	0.465
2789m10	3A	7/94	7.2	7.17	2.53	58	1.0	535.4	28.13	51.57	10.32	33	2	4	0.415
2791m10	3A	7/94	7.3	7.14	2.60	96	2.7	532.0	27.88	51.39	10.43	36	3	6	0.509
2947m14	hdv	7/95	16.7	6.66		138	2.3	540.3	27.96	51.49	10.69	46	2	30	0.641
2949d6	26	7/95	5.6	7.33	2.58	47	0.8	539.0	28.55	52.32	10.11	26	2	3	0.288
2949d4	26	7/95	5.6	7.30		78	1.2	539.1	28.52	52.14	10.17	26	2	4	0.330

Table 1. Cont.

sample#	vent name or marker	date	temp °C	pH at 25 °C	Alk	H₂S	NH₃	Cl	SO₄	Mg	Ca	Li	Fe	Mn	Si
Floc area															
2947d6	hdv	7/95	16.7	6.61		231	3.3		27.85	50.97	10.91	53	3	38	0.793
2947d4		7/95	2	7.69		0		537.7	28.42	52.44	10.00	25	0.3	<0.08	0.168
2947m11	hdv	7/95	16.7	6.66		127	2.1	540.7	27.97	51.35	10.64	47	1	30	0.611
2949m15	18	7/95	5.9	7.24		2	<0.34	538.7	28.73	51.51		29			0.320
2951m11	hdv	7/95	13	7.00	2.56	72	1.7	538.5	28.19	51.51	10.49	38	2	19	0.565
2951m14	hdv	7/95	13	6.92	2.58	32	0.9		28.35	51.60	10.44	38	2	16	0.534
2949m11	15	7/95	5	7.36		1	<0.34	539.0	28.51	52.38	10.07	26	2	2	0.265
2949m13	18	7/95	5.9	7.28		1		538.1	28.52			27	6	2	0.319
2949m14	15	7/95	5	7.42		3	0.4	538.9	28.54	52.33	10.04	27	6	2	0.266
2946m15	3A/17	7/95	8	7.19		2	0.4	539.3	28.44	52.02	10.40	28	2	3	
2946d6	3A/17	7/95	2			0	<0.3	540.5	28.58			27	5	2	0.196
2946d5	3A/17	7/95	2			0		540.3	28.50	52.58	10.04	24			
2946d4	3A/17	7/95	8	7.10		11	1.0	538.6	28.18	51.91	10.47	35	2	6	
2946m14	11	7/95	2.5	7.42		0	<0.3	540.1	28.59	52.51	10.13	24	5	1	0.193
2946m13	3A/17	7/95	8			1	0.3	538.8	28.32			30	2	3	0.357
2946m11	11	7/95	2.5	7.35		0	<0.3	539.5	28.56	52.40	10.08	24	4	1	0.187
background seawater															
2788d9	seawater	7/94			2.58	0	tr	538.4	28.40	52.57	10.19		0	0	0.177
2792m13	seawater	7/94	2	7.89	2.58	0		538.9	28.39	52.22	10.21		2	1	0.176
2792m10	seawater	7/94	2	7.88	2.52	0	0.1	538.8	28.39	52.29	10.20		1	1	0.175

(b) Flow site

The Flow site vents appeared as shimmering water exiting cracks and interstices of a fresh lobate/pillow lava mound at ca. 2390 m depth. Extensive surface alteration and precipitation of hydrothermal sediment was already apparent within one week of the eruption, and globules of orange material could be seen coming directly out of cracks in some places. The initial samples recovered by ROPOS were depleted in Mg by only 0.5% relative to ambient bottom seawater (indicative of substantial seawater entrainment during sampling), did not differ significantly from seawater in chlorinity, but were significantly enriched in Li, H_4SiO_4, Ca, Fe and Mn. H_2S was undetectable (below 0.5 µmol kg^{-1}) in all samples collected by ROPOS and Alvin at the Flow site.

The Flow site was revisited approximately nine weeks after the first samples were taken, and some subtle changes in chemical composition had occurred. Recovered samples had Mg depletions ranging from 0.7–2.3% relative to ambient seawater, and temperatures ranged from 4–36 °C. Though the October samples were more depleted in Mg, they had lower iron concentrations than the initial ROPOS samples, and the chlorinities were now clearly higher than seawater. Measured ammonia concentrations ranged from 1–11 µmol kg^{-1}, compared to background seawater at less than 0.3 µmol kg^{-1} (greater than 30 µmol kg^{-1} nitrate in bottom seawater is more than sufficient as a nitrogen source for the ammonia). Slightly different element–magnesium trends are apparent in the different vents sampled (figure 2); vent 6a (46° 31.609' N, 129° 34.713' W) has higher Fe, Cl and temperature than vent 7a, located about 330 m to the southwest (46° 31.451' N, 129° 34.811' W). Two samples from lower temperature vents in the graben just south of the lava flow (46° 30.581' N, 129° 35.33' W) were closer to seawater in composition.

The Flow site was visited again by Alvin in July 1994. Venting had nearly ceased one year after the event. One vent (46° 31.444' N, 129° 34.804' W, marker 0) with very little flow and temperature of 9 °C was sampled, and the samples did not differ significantly from ambient seawater in Mg or Cl. The fluids did have very slightly elevated silica, iron, manganese and alkalinity, indicating that water–rock reaction and, possibly, microbial activity were still taking place at greatly reduced levels relative to the immediate post-eruption period. Flow site vents were not sampled in 1995, when water column surveys indicated no detectable thermal anomalies over Flow (E. T. Baker, personal communication 1995).

(c) Floc site

The seafloor vents giving rise to the water column 'Floc snowstorm' observed with ROPOS on 2 August were found with Alvin in October 1993, when fluids were collected from vents near marker 3A (16–18 °C) and marker 11 (22 °C), separated by a distance of ca. 580 m. The vents differed markedly in appearance from the Flow site, as they were associated with fissures in older basalt between 2220 and 2290 m depth, and were emitting white flocculent material thought to be bacterial in origin (Juniper et al. 1995). Also in contrast to the Flow site, Floc vent fluids had elevated levels of H_2S and were lower than seawater in chlorinity (figure 2), indicative of a vapour component. In contrast to the rapid exhaustion of heat and diminution of fluid flux seen at Flow site after one year, significant fluid fluxes were maintained at the Floc site up to two years after the volcanic event. Observations of the seafloor in the Floc area support a waning of venting at particular sites and a general decrease in the area of active venting over time, but our submersible surveys do not allow

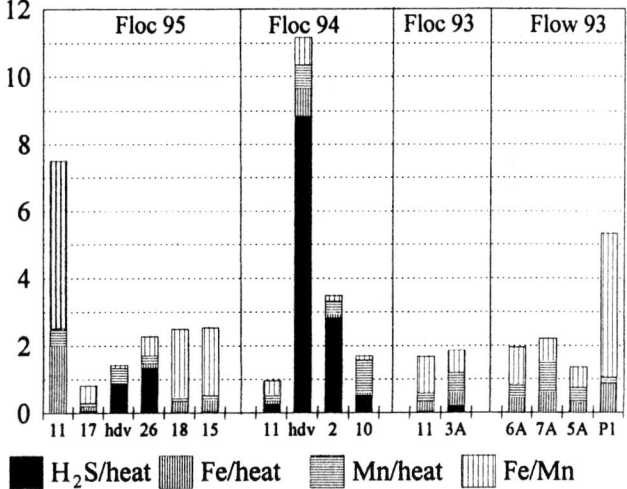

Figure 3. Temporal and spatial variation in diffuse fluids. Stacked bar graph of average element to heat ratios in nmol J^{-1} (H$_2$S/heat values have been divided by three to match scales) and Fe/Mn ratio. Vent marker names (see figure 1 and §2) are given below the bars, which are ordered from N (left) to S (right) for each area. Absence of H$_2$S at Flow site indicates that the degree of subseafloor mixing caused the redox potential to be dominated by seawater. Iron oxidation is significant at the Flow site, as plentiful amorphous iron oxides and ferric phosphate coatings on bacterial sheaths (Juniper *et al.* 1995) attest. Note H$_2$S pulse in 1994 at HDV. Spatial variation in H$_2$S/heat (0.8–31 nmol J^{-1}) and H$_2$S/Fe (10–340) in Floc vent fluids is enormous, with highest values near HDV and marker 26. Lower values at the ends of observed venting may be due to a combination of precipitation and oxidation of H$_2$S toward the periphery of the Floc upflow zone. H$_2$S may be the biolimiting energy source for micro-organisms in some Floc vents (e.g. markers 11, 18 and 15 in 1995).

quantitative assessment of changes in fluid or heat flux. Fluid chemistry exhibits detectable temporal and spatial variation at Floc (figures 2 and 3). In October 1993, chlorinities were lower than seawater; in July 1994, more samples were recovered showing a range of chlorinities both higher and lower than the initial three samples indicated. There were slight changes in composition at reoccupied marker 11 and 3*a* sites. In July 1995, chlorinities of all the recovered samples from Floc were very close to seawater.

In October 1993, Floc samples had significant levels of H$_2$S and Fe (up to 55 and 27 µmol kg^{-1}, respectively). In 1994, H$_2$S increased to a maximum of 380 µmol kg^{-1}, while iron was below 3 µmol kg^{-1} in all but one sample. In 1995, maximum H$_2$S was 230 µmol kg^{-1} and iron was below 6 µmol kg^{-1}. While iron was virtually unchanged at HDV vent from 1994–1995, hydrogen sulphide concentration apparently reached a peak in 1994, which coincides with maximum light attenuation in the plume over Floc (E. T. Baker, personal communication 1996).

The continued presence of H$_2$S, Li, Fe and Mn in the fluids indicates that high-temperature reactions continue beneath the Floc site. If the reaction zone produces a high-temperature fluid that has zero sulphate as well as zero magnesium, then sulphate reduction in the near-seafloor upflow zone is not important enough to cause significant depletion of sulphate below a conservative mixing line in the sulphate versus magnesium plot. Nearly all of the samples are slightly above the conservative mixing line, suggesting that oxidation of sulphide or dissolution of sulphate minerals (e.g. anhydrite or caminite) is quantitatively more important than sulphate reduction.

(d) Source site

There are four known high-temperature vents (Beard, Church, Mongo and Twin Spires) at the Source site spread out over *ca.* 100 m along a fissure running *ca.* 020°NNE near the crest of a pillow lava ridge at *ca.* 2055 m depth. The Source site is at the southern end of the CoAxial neovolcanic zone, collinear with Flow and Floc sites. The chemistry of the vent fluids from this site indicates that it was unaffected by the 1993 eruptive activity and probably existed prior to the eruption.

All four of the Source vents have virtually the same major element composition, which did not change significantly over the two years following the eruption (figure 4). Source fluids are moderate temperature (223–294 °C) brines (695 mmol kg^{-1} Cl) with low H_2S (1.0–1.7 mmol kg^{-1}), iron (45–120 µmol kg^{-1}), Fe/Mn (0.4–0.5), and zinc (6–17 µmol kg^{-1}) relative to other MOR vents with similar temperature and chlorinity. The combination of low Fe and a pH of 4.5–4.8 means that significant iron sulphide precipitation has not occurred in the upflow zone near the vents (otherwise the pH would be lower) and implies that the fluids last equilibrated near the temperature of venting (Seyfried & Mottl 1995). These fluids have very high Ca (67 µmol kg^{-1}), high Li (900 µmol kg^{-1}) and low Na/Cl ratio (0.76 compared to 0.86 in seawater), indicative of extensive water–rock reaction (Li-derived water–rock ratio is 1.0, similar to many high-temperature MOR vents). These characteristics are consistent with a brine that formed at high temperatures (greater than 400 °C), then remained in the crust, cooled or mixed with cooler seawater-derived fluid, and equilibrated at temperatures closer to 300 °C before venting.

4. Discussion

The CoAxial segment eruption illustrates how geological, chemical, and microbiological processes are intimately linked in seafloor hydrothermal systems. Composition of high-temperature hydrothermal fluids is controlled primarily by the pressure–temperature conditions, reaction kinetics and rock composition in the high-temperature reaction zone. Diffuse fluids on the ridge axis result from a subseafloor dilution of hotter hydrothermal fluids, and the range of compositions of diffuse fluids can be influenced by the degree of dilution, the time between dilution and venting, and by microbially-mediated and inorganic chemical reactions occurring throughout the upflow–mixing zone. The prime hydrothermal environment for microbiological colonization is within the redox and temperature gradient below the seafloor, where electron donors generated from the dyke (H_2, H_2S, Fe^{2+}, etc.) can mix and react with electron acceptors in circulating seawater (O_2, SO_4^{2-}, NO_3^-), and it is within this zone that fluid–microbe interactions will be most important (see discussions in Jannasch (1995) and Karl (1995)). If heat, volatiles and metals are rapidly exchanged in the very shallow subseafloor, then size of habitat, opportunity for microbiological growth, and impact of microbes on fluid chemistry are all diminished. The chemical composition of the fluids during the post-eruption period is an important factor influencing microbial growth and colonization (Holden 1996), and is ultimately linked to the details of the volcanic event.

The two very different types of diffuse fluids seen at CoAxial (oxidized, Fe-rich fluids at Flow, and more reducing H_2S-rich fluids at Floc) can be explained by a

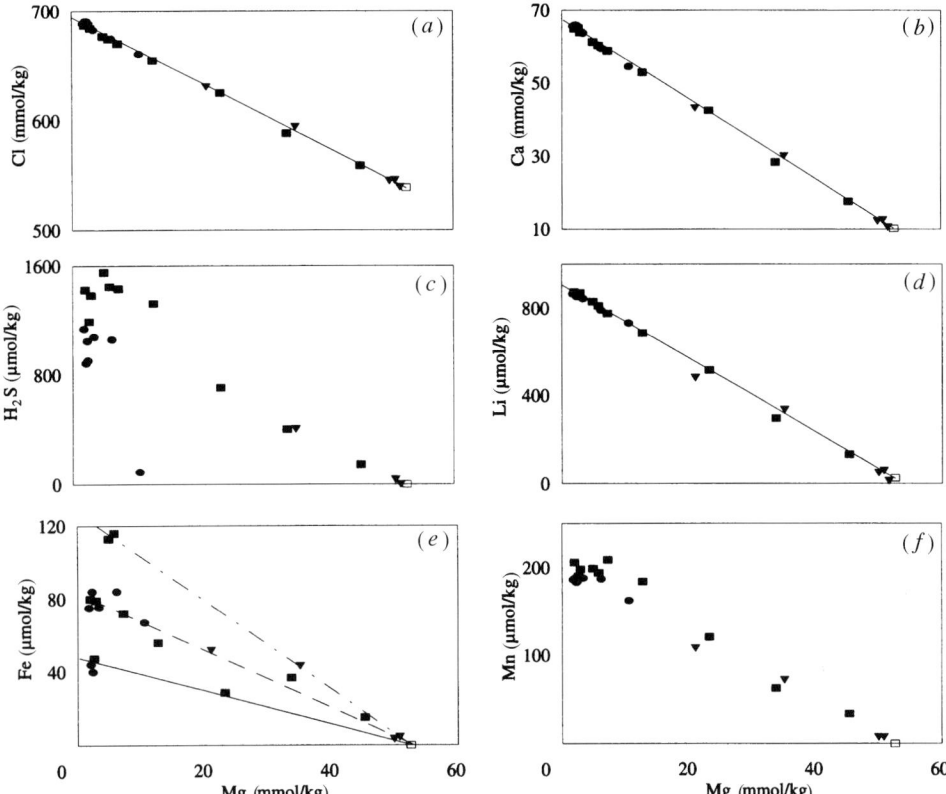

Figure 4. Plots of (a) Cl, (b) Ca, (c) H_2S, (d) Li, (e) Fe and (f) Mn versus Mg for source vent fluids. Symbols: triangle, 1993, rectangle, 1994, circle, 1995, open box, ambient seawater. Cl, Ca, and Li data show that there is a single primary endmember for this site that is not changing significantly over time. Lines drawn in (e) to show temperature dependence of Fe concentration: solid line for Twin Spires at 223 °C, dashed line for several vents in range of 250–275 °C and dot-dashed line for a 294 °C orifice at Church vent.

general model of how vent fluids evolve following a volcanic event (figure 5), taking into account the likely differences in the location of the heat source within the oceanic crust (figure 6). Immediately following a volcanic eruption or dyke injection, heat flux increases greatly, triggering phase separation and preferential venting of the more buoyant vapour phase. Continuous venting of brines observed in some systems means the conjugate brine phase must be temporarily retained around the heat source (by virtue of higher density or other physical properties), while higher-enthalpy vapour-rich fluids with low chlorinities remove heat, volatiles (H_2S, CO_2, He and H_2 from magmatic degassing and water-rock interaction partition into the vapour phase) and metals (figure 6). Iron concentration during the vapour-dominated period will be a function of temperature, pH, Cl concentration and possibly reaction kinetics, and cannot be assumed to be simply proportional to Cl (Seyfried & Ding 1995). As heat is removed and the system cools, phase separation slows and stops, and the fluids go through a transition from vapour-dominated to brine-dominated (chloride and metals increase while volatiles decrease). Brine content in the vented fluids may reach a peak and then decline as heat from the system runs out and fluid compositions decay back toward seawater.

The time scale in figure 5 is relative to the size of the volcanic event. Vents at

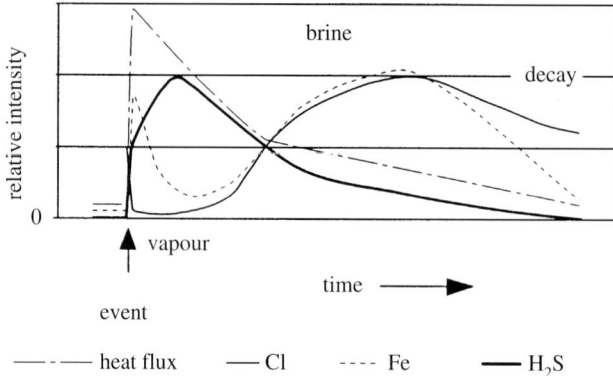

Figure 5. Response of hydrothermal systems to a volcanic event. Relative intensity of heat flux (dot-dashed line) and vent fluid concentrations of chloride (thin line), iron (dotted line) and hydrogen sulphide (thick line) over time. Systems evolve from a vapour-dominated, high heat flux stage accompanied by phase separation, through a transition to brine-dominated discharge, and eventually decay back toward zero heat flux and seawater composition. Our observations suggest high Fe concentrations in immediate post-eruptive fluids. Response model is based on this work, Butterfield & Massoth (1994), Von Damm et al. (1995), Lupton (1995), and Baker (1995).

9° 46.5′ N on the EPR showed an increase in chlorinity but were still below seawater chlorinity three years after a seafloor eruption (Von Damm et al. 1995). North Cleft segment diffuse and high-temperature fluids reached the brine emission stage at least three years after the volcanic event (Butterfield & Massoth 1994). High-temperature fluids are still evolving at North Cleft (Butterfield, unpublished data) and it is not clear how long the decay phase may last. We propose that cooling and evolution of fluid chemistry was so fast at the Flow site that we missed the vapour-dominated venting period entirely. It took only 3–12 weeks to reach the brine stage at Flow, and venting was virtually exhausted after one year, while venting of low-chlorinity fluids continued for at least a year at Floc.

We attribute this post-eruptive fluid evolution to cooling of magma injected into the permeable upper layer of oceanic crust. In the absence of recent volcanic perturbations, it appears that low-chlorinity volatile-rich fluids are associated with high heat flux systems driving deeper phase separation (e.g. Endeavour Main Field), while brine-dominated fluids are associated with lower heat flux systems (e.g. Cleft segment) (see heat flux estimates in Baker 1995). The model depicted in figure 5 may apply in a general sense to all hydrothermal systems insofar as intensity of heat output over time should correlate with fluid chemistry. The Source vent area at CoAxial could be interpreted to be in the decay phase in figure 5. So far, no hydrothermal system has been sampled before and after a known volcanic event, but we propose that the model for post-eruptive fluid evolution would apply in an established hydrothermal system as well as in areas with no pre-existing, active hydrothermal system.

The Flow site is located at the distal end of the dyke injection (Dziak et al. 1995) and was characterized by shallow crustal activity relative to the Floc site (Schreiner et al. 1995) and a pillow–lobate eruption (Embley et al. 1995). The chemistry of the event plumes, vent fluids and basalt alteration products at Flow site is consistent with an initial high-temperature venting period lasting less than three weeks after the end of the eruption, followed by rapid cooling of the lava mound and underlying dyke. The large inventory of Fe and Mn and the presence of ZnS particles in the event

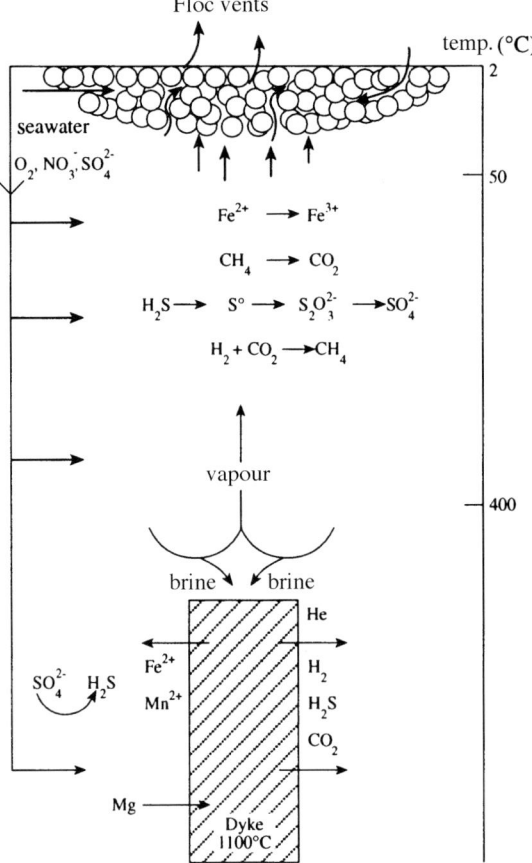

Figure 6. Chemical and microbial processes in a diffuse upflow zone. Injection of a dyke results in delivery of reduced volatiles and metals through outgassing and water–rock interaction. Phase separation partitions volatiles and some metals into vapour phase, and brines accumulate around heat source. Thermal and redox gradient provides a zone for chemical and microbial oxidation of reduced gases and metals as circulating seawater is entrained (microbial methanogenesis and sulphur reduction are known to occur at 110 °C).

plumes (Massoth *et al.* 1995) and the high particulate Cu/Fe ratios in event plume A (0.002–0.005, compared to less than 0.0016 in the diffuse fluid-dominated chronic plumes) indicate that a high-temperature fluid contributed to event plume formation. (Plume particle chemistry determined by XRF.) Many of the freshly erupted basalt samples recovered showed evidence of an initial high-temperature reaction period: halite coatings precipitated by direct contact of seawater with very hot rock; fresh needles of sphalerite and small crystals of pyrite and chalcopyrite lodged on top of halite and other surfaces (figure 7), indicating flow of fluids from a hot (greater than 350 °C) reaction zone; and pervasive staining consisting of oxides (amorphous iron oxide, boehmite, anatase), kaolinite and chlorite, indicating temperatures of 250 °C and up to greenschist conditions. The precipitation of halite coatings on altered glass surfaces is consistent with heating seawater to temperatures of 440 °C at 235 bar local seafloor pressure (Bischoff & Pitzer 1989) and represents a short-term storage of chloride. Vapour-dominated fluids must have vented when halite was precipitating. Continued circulation of warm fluids through the lava flow should dissolve the halite

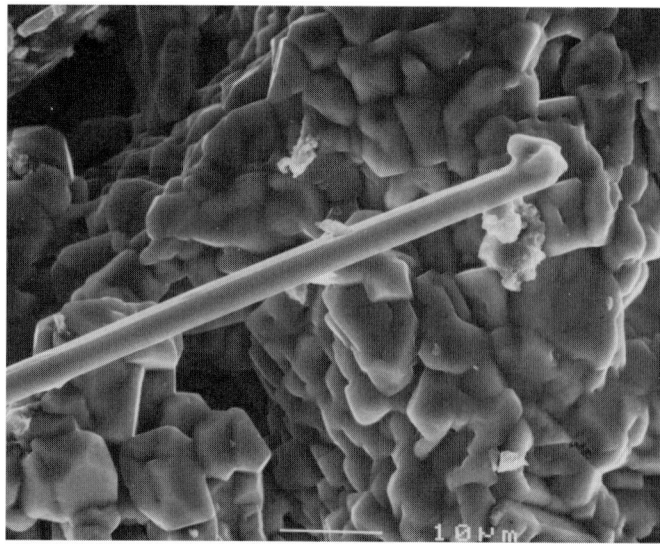

Figure 7. Photograph of halite coating on basalt within a cavity from Alvin rock sample 2672–7, recovered in October from the crest of the July 1993 lava flow. Halite coating is intergrown with anatase (TiO_2), boehmite ($AlO(OH)$), and rare sphalerite (ZnS) needles (prominent in this photograph). Interpretation is that eruption caused immediate phase separation and halite precipitation, followed by a high-temperature reaction period. Mineral compositions were determined by XRD and EDS-SEM at the Geological Survey of Canada in Ottawa.

and increase the chlorinity of vented fluids. Some of the halite coatings on basalts collected in October 1993 showed surface textures indicative of partial dissolution.

At the Floc site, we propose that the heat source was deeper, larger and not directly exposed to ambient seawater (figure 8), allowing the vapour-dominated period to last for at least one year. At two years after the event, fluid chlorinities were nearly identical to seawater, suggesting that this site was making the transition toward brine-like composition. We predict that Floc site fluids will evolve to become brine-dominated, eventually becoming totally depleted in H_2S and dominated by iron, as seen at North Cleft and at Flow.

Subseafloor chemical and microbial oxidation of sulphide has the potential to generate significant quantities of particulate elemental sulphur, which we have observed in direct association with biogenic particles in the plume over Floc. Our data and general model strongly suggest that methane (possibly from subseafloor methanogens) and biogenic particles vented from post-eruptive vapour-dominated diffuse vents contribute significantly to the high CH_4/Mn and particulate S/Fe ratios observed in hydrothermal plumes over magmatically active ridge segments (Lupton et al. 1993).

We hypothesize that nearly all of the potential heat to be extracted at the Flow site is contained in the seafloor lava flow and shallow underlying dyke and conduits, and that a significant part of this heat was extracted and formed event plumes during the eruptive event. The decline in observable fluid flux and maximum measured vent fluid temperature from 51 °C on 1 August 1993, to 36 °C in mid-October, to only 9 °C in July 1994 shows that the heat source at the Flow site was nearly exhausted within a year. Using the volume of extruded lava (5.4×10^6 m^3) estimated by Chadwick et al. (1995) and the basalt properties found in Sleep et al. (1983), we calculate that the heat available in the lava mound through latent heat of crystallization and cooling to ambient seawater temperatures is 2.7×10^{16} J. The estimate of Baker et al. (1995)

Seafloor eruptions and evolution of hydrothermal fluid chemistry 167

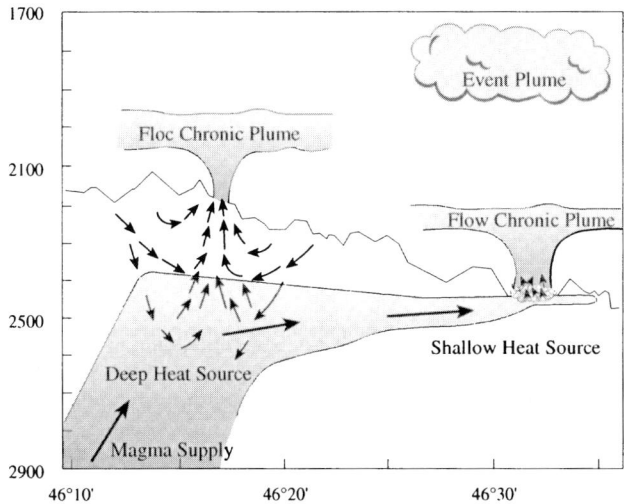

Figure 8. Geologic interpretation of CoAxial hydrothermal evolution, with depth below sea level in meters on vertical axis and latitude (°N) on horizontal axis. At the distal end of the dyke injection, heat is removed rapidly by formation of event plumes and circulation of seawater through the permeable lava mound, while a larger and deeper heat source near the magma supply continues to discharge for several years and provides a habitat for microbial communities. See discussion.

for the heat contained in the three event plumes (1.8×10^{16} J) is equivalent to $\frac{2}{3}$ of the heat within the lava mound. (Diffuse venting was seen north and south of the lava mound, and consideration of the potential volume of dyke beneath the entire area of known venting at Flow site. (1 m $\times$ 8 km $\times$ 2 km) increases the available heat by a factor of three.) There is no evidence for hydrothermal discharge at the Flow site immediately preceding the 1993 eruption, so it is difficult to invoke a significant shallow crustal hydrothermal fluid reservoir, as required in some event plume models (Cann & Strens 1989). High-temperature sulphide minerals clearly precipitated from the early fluids exiting the lava mound (figure 7) and therefore provide a source for the sulphide minerals observed in the event plumes. Because the CoAxial volcanic event was pulsed and episodic over several weeks, the initial penetration of a dyke in the eruption area creates a high-temperature reaction zone which could be disrupted and emptied during later eruptive pulses. The very large permeability required to generate event plumes by dyke injection (Lowell & Germanovich 1995) is provided here by the presence of lava at, and directly below, the seafloor. The published plume data (Baker et al. 1995; Cannon et al. 1995; Lavelle 1995; Lupton et al. 1995; Massoth et al. 1995) and our evidence for very short-lived, high-temperature venting support the hypothesis that the observed CoAxial event plumes resulted directly from rapid high-temperature water–rock interaction during the seafloor eruption and shallow dyke injection. The presence of erupted lava mounds near the North Cleft megaplume site provides some support that this is a general mechanism of event plume formation but questions remain regarding the relationship of event plume chemistry to event plume formation mechanisms.

We thank the ROPOS and Alvin Deep Submergence teams, Ed Baker and John Lupton for comments on the manuscript, Joe Cann for editing, Ryan Whitney and Martha Jackson for word processing, Aries Galindo for graphics, and My Nguyen for assistance with chemical analysis. This research was supported by the NOAA VENTS Program, the National Undersea Research

Program, and the U.S. National Science Foundation. PMEL contribution number 1731. JISAO contribution number 362.

References

Auzende, J.-M., Ballu, V., Batiza, R., Charlou, J.-L., Cormier, M.-H., Fouquet, Y., Geistdorfer, P., Lagabrielle, Y., Sinton, J. & Spadea, P. 1996 Recent tectonic, magmatic and hydrothermal ctivity on the East Pacific Rise between 17° S and 19° S: submersible observations. *J. Geophys. Res* **101**, 17 995–18 010.

Berndt, M. E. & Seyfried W. E. Jr 1990 Boron, bromine, and other trace elements as clues to the fate of chlorine in mid-ocean ridge vent fluids. *Geochim. Cosmochim. Acta* **54**, 2235–2245.

Baker, E. T. 1995 Characteristics of hydrothermal discharge following a magmatic intrusion. In *Hydrothermal vents and processes* (ed. L. M. Parson, C. L. Walker & D. R. Dixon), pp. 65–76. Special Publication No. 87. London: Geological Society.

Baker, E. T., Massoth, G. J., Feely, R. A., Embley, R. W., Thomson, R. E. & Burd, B. J. 1995 Hydrothermal event plumes from the CoAxial seafloor eruption site, Juan de Fuca Ridge. *Geophys. Res. Lett.* **22**, 147–150.

Butterfield, D. A., Massoth, G. J., McDuff, R. E., Lupton, J. E. & Lilley, M. D. 1990 The geochemistry of hydrothermal fluids from ASHES vent field, Axial Seamount, Juan de Fuca Ridge: subseafloor boiling and subsequent fluid–rock interaction. *J. Geophys. Res.* **95**, 12 895–12 922.

Butterfield, D. A. & Massoth, G.J. 1994 Geochemistry of north Cleft segment vent fluids: temporal changes in chlorinity and their possible relation to recent volcanism. *J. Geophys. Res.* **99**, 4951–4968.

Cann, J. R. & Strens, M. R. 1989 Modeling periodic megaplume emission by black smoker systems. *J. Geophys. Res.* **94**, 12 227–12 237.

Cannon, G. A., Pashinski, D. J. & Stanley, T. J. 1995 Fate of event hydrothermal plumes on the Juan de Fuca Ridge. *Geophys. Res. Lett.* **22**, 163–166.

Charlou, J.-L., Fouquet, Y., Donval, J. P., Auzende, J. M., Jean-Baptiste, P., Stievenard, M. & Michel, S. 1996 Mineral and gas chemistry of hydrothermal fluids on an ultrafast spreading ridge: East Pacific Rise, 17° to 19° S (Naudur cruise, 1993) phase separation processes controlled by volcanic and tectonic activity. *J. Geophys. Res.* **101**, 15 899–15 919.

Deming, J. W. & Baross, J. A. 1993. Deep-sea smokers: window to a subsurface biosphere? *Geochim. Cosmochim. Acta* **57**, 3219–3230.

Dziak, R. P., Fox, C. G. & Schreiner, A. E. 1995 The June–July 1993 seismo-acoustic event at CoAxial segment, Juan de Fuca Ridge: evidence for a lateral dyke injection. *Geophys. Res. Lett.* **22**, 135–138.

Edmond, J. M., Von Damm, K. L., McDuff, R. E. & Measures, C. I. 1979 Ridge crest hydrothermal activity and the balance of the major and minor elements in the ocean: the Galapagos data. *Earth Planet. Sci. Lett.* **46**, 1–18.

Embley, R. E. & Chadwick, W. W. Jr 1994 Volcanic and hydrothermal processes associated with a recent phase of seafloor spreading at the southern Cleft Segment, Juan de Fuca Ridge. *J. Geophys. Res.* **99**, 4741–4760.

Embley, R. W., Chadwick, W. W., Jonasson, I. R., Butterfield, D. A. & Baker, E. T. 1995 Initial results of the rapid response to the 1993 CoAxial event: relationships between hydrothermal and volcanic processes. *Geophys. Res. Lett.* **22**, 143–146.

Fox, C. G. 1990 Consequences of phase separation on the distribution of hydrothermal fluids as ASHES vent field, Axial Volcano, Juan de Fuca Ridge. *J. Geophys. Res.* **95**, 12 923–12 926.

Fox, C. G. 1995 Special collection on the June 1993 volcanic eruption on the CoAxial segment, Juan de Fuca Ridge. *Geophys. Res. Lett.* **22**, 129–130.

Fox, C. G., Radford, E., Dziak, R. P., Lau, T.-K., Matsumoto, H. & Schreiner, A. E. 1995 Acoustic detection of a seafloor spreading episode on the Juan de Fuca Ridge using military hydrophone arrays. *Geophys. Res. Lett.* **22**, 131–134.

Goldfarb, M. S. & Delaney, J. R. 1988 Response of two-phase fluids to fracture configurations within submarine hydrothermal systems. *J. Geophys. Res.* **93**, 4585–4594.

Haymon, R. M. et al. 1993 Volcanic eruption of the mid-ocean ridge along the East Pacific Rise crest at 9° 45–52′ N, direct submersible observations of seafloor phenomena associated with an eruption event in April 1991. *Earth Planet. Sci. Lett.* **119**, 85–101.

Holden, J. F. 1996 Ecology, diversity, and temperature-pressure adaptation of the deep-sea hyperthermophilic archaea Thermococcales. Ph.D. dissertation, University of Washington, Seattle.

Jannasch, H. W. 1995 Microbial interaction with hydrothermal fluids. In *Seafloor hydrothermal systems: physical, chemical, biological, and geological interactions* (ed. S. E. Humphris et al.), pp. 273–296. Washington, DC: AGU.

Juniper, S. K., Martineau, P., Sarrazin, J. & Gelinas, Y. 1995 Microbial-mineral Floc associated with nascent hydrothermal activity on CoAxial segment, Juan de Fuca Ridge. *Geophys. Res. Lett.* **22**, 179–182.

Karl, D. M. 1995 Ecology of free-living, hydrothermal vent microbial communities. In *The microbiology of deep-sea hydrothermal vents* (ed. D. M. Karl), pp. 35–125. Boca Raton, FL: Chemical Rubber Company.

Lavelle, J. W. 1995 The initial rise of a hydrothermal plume from a line segment source-results from a three-dimensional numerical model. *Geophys. Res. Lett.* **22**, 159–162.

Lowell, R. P. & Germanovich, L. N. 1995 Dike injection and the formation of megaplumes at ocean ridges. *Science* **267**, 1804–1807.

Lupton, J. E. 1995 Hydrothermal plumes: Near and far field. In *Seafloor hydrothermal systems: physical, chemical, biological, and geological interactions* (ed. S. E. Humphris et al.), pp. 317–346. Washington, D.C.: American Geophysical Union.

Lupton, J. E., Baker, E. T., Massoth, G. J., Thomson, R. E., Burd, B. J., Butterfield, D. A., Embley, R. W. & Cannon, G. A. 1995 Variation in water-column 3He/heat ratios associated with the 1993 CoAxial event, Juan de Fuca Ridge. *Geophys. Res. Lett.* **22**, 155–158.

Lupton, J. E., Baker, E. T., Mottl, M. J., Sansone, F. J., Wheat, C. G., Resing, J. A., Massoth, G. J., Measures, C. I. & Feely, R. A. 1993 Chemical and physical diversity of hydrothermal plumes along the East Pacific Rise, 8° 45′ N to 11° 50′ N. *Geophys. Res. Lett.* **20**, 2913–2916.

Massoth, G. J., Butterfield, D. A., Lupton, J. E., McDuff, R. E., Lilley, M. D. & Jonasson, I. R. 1989 Submarine venting of phase-separated hydrothermal fluids at Axial Volcano, Juan de Fuca Ridge. *Nature* **340**, 702–705.

Massoth, G. J., Baker, E. T., Feely, R. A., Butterfield, D. A., Embley, R. W., Lupton, J. E., Thomson, R. E. & Cannon, G. A. 1995 Observations of manganese and iron at the CoAxial seafloor eruption site, Juan de Fuca Ridge. *Geophys. Res. Lett.* **22**, 151–154.

Palmer, M. R. & Edmond, J. M. 1989 The strontium isotope budget of the modern ocean. *Earth Planet. Sci. Lett.* **92**, 11–26.

Seyfried, W. E. Jr & Ding, K. 1995 Phase equilibria in subseafloor hydrothermal systems: a review of the role of redox, temperature, pH, and dissolved Cl on the chemistry of hot spring fluids at mid-ocean ridges. In it Seafloor hydrothermal systems: physical, chemical, biological, and geological interactions (ed. S. E. Humphris et al.), pp. 248–272. Washington DC: AGU.

Seyfried, W. E. Jr. & Mottl, M. J. 1995 Geologic setting and chemistry of deep-sea hydrothermal vents. In *The microbiology of deep-sea hydrothermal vents* (ed. D. M. Karl), pp. 1–34. Boca Raton, FL: Chemical Rubber Company.

Sleep, N. H., Morton, J. L., Burns, L. E. & Wolery, T. J. 1983 Geophysical constraints on the volume of hydrothermal flow at ridge axes. In *Hydrothermal processes at seafloor spreading centers* (ed. P. A. Rona, K. Bostrom, L. Laubier & K. L. Smith Jr), pp. 53–69. New York: Plenum.

Tunnicliffe, V., Embley, R. W., Holden, J. F., Butterfield, D. A., Massoth, G. J. & Juniper, S. K. 1997 Biological colonization of new hydrothermal vents following an eruption on Juan de Fuca ridge. *Rev. Deep Sea Res.* (In preparation.)

Von Damm, K. L. 1995 Controls on the chemistry and temporal variability of seafloor hydrothermal fluids. In *Seafloor hydrothermal systems: physical, chemical, biological, and geological interactions* (ed. S. E. Humphris, R. A. Zierenberg, L. S. Mullineaux & R. E. Thomson), pp. 222–247. Washington, DC: AGU.

Von Damm, K. L., Edmond, J. M., Grant, B., Measures, C. I., Walden, B. & Weiss, R. F. 1985 Chemistry of submarine hydrothermal solutions at 21° N, East Pacific Rise. *Geochim. Cosmochim. Acta* **49**, 2197–2220.

Von Damm, K. L., Oosting, S. E., Kozlowski, R., Buttermore, L. G., Colodner, D. C., Edmonds, H. N., Edmond, J. M. & Grebmeir, J. M. 1995 Evolution of East Pacific Rise hydrothermal vent fluids following a volcanic eruption. *Nature* **375**, 47–50.

Controls on the physics and chemistry of seafloor hydrothermal circulation

By Adam Schultz and Henry Elderfield

University of Cambridge, Department of Earth Sciences, Downing Street, Cambridge CB2 3EQ, UK

Low temperature diffuse hydrothermal circulation is a natural consequence of the cooling of the oceanic lithosphere. Diffuse flow is expected to be ubiquitous, and will be present both within mid-ocean ridge crest axial zones of young age (0–1 Ma), and also on the older ridge crest flanks and limbs. If underlying thermal models are correct, hydrothermal circulation should persist for oceanic lithosphere of age 0–65 Ma, and is present over half the total area of the ocean basins. By using numerical models of hydrothermal circulation in cracked permeable media, we show qualitatively how diffuse flow is an intrinsic feature of high temperature axial (0–1 Ma) hydrothermal systems, and is not restricted to older (more than 1 Ma) lithosphere. This is in agreement with our field observations which suggest that in such high temperature vent fields the greatest part of the heat and volume flux is due to lower temperature diffuse flow, rather than high temperature black smoker venting.

By combining direct measurements of the physical properties of diffusely flowing effluent within axial hydrothermal systems with concurrent sampling of the chemical properties of that effluent, and by considering also the chemistry of unmixed black smoker endmember fluids from the same hydrothermal systems, the processes of mineral deposition and dissolution can be studied directly. By referring to the present-day lithology of such areas, it is possible to examine the balance between concurrent mineral deposition and dissolution processes, and the retention rate of specific mineral assemblages integrated over the history of the hydrothermal system. Thus details of the episodicity of hydrothermal venting within the system may be revealed. An example of this method of combining a variety of direct measurements of diffuse and high temperature effluent properties is given from the TAG hydrothermal field, Mid-Atlantic Ridge.

Long time series observations of the physical properties of diffuse and high temperature effluent reveal the importance both of tidal variability and also the response to changes in the permeability structure of the system brought about by natural and anthropogenic processes. Several mechanisms are considered to explain the relationship between ocean tidal loading, solid Earth tidal deformations, and the observed changes in flow within axial hydrothermal systems.

1. Introduction

The dominant means by which the newly emplaced axial (age 0–1 Ma) oceanic lithosphere is cooled is through the advection of seawater into, and circulation through, the cracked permeable seafloor. The higher temperature part of that process has received the greatest theoretical and experimental interest since the initial discovery of high temperature venting in the Galapagos (Corliss et al. 1977). In the present paper we consider the lower temperature more diffuse component of that circulation. Diffuse hydrothermal fluids circulating through the oceanic lithosphere and subsequently emitted into the water column present an observational challenge. This mode of flow is associated with low rates of fluid transport. Diffuse effluent temperature may be elevated only slightly above that of ambient seawater, and diffuse effluent velocity may span the range of centimetres per second (Schultz et al. 1992) down to rates just in excess of molecular diffusion. Furthermore, plumes of hydrothermal effluent resulting from diffuse sources may not necessarily be entrained into local high temperature plumes, and thus might not be carried into the water column efficiently (Schultz et al. 1992; Rosenberg et al. 1988).

Under certain conditions diffuse plumes may be sufficiently saline when mixed with seawater such that they are non-buoyant (Turner & Gustafson 1978), and thus may have little signature in the overlying water column. It is possible that effluent may pool along the seafloor, or under the appropriate range of physical and chemical conditions, may form warm briny submarine aquifers. This complicates efforts to quantify hydrothermal fluxes within such systems.

It is therefore appropriate to consider further the physics and chemistry of the lower temperature component of seafloor hydrothermal circulation, and to give special attention to the observational problems these fluids entail. Constraints on the total heat flux available to drive hydrothermal circulation globally and the distribution of that flux with lithospheric age are based largely on parametric thermal models. We review in some detail the assumptions behind these underlying models and discuss the uncertainties that arise in estimates of the total volume of seawater circulating through and interacting with the oceanic lithosphere.

We present data obtained from direct seafloor measurements of the physical and chemical properties of diffuse hydrothermal effluent. These were obtained during the joint British–Russian BRAVEX/94 MIR submersible expedition to the TAG (26°08′ N, 44°49′ W) and Broken Spur (29°10′ N, 43°10′ W) vent fields, Mid–Atlantic Ridge, during the period August–October 1994. Further data were recovered during the joint US-British Atlantic II Alvin Leg 132–02 return expedition to TAG in February–March 1995. The samples were obtained from the 'Medusa' hydrothermal monitoring system, a set of instruments that capture fluid samples and measure effluent velocity and temperature directly at the point of emission on the seafloor (Schultz et al. 1996). These instruments were used to monitor the hydrology of the TAG mound before, during and after Ocean Drilling Program (ODP) leg 158, and also to map the advective component of heat flux density during those times.

2. Underlying thermal models

It is the cooling of the newly formed oceanic lithosphere that provides the source of heat to drive hydrothermal circulation. In the following section we review the underlying theoretical and observational basis of our understanding of that process.

We consider two methods of establishing the total budget of heat available to drive global seafloor hydrothermal circulation. The first method is based on considering the thermal behaviour of the oceanic lithosphere as equivalent to that of a conductively cooled plate. The second method considers the total global ridge crest magmatic

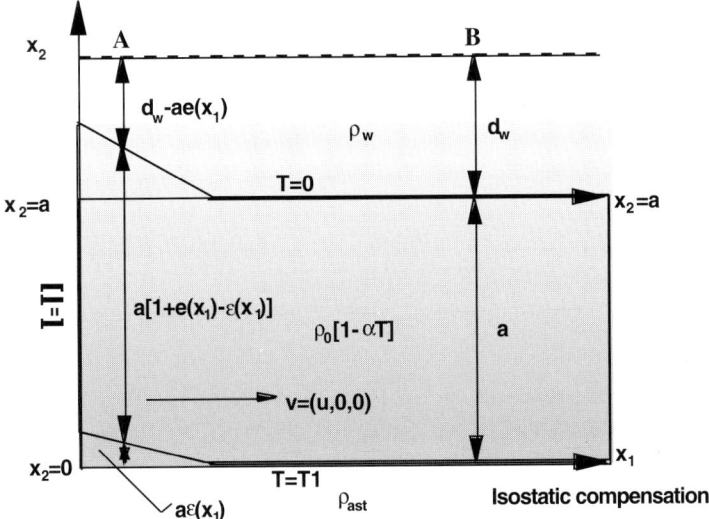

Figure 1. Schematic diagram of lithospheric cooling represented by conduction through a two-dimensional plate heated from the side and below.

budget and uses this, as well as knowledge of the latent heat of crystallization and the specific heat of seawater to place bounds on the heat available to drive hydrothermal circulation.

(a) Conductively cooled plate model

The difference between the heat flow predicted from a model representing the purely conductive cooling of the oceanic lithosphere (McKenzie 1967) and that observed globally from seafloor heat flow measurements is taken to be due to the heat carried away from the slab through advection, i.e. seafloor hydrothermal circulation and venting. This follows a tradition dating back at least as far as Parsons & Sclater (1977).

In the Parsons & Sclater model (figure 1), T is temperature, T_1 is a constant temperature applied to the base of the model and to the vertical planar origin at horizontal distance $x_1 = 0$, α is the thermal expansion coefficient, ρ_w the density of seawater, ρ_0 the density of 0 °C lithosphere, ρ_{ast} the density of aesthenosphere, d_w the asymptotic depth to the seafloor at large distances from the ridge, a the asymptotic thickness of the lithosphere, and $a\epsilon(x_1)$ is the displacement of the bottom of the lithosphere above a somewhat arbitrary level of compensation. The ridge is elevated above such a reference level due to the thermal expansion of the hotter materials near the ridge crest relative to the thermal contraction of cooler materials at distance from the ridge.

The topographic anomaly due to thermal expansion effects is (Davis & Lister 1974)

$$h = \alpha_{\text{eff}} \int_0^\infty \{T_1 - T(z)\}\, dz = (2/\sqrt{\pi})\alpha_{\text{eff}} T_1 \sqrt{\kappa t}, \qquad (2.1)$$

where α_{eff}, the effective thermal expansion coefficient, is modified for isostatic equilibrium (i.e. the material expands thermally but then is further compressed or dilated through changes in hydrostatic pressure due to changes in the height of the overlying water column) thus

$$\alpha_{\text{eff}} = \frac{\rho_0}{\rho_0 - \rho_w} \alpha. \tag{2.2}$$

The assumption of isostatic equilibrium is justified on the basis of marine gravity data. These reveal that on a gross scale mid-ocean ridges are not associated with significant free-air gravity anomalies. By assuming perfect isostasy we take the lithosphere to float atop a semi-fluid aesthenosphere, which is equivalent to maintaining a constant-pressure zone at depth. There is therefore no mechanical impediment to the subsidence of oceanic lithosphere after cooling.

The non-dimensional heat transport equation is given for the present example by Parsons & Sclater (1977),

$$\frac{\partial^2}{\partial x_1'^2} T' - 2Pe \frac{\partial}{\partial x_1'} T' + \frac{\partial^2}{\partial x_2'^2} T' = 0, \tag{2.3}$$

where the primes denote non-dimensional variables $T = T_1 T'$, $x_1 = ax_1'$, and $x_2 = ax_2'$. The Peclet number (i.e. the thermal Reynolds number, e.g. the ratio of heat convection to heat conduction) is given by $Pe = ua/2\kappa$, where κ is the thermal diffusivity and u is the ridge crest half-spreading rate.

The solution for the non-dimensional heat flow equation which satisfies the boundary conditions given in figure 1 is

$$T = (1 - x_2) + 2 \sum_{n=1}^{\infty} \frac{(-1)^{n+1}}{n\pi} \sin(n\pi x_2) e^{-\mu_n x_1}, \tag{2.4}$$

where

$$\mu_n = [(Pe^2 + n^2\pi^2)^{1/2} - Pe]. \tag{2.5}$$

When the product of the age of the lithosphere and the thermal diffusivity is much less than the square of the asymptotic plate thickness, the above relationship assumes the form

$$T \sim \text{erf}[1 - x_2'/2t^{1/2}]. \tag{2.6}$$

Lister (1972) obtained the above expression for the temperature inside a semi-infinite halfspace cooled from a uniform temperature through vertical conduction. Parsons & Sclater (1977) find that the general solution to the two-dimensional heat flow equation reduces to Lister's halfspace form also when $x_1' Pe \gg 2$. For a reasonable set of thermal conductivities this condition holds for all lithosphere older than 1 Ma regardless of the spreading rate.

(i) Predictions of bathymetry and heat flow

From the Parsons & Sclater model one can predict simultaneously the bathymetric height and the heat flow due to thermal conduction for seafloor of a given age. There is good agreement, to first order, between observed broad-scale seafloor bathymetry and that predicted from the cooling plate model (Parsons & Sclater 1977; Sclater et al. 1980; Stein & Stein 1992, 1994), i.e. the height of the seafloor scales proportionally with the square-root of age. While Carlson & Johnson (1994) suggest that the cooling plate model fails to fit seafloor bathymetry systematically for lithospheric ages greater than 81 Ma, this is a second order effect, and is observed for ages at which there is no longer direct evidence of hydrothermal circulation.

The general concurrence between the predicted and observed seafloor bathymetry suggests strongly that the underlying thermal model approximates to a reasonably accurate degree the product of the true temperature distribution and the effective coefficient of thermal expansion. This is in dramatic contrast to predictions of heat flow (rather than bathymetry) from the cooling plate model which overestimate the true heat flow for lithospheric ages less than approximately 50 Ma. Furthermore, the Parsons & Sclater model underestimates the true heat flow and overestimates the true seafloor bathymetric depths for ages greater than 70 Ma.

(ii) *Vertical extent of hydrothermal circulation*

There is considerable evidence that the 'reaction zone', i.e. the region at which there is direct interaction between newly injected magma and hydrothermal fluids is found at 2.5 km (± 2.0 km) depth throughout much of the global ridge crest system. Seismic data from a variety of ridge crest segments show the presence of a low velocity layer consistent with the existence of magma chambers. Two-way travel time data for a multichannel wide aperature profile obtained from $9°30'$–$10°$ N on the East Pacific Rise (Vera & Diebold 1994) show a low velocity zone at depths varying between 1390–1600 mbsf along the strike of the ridge axis. Other studies reveal similar structures at depths spanning 2300–2400 mbsf beneath the Juan de Fuca Ridge (Morton 1984), 3500 mbsf beneath the Lau Basin (Morton & Sleep 1985), and elsewhere.

Silica geobarometry reinforces further the extent of the depth of high temperature water–rock interactions. Von Damm *et al.* (1985) note that dissolved SiO_2 concentration is often in equilibrium with solid quartz. Quartz solubility is a function of both temperature and pressure. If the silica content of the hydrothermal effluent is measured, and the temperature can be determined separately, then the remaining unknown – pressure – can also be determined. Curves of silica dissolution *vs* temperature for various pressures are given, e.g. in Kennedy (1950) and in various solutions by Chen & Marshall (1982).

The temperature of black and white smoker vent fluids is measured routinely at the base of smoker chimney orifices. The pressure can be assumed to be hydrostatic in a crack-dominated hydrothermal system, and the depth of the water column is known. The pressure calculated from the silica concentration can therefore be transformed to depth within the hydrothermal system, i.e. the depth where the silica dissolution took place. Von Damm *et al.* (1985) report that depths determined in this way span 0.5–3.5 km at sites in their original study. Subsequent work tends to confirm this depth range.

Hydrothermal circulation is therefore likely to be concentrated within the upper 2–3 km of the oceanic lithosphere. As a result, only the uppermost few percent of any vertical lithospheric section will be affected directly by advective heat loss. Hydrothermal circulation will therefore play little role in changing the broad-scale bathymetry of the seafloor. In contrast, the heat flow will be influenced directly by the removal of heat from the upper part of the lithosphere through the advective losses due to hydrothermal discharge.

Before taking the existing framework and using it to place bounds on the global hydrothermal heat flux budget, it is necessary to refine the underlying thermal model. We consider initially the Stein & Stein (1992, 1994) reformulation of the Parsons & Sclater model. This differs from the original in that it is expressed in terms of bathymetric depth, rather than height above a hard-to-define compensation depth,

$$T(x,z) = T_1\left[\frac{z}{a} + \sum_{n=1}^{\infty} c_n e^{-\mu_n(x/a)} \sin\left(n\pi\frac{z}{a}\right)\right], \quad (2.7)$$

where $c_n = 2/(n\pi)$ and the other quantities match those in Parsons & Sclater (1977). The equation for surface heat flux density may be written

$$q(x) = k\left(\frac{\partial T}{\partial z}\right)_{s=0} = q_s\left[1 + 2\sum_{n=1}^{\infty} e^{-\mu_n(x/a)}\right], \quad (2.8)$$

where $q_s = kT_1/a$ is the asymptotic heat flow for old lithosphere in which the thermal gradient is linear (Stein & Stein 1992).

The bathymetric depth to the seafloor vs. distance from the spreading centre is

$$d(x) = d_w\left[2 - \frac{8}{\pi^2}\sum_{n=1,\text{odd}}^{\infty} n^{-2} e^{-\mu_n(x/a)}\right] - ae(0), \quad (2.9)$$

where $d_w - ae(0)$ is the bathymetric depth to the top of the ridge axis as seen in figure 1, and d_w (the depth to the top of the asymptotically old lithosphere) is given by

$$d_w = \frac{\alpha \rho_{\text{ast}} T_1 a}{2(\rho_{\text{ast}} - \rho_w)}. \quad (2.10)$$

Stein & Stein (1992, 1994) conducted a model search where the free parameters were the asymptotic thickness of the lithosphere, the base temperature, and the coefficient of thermal expansion. The best fitting coefficient of thermal expansion was found to be $\alpha = 3.1 \times 10^{-5}$ K^{-1}, the best fitting plate thickness was 95 km, and the best fitting basal temperature was 1450 °C. The asymptotic depth to the top of the lithosphere in this model is 5651 m, and the asymptotic heat flow is 48 mW m^2.

The Stein & Stein reformulation of the Parsons & Sclater conductively cooled plate model is an improvement over its direct ancestor as it fits gross bathymetry data and heatflow data exceptionally well (except for heat flow for ages less than 50–75 Ma). The RMS misfit between calculated and observed bathymetric depths was 0.42, and for heat flow values, 0.40. Corresponding values for the Parsons & Sclater model were 1.0 and 0.83. From a purely statistical point of view, both models are acceptable, but the reliability of these models is difficult to appraise in the absence of a sensitivity analysis.

An additional caveat is that the cooling plate model is two-dimensional and based on fitting data sets that have been grouped according to age and spreading rate and thus represent global averages. This ignores smaller-scale heterogeneities in heat flow which can accommodate (locally) much larger components of hydrothermal cooling, as well as areas where there may be no appreciable misfit between the conductive cooling model and the local observations and thus no direct surface evidence for hydrothermal circulation.

Stein & Stein find that the conductively cooled plate model fits observed heat flow adequately at ages greater than 65 ± 10 Ma. This is often referred to as the 'sealing age', and has been interpreted in the past to represent a critical age in oceanic lithosphere where significant levels of hydrothermal circulation are no longer supported due to a drop in the permeability of the basement rock. Alternatively, it has been interpreted as the age of lithosphere associated with sufficient sedimentary cover such that underlying circulation cells are contained entirely within an impermeable cap, and have no surface heat flow expression.

The heat flow deficit at younger ages is attributed to hydrothermal circulation in the oceanic lithosphere. Elder first speculated on this in 1965. This was followed up by LePichon & Langseth (1969). Finally Lister (1972) provided a theoretical basis for predicting that the heat deficit could be explained best by hydrothermal circulation, and it is arguable that it is to Clive Lister we owe the credit for the first prediction of concentrated hydrothermal circulation at ridge crests, i.e. smoker vent fields and the like.

The conductive heat flux density may be represented in this model by

$$\left.\begin{array}{l} q'(t) = 510 t^{-1/2} \text{ MW m}^{-2}, \quad t \leqslant 55 \text{ Ma}, \\ q'(t) = 48 + 96 e^{-0.0278 t} \text{ MW m}^{-2}, \quad t > 55 \text{ Ma}. \end{array}\right\} \quad (2.11)$$

where t is lithospheric age in Ma. In order to determine the total heat flux for the oceanic lithosphere, i.e. the heat flux density integrated over the upper surface area of the lithosphere, it is necessary to divide the lithosphere into a number of bands of different age, determine the average heat flux density for each of these bands, and their respective surface areas.

The expressions for heat flux density (2.11) take the form of the Lister (1972) halfspace cooling approximation. Stein & Stein report that use of the full series expansion rather than this approximation changes the result by 0.4%, even when a modified side boundary condition due to Davis & Lister (1974) is used which conserves energy by balancing the total heat flux, convective plus conductive, from the origin with that convected at the supply temperature T_1, ignoring the latent heat of the small melt volume (i.e.)

$$-k\frac{\partial T}{\partial x} + \rho c_{\mathrm{p}} u T = \rho c_{\mathrm{p}} u T_1, \quad (2.12)$$

or in non-dimensional terms

$$\frac{-1}{2Pe}\frac{\partial}{\partial x'}T' + T' = 1\big|_{x'=0} \quad (2.13)$$

rather than the original Parsons & Sclater (1977) isothermal boundary condition, i.e. $T = 1$ at $x = 0$. Note here that c_{p} is the specific heat. This new boundary condition equates the heat introduced at the origin by magma injection to balance that lost to horizontal conduction and to advection by horizontal motion of the plate.

It is intuitively appealing that there should be a heat balance in the steady state, but that imposed balances only horizontal advection/diffusion. An additional term involving advection of heat due to hydrothermal circulation must be considered. Certain aspects of Sleep's (1975) boundary conditions are also appealing, as they attempt to bring in the behaviour of melt including energy in the form of latent heat into the balance, which is ignored here. It is not clear that the approximation that latent heat may be ignored due to small melt volumes is necessarily true, particularly at faster spreading ridges where larger quantities of melt are expected.

Stein & Stein (1994) calculate the integrated heat flux for oceanic lithosphere of ages less than t_n as

$$Q'_n = \sum_{i=1}^{n} A_i q'_i, \quad (2.14)$$

Table 1. *Global heat budget*

(Observed and model conductive heat flows, and residual cumulative heat flow presumed due to hydrothermal advection, adapted from Stein & Stein (1994).)

cumulative ages (Ma)	predicted	observed	hydrothermal
1	3.6	0.4 ± 0.3	3.2 ± 0.3
2	5.1	1.0 ± 0.7	4.1 ± 0.7
4	7.2	1.8 ± 1.4	5.4 ± 1.4
9	11.3	3.8 ± 2.1	7.4 ± 2.1
20	15.6	6.5 ± 2.7	9.1 ± 2.7
35	19.8	9.2 ± 3.2	10.5 ± 3.2
52	22.7	11.5 ± 3.4	11.2 ± 3.4
65	24.6	13.3 ± 3.5	11.3 ± 3.5
80	26.9	15.6 ± 3.7	11.3 ± 3.7
95	28.5	17.3 ± 3.9	11.2 ± 3.9
110	29.8	18.7 ± 3.9	11.1 ± 3.9
125	30.6	19.5 ± 3.9	11.1 ± 3.9
140	31.5	20.4 ± 3.9	11.1 ± 3.9
160	31.9	20.8 ± 3.9	11.1 ± 3.9
180	32.0	21.0 ± 3.9	11.0 ± 3.9

where A_i is the total surface area for lithosphere spanning ages t_{i-1} and t_i, and q'_i is the average heat flux density for lithosphere spanning those ages as predicted by the cooling plate model. Similarly, the observed integrated heat flux is

$$Q_n = \sum_{i=1}^{n} A_i q_i, \qquad (2.15)$$

where q_i is the observed average heat flux density for that age range.

The difference between the integrated heat flux for the lithosphere of age less than t_n predicted by the thermal model and the observations is

$$Q_n^h = Q'_n - Q_n = \sum_{i=1}^{n} A_i(q'_i - q_i), \qquad (2.16)$$

which is taken to be the total heat flux due to hydrothermal circulation integrated over all oceanic lithosphere younger than t_n.

The results of these calculations are summarized in table 1 which contains the cumulative ages of the seafloor from zero age, the heat flow in TW (1 TW = 10^{12} W) predicted from this model, the heat flow actually observed as a global average for oceanic lithosphere of the specified cumulative age, and the difference which is taken to be the total budget available to drive hydrothermal circulation of cool seawater into the cooling oceanic lithosphere.

Stein & Stein (1994) calculate from their catalogue of heat flow observations that the total ocean lithospheric heat flux is 32 TW. From the misfit to the cooling plate model, they predict that of this total budget, 11 ± 4 TW, or 34 ± 12% is due to

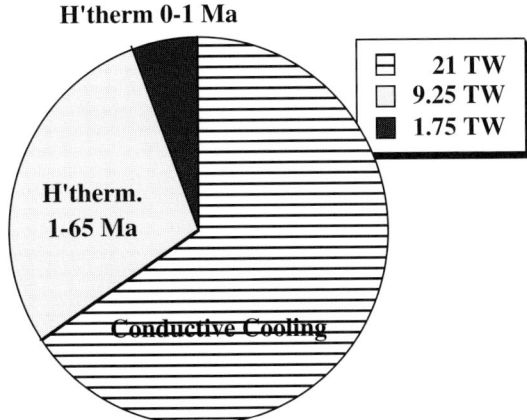

Figure 2. Total heat budget of the oceanic lithosphere showing relative partitioning between heat lost through conductive cooling and heat lost through advection of hydrothermal fluids. The advective component is subdivided into ridge crest axial hydrothermal flow for lithosphere no older than 1 Ma, and the remaining 64 Ma of lithosphere where hydrothermal circulation persists unambiguously.

hydrothermal circulation processes (this includes ridge crests and ridge flanks out to the 'sealing age' of 65 ± 10 Ma). It is clear that the greatest part of hydrothermal circulation occurs in older crust far from the 'active' circulation found at ridge crests. The division between conductive cooling, off-axis hydrothermal advective cooling, and on-axis hydrothermal cooling is seen in figure 2.

(b) Persistence of hydrothermal circulation in older lithosphere

There is evidence from K/Ar dating of celadonites (a low-temperature, e.g. less than 30 °C) alteration mineral) in the Troodos ophiolite complex that low-temperature hydrothermal alteration persisted there for at least 40 Ma after crustal formation (Gallahan & Duncan 1994). Localized depressions in predicted heat flow values from areas such as the equatorial Pacific basin are consistent with thermal circulation and large-scale lateral advection of water through basaltic crust. Analysis of major elements in fluids extracted from sediment pores in this region appear to corroborate this view, and to show evidence for hydrothermal circulation for ocean lithospheric ages spanning 15–70 Ma (Baker et al. 1991).

Direct evidence of off-axis hydrothermal flow at elevated temperatures has been obtained in recent years on the eastern flank of the Juan de Fuca Ridge (Davis et al. 1992; Wheat & Mottl 1994). There is additional evidence for off-axis hydrothermal circulation from a variety of theoretical and observational studies. These include numerical solutions for broad-scale flow within permeable media heated from below (Fehn et al. 1983), tank experiments using Hele-Shaw cells (Hartline & Lister 1981), and from observations of the large-scale fabric of the seafloor in the form of ridge-parallel faults (Johnson et al. 1993). Persistent zones of in-flow and out-flow, either channelled through permeable pathways following seafloor faults, or as an intrinsic feature of convection in permeable media (or both) are therefore expected over broad areas of the flanks and limbs of the ridge crests and older oceanic lithosphere. The heat flux models described thus far would suggest that the occurrence of off-axis systems of the Juan de Fuca–flank type should not be uncommon.

(i) *Lithospheric cooling models accommodating hydrothermal sinks and magmatic sources*

The cooling plate model does not contain hydrothermal circulation explicitly, thus the hydrothermal component is constrained only weakly. Morton & Sleep (1985) re-examined Sleep's (1975) model of a dyke annealing to the vertical boundary at the origin of a ridge crest spreading centre. A new term is added to the original two-dimensional heat flow equation,

$$u\frac{\partial T_e}{\partial x} = \frac{k}{\rho c_p}\left(\frac{\partial^2 T_e}{\partial x^2} + \frac{\partial^2 T_e}{\partial z^2}\right) + \frac{H}{\rho c_p}, \tag{2.17}$$

where T_e is the excess temperature in °C (i.e. the temperature at a given position within the model lithosphere less the steady state solution for the asymptotic temperature at $x = \infty$), z is the depth below the top of the oceanic lithosphere, k is the thermal conductivity, ρc_p is the volume specific heat, and H represents all excess sources and sinks of heat including heat brought in by intrusion, latent heat of crystallization, and heat removed by hydrothermal circulation.

The appropriate boundary conditions are:

$T_e = 0|_{z=0}$ (i.e. no excess temperature at the seafloor);

$T_e = 0|_{z=\lambda}$ (i.e. no excess temperature at the base of the lithosphere);

$T_e = 0|_{x=\infty}$;

$$\frac{\partial T_e}{\partial x} = \frac{u\phi c_p T_e}{k}\bigg|_{x=0}.$$

where λ is the depth to the base of the lithosphere. The final boundary condition balances advection and diffusion across the point of dyke injection at the origin.

This solution to the heat flow equation does not reduce to the simple form found in Sleep (1975), but it does remain tractable if one assumes a simple form for the heat sources and sinks. Morton & Sleep (1985) take this to be a Dirac delta function located a distance x_0 from the origin (i.e. an infinitesimally thin spike of unit area centred at $x = x_0$). Therefore, hydrothermal circulation is modelled as a set of infinitesimally thin heat sinks. This is inadequate to represent the details of flow, or of any chemical reactions taking place within such a flow regime, but it is adequate to understand the impact of localized hydrothermal convection on the broad-scale heat budget of a cooling plate.

Morton & Sleep (1985) take the top of a magma chamber to represent the 1185 °C isotherm, which they associate with the 70% solid temperature of MORB (Mid-Ocean Ridge Basalt). Morton & Sleep note that in magma chamber models based on ophiolite studies and thermal models, most of the latent heat of crystallization is released at the top of the chamber, resulting in plated gabbros and crystals that settle to form cumulate layers. A small amount of melt may be retained in the crystal mush and may crystallize at the sides of the magma chambers. This is used as justification for placing 70% of the latent heat resulting from magma chamber crystallization along the top of the magma chamber, and the remainder along the sloping sides. Latent heat resulting from crystallization of dykes and flows is placed directly above the chamber, on-axis.

The model is constructed by placing all of the latent heat initially at the ridge axis in order to determine the temperature distribution in the plate (as before). By so doing, it is possible to identify the 1185 °C isotherm. The latent heat is then redistributed on the top and sides of the magma chamber (i.e. along the 1185 °C isotherm), thereby changing the temperature field and shifting the 1185 °C isotherm. The final model is constructed by carrying out this process iteratively until the isotherms converge to a stable geometry such that the location of the 1185 °C isotherm corresponds to the latent heat distribution.

Morton & Sleep find that the depths to the 1185 °C isotherm tend to be shallower than the depths to the tops of the corresponding axial magma chambers, at least as inferred from seismic observations. This situation is rectified if one introduces a set of heat sinks representing heat loss due to hydrothermal flow. By introducing a (non-unique) set of sinks, the isotherm may be depressed such that its contour matches that of the presumed magma chamber.

In order to quantify the total budget of heat available to drive hydrothermal circulation, we make use of the best features of the Stein & Stein (1992, 1994) model and the Morton & Sleep (1985) boundary conditions. The details of such a model may be found in Stein et al. (1995).

The constraints on this model are the depth and heat flow data for older ages (which constrain the overall thermal model), earthquake and magma chamber depths (which constrain the near-axial geotherm), and the heat flow data for 10–50 Ma (which constrain the surface temperature gradient for these ages). Furthermore, two sets of models are produced for two different classes of spreading rates. For lithospheric ages greater than 0.5 Ma, the difference between the two spreading rate models is small, and they may be combined. For ages between 0.5 Ma and 50 Ma, the predicted model heat flow is approximated by $q_c = 308 t^{-1/2}$, where q is in mW m^{-2}, and t is in Ma.

The Stein et al. composite model predicts a higher heat loss for ages less than 10 Ma than the simple conductively cooled plate model. This adjusts the hydrothermal component of heat loss (i.e. the mismatch between the model prediction and the observations) by lowering slightly the hydrothermal component relative to the simpler model. Regardless of this detail, the majority of cumulative hydrothermal heat loss remains associated with ages greater than 1 Ma. Another result of these calculations is that higher rates of hydrothermal heat loss are associated with faster spreading rates than slow spreading rates.

(ii) *Predicted hydrothermal heat loss per unit length ridge axis*

The Stein et al. (1995) model permits us to estimate the predicted heat flux density at each point on the oceanic lithosphere, and to integrate over surface areas of a given age to determine the cumulative hydrothermal heat flux due to crust of a given age range. Stein et al. (1995) assume that sediment cover does not control the 'sealing age' but Davis et al. (1992) dispute this view, at least on a local scale. Therefore, it should be remembered that if the sediment cover on a ridge is unusually thick for its age, the surface heat expression may be reduced or suppressed.

For slow-spreading plate boundaries (e.g. half rates of 10 mm yr^{-1}), the model predicts the hydrothermal heat flux would be the product of 1 W m^{-2} average heat flux density with the total area on both sides of the ridge axis within 0.1 Ma, or 2 MW km^{-1} of ridge axis for lithosphere younger than 0.1 Ma. For half rates of 30 mm yr^{-1} (e.g. Juan de Fuca Ridge), the equivalent value is 15 MW km^{-1} of ridge axis.

For older ocean lithosphere (older than 0.5 Ma), the slow spreading and fast spread-

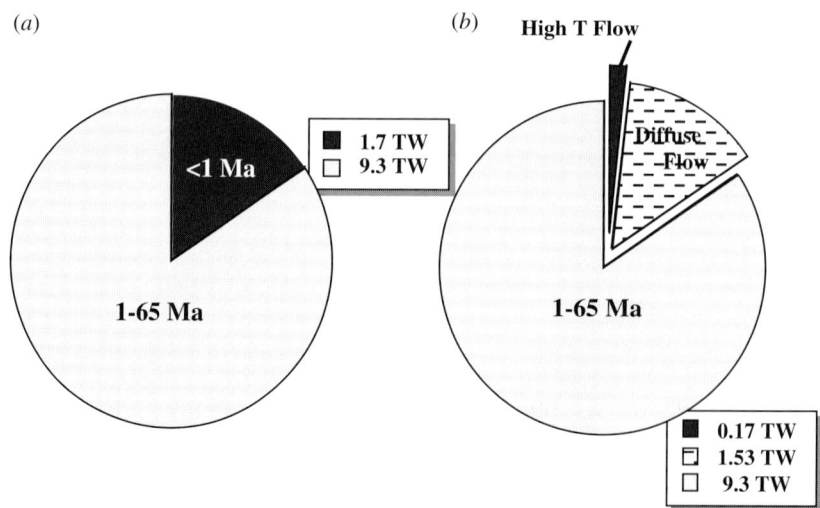

Figure 3. Total hydrothermal heat budget of the oceanic lithosphere (left). The advective component is subdivided into ridge crest axial hydrothermal flow for lithosphere no older than 1 Ma, and the remaining 64 Ma of lithosphere where hydrothermal circulation persists unambiguously (right).

ing models are almost indistinguishable. Stein et al. (1995) estimate that the total heat flux density budget for hydrothermal circulation averaged over ages of 0–1 Ma is ca. $0.5 \text{ W m}^{-2} \times (3.5 \times 10^6 \text{ km}^2$ surface area for crust of that age) (Stein & Stein 1994) yielding a total heat flux of 1.75 TW.

Recall that Stein & Stein (1994) estimated that the total ocean lithospheric heat flux is 32 TW, of which 11 ± 4 TW, or $34 \pm 12\%$ is due to hydrothermal circulation processes (including ridge crests and ridge flanks out to the 'sealing age' of 65 ± 10 Ma). While Stein et al. (1995) have made no similar calculation of the total budgets integrated over all crustal ages for the improved composite model, the cumulative budgets will be qualitatively the same as the earlier model. If so, then hydrothermal heat flux in the near-axial zone (0–1 Ma) accounts for only ca. 16% of the total hydrothermal heat flux budget for ages (0–65 Ma). The partitioning of the total budget of hydrothermal heat flux between on-axis (0–1 Ma) and off-axis (1–65 Ma) is seen in figure 3. On the right-hand-side of this figure we permit approximately 90% of the total near-axial hydrothermal heat flux to be expressed in the form of low temperature flow, and 10% in the form of high temperature (i.e. 'smoker') flow. This division is in rough agreement with observations of seafloor hydrothermal systems (Schultz et al. 1992; Rona & Trivett 1992). We shall make use of this subsequently in calculating total global water fluxes. A theoretical underpinning for the generation of diffuse flow within high temperature hydrothermal systems is given later in this paper.

(c) Magmatic budget

An independent method of bounding the global hydrothermal heat budget comes from consideration of the total volume of new magma emplaced annually along the global ridge system. Kadko et al. (1994) and Elderfield & Schultz (1996) considered the latent heat of crystallization of basaltic magma (ca. $0.676 \times 10^6 \text{ J kg}^{-1}$) which takes place at temperatures near 1200 °C, and then the residual heat of cooling the newly crystallized rock from these temperatures to the temperatures of the fluids found to exit from high temperature hydrothermal vents (1.130 MJ kg^{-1} for cooling from 1200 to 350 °C typical black smoker vent temperature).

It may be argued that a more appropriate temperature for hydrothermal end-member fluids at the reaction zone (typically 2 km below seafloor) would either be 465 °C, i.e. by extrapolating surface hydrothermal temperatures adiabatically to the reaction zone, or even as high as 600 °C, (Morton & Sleep 1985; Phipps Morgan & Chen 1993). We shall therefore take 350 °C as a lower bound on the temperature (i.e. the upper bound on the hydrothermal component of lithospheric cooling), and 600 °C as an upper bound.

The total heat available for hydrothermal circulation based entirely on cooling magma from liquidus temperatures to the temperature of vent fluids is therefore 1.806 MJ kg^{-1} (Kadko *et al.* 1994). We compare this to the heat required to elevate a given mass of seawater at 2 °C (typical benthic temperature) to hydrothermal temperatures, or 350 °C. This is of course dependent upon depth, e.g. pressure. We expect a balance between magmatic cooling and seawater heating, since in the area of 'active' hydrothermal circulation, it is cooling of the magmatic intrusions by heating of seawater which leads to the production of high temperature endmember hydrothermal effluent. We have already concluded from the thermal models that conductive cooling in the near-axial zone contributes only a small part to the total cooling budget.

Bischoff & Rosenbauer (1985) have investigated the equation of state for a hydrothermal fluid analogue, i.e. 3.2% NaCl seawater by determining experimentally the pressure dependence of the specific volume of seawater. Temperatures of the isotherms were then fit to a compressibility equation, and an empirical equation of state was determined that relates the specific volume, compressibility and expansivity of seawater to the temperature and pressure.

For typical high temperature vent fluid exit temperatures of 350 °C, and for a typical depth of medium and fast-spreading mid-ocean ridges, i.e. 2500 mbsl, Bischoff & Rosenbauer's relationship shows the heat capacity of the hydrothermal fluids is approximately 6.5×10^{-3} MJ kg^{-1}K^{-1} (figure 4). There is a complication in that values of the specific heat come close to the two-phase boundary, particularly near the critical point of *ca.* 405 °C and 300 bar where values of specific heat are known only approximately.

There is petrological evidence from fluid inclusions and from chemical analysis of quartz-cemented breccias (Delaney *et al.* 1987), and from direct measurements of high temperature vent fluid at the Juan de Fuca Ridge (Butterfield *et al.* 1994) that phase separation has been observed within certain hydrothermal systems. Butterfield & Massoth (1994) report on significant temporal variation of vent chemistry at the North Cleft segment of the Juan de Fuca Ridge between 1988 and 1992, suggesting transition from vapour-enriched to brine-enriched effluent during the course of volcanic activity. This is in general agreement with the Edmonds & Edmond (1995) model of mixing of phase-separated brine, vapour and hydrothermally altered seawater within the hydrothermal system, which in turn is consistent with Bishoff & Rosenbauer's (1989) double-diffuse convection model of the reaction zone. Phase separation effects will be considered later.

To heat 1 kg of seawater at 2 °C to 350 °C therefore is equivalent to *ca.* 2.262 MJ. The amount of heat released in cooling 1 kg of 1200 °C magma to 350 °C (which we presume is due to hydrothermal processes) is in the ratio of 1806/2262 the amount of

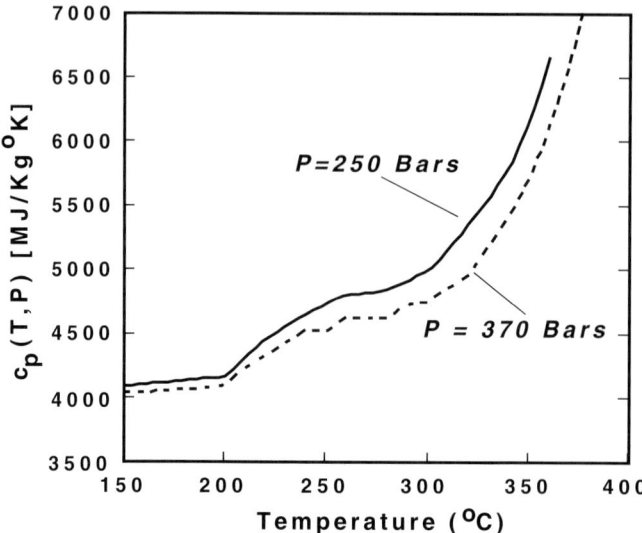

Figure 4. Specific heat of seawater *vs* temperature (°C) at 250 bar (solid curve) and 370 bar (dashed curve) pressure, roughly equivalent in the first case to the ambient seafloor hydrostatic pressure at vent fields throughout the East Pacific Rise and Juan de Fuca Ridge, and in the second to that at the TAG hydrothermal mound, Mid-Atlantic Ridge. The values shown have been interpolated from those found in Bishoff & Rosenbauer (1985).

heat necessary to heat 1 kg of ambient seawater to 350 °C; i.e. the heat liberated by cooling 1 kg of magma is sufficient to heat 798 g of water. Of greater interest perhaps is determination of the relative volume of magma involved in heating the equivalent volume of water. It is also of interest to compare the estimates of the total heat available to drive hydrothermal circulation determined from the mismatch between the Stein *et al.* (1995) lithospheric cooling model and observed heat flow, and the budget as estimated from the heat content of newly emplaced magma, integrated over the global ridge crest system.

Given a total volume of newly emplaced magma of 17 km^3 yr^{-1}, and taking the density of seafloor basalt to be *ca.* 3000 kg m^{-3}, the total mass-per-year of newly emplaced magma is 5.1×10^{13} kg yr^{-1}. Given the figure above of 1.806 MJ kg^{-1} liberated from cooling uncrystallized magma at 1200 °C to 350 °C, the total heat released by this process, globally, is 9.21×10^{19} J yr^{-1}, for a total heat flow of *ca.* 2.9×10^{12} W, or *ca.* 3 TW. This figure is in close agreement with that predicted by the mismatch between heat flow observations and the Stein & Stein (1994) and Stein *et al.* (1995) thermal models.

The estimated total heat budget can be more than doubled by increasing the reaction temperature from 350 °C to the upper bound of 600 °C. Even greater values are possible if the dramatic increase in c_p near the point of phase separation is taken into account. The aforementioned close agreement between the calculations appropriate for 350 °C with those obtained from the underlying thermal models suggests that this lower bound may be closer to the globally averaged truth, although locally there may be large variations from the global average.

Bishoff & Rosenbauer's calculations of specific volume for seawater at 250 bar and 350 °C yield a density of approximately 680 kg m^{-3}, compared with the density of ambient (*ca.* 2.7 °C) seawater at the same pressure of 1028 kg m^{-3}. In volumetric terms therefore, the amount of heat released in cooling 1 m³ of magma is sufficient to heat 3.52 m³ of seawater. In mass heat exchange terms, the water/rock ratio for seafloor hydrothermal systems is nearly unity, while in volumetric terms it is about 3.5:1, i.e. each litre of magma will heat *ca.* 3.5 litres of high temperature hydrothermal effluent.

(d) Global hydrothermal water budget

Given knowledge of the specific heat of seawater under hydrothermal conditions, and given estimates of the total heat budget available to drive hydrothermal circulation through an area of seafloor at a particular temperature and pressure, it is possible to place bounds on the total volume of seawater necessary to transport that heat, i.e. the volume of seawater that must circulate through the global hydrothermal system each year. Such calculations are strongly parametric; they depend not only on the underlying thermal model (which we have seen is dependent itself on numerous assumptions), but also on estimates of the temperature to which those fluids are heated during their residence time within the lithosphere. Accurate estimation of the water budget would require full knowledge of the total path length over which the fluid circulates, and then integration of all pressure and temperature effects over that path length.

By ignoring considerations of variable P–T conditions over the path-length of the flow, one can transform the known quantities into water mass flux,

$$Q_m = \frac{Q_a}{\Delta T c_p}, \tag{2.18}$$

where Q_m is the water mass flux (kg yr^{-1}), Q_a is the heat flux (TW), and ΔT is the temperature by which the ambient seawater is raised during its circulation within the lithosphere. An example of this naive means of calculating the global water flux is seen in figure 5. Here we have assumed that all of the axial hydrothermal heat flux is removed by hydrothermal flow at 350 °C, and that these fluids are heated at a depth equivalent to 250 bar. We assume further that all off-axis flow takes place at 5 °C and 250 bar. This ignores entirely the partitioning, on-axis, between high and low temperature flow, and also ignores the details of the reaction zone depths, the integrated path lengths of fluid flow, and seafloor bathymetric effects off-axis. The total water flux due to off-axis flow for lithospheric ages of 1–65 Ma (taken to be uniformly 5 °C above ambient temperatures) is 2.4×10^{16} kg yr^{-1}, while that due to high temperature axial flow within ages 0–1 Ma (uniformly 350 °C) is 2.4×10^{13} kg yr^{-1}.

(i) Temperature and pressure effects on water volume calculations

The accuracy of these estimates depends greatly on the underlying assumption that on average this circulation results in heating of seawater by 5 °C above ambient temperatures during the time it is resident within the lithosphere. This is an arbitrary assumption, and it may equally well be that most of the circulation takes place at cooler temperatures. It seems certain that the bulk of off-axis flow takes place at temperatures below 30 °C (Honnorez 1981; Gallahan & Duncan 1994) but the lower bound on, and spatial distribution of off-axis upper lithospheric temperatures are as yet constrained poorly. The effect of this would be that the total transport of seawater into the lithosphere would scale, almost linearly at these temperatures, with the reciprocal temperature increase above ambient.

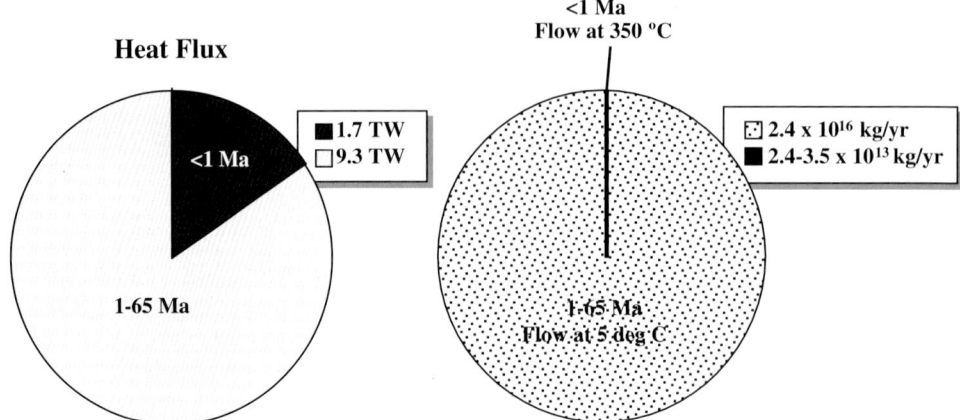

Figure 5. Total budget of hydrothermal fluids circulating through the oceanic lithosphere. The heat flux which drives the circulation (left). The mass of water necessary to cool the lithosphere, divided into axial (0–1 Ma) and off-axis (1–65 Ma) components (right). The axial component is calculated by making two different assumptions (described in the text), leading to a 46% difference in the estimates of total water circulation.

The effects of variable temperature on the specific heat of seawater may be taken into account by integrating with respect to temperature the relationship between specific heat and volume flux,

$$Q_m = Q_a \bigg/ \int_{2\,°C}^{350\,°C} c_p(P,T)\,\mathrm{d}T. \tag{2.19}$$

If we consider that the fluid has been heated at a constant pressure of 250 bar, the water volume flux, integrating over all temperatures from 2 to 350 °C becomes 3.5×10^{13} kg yr^{-1}, an increase of 46% over the more naive estimate above.

The effects of pressure on c_p of seawater are less than, and act in opposition to, those due to temperature (Bischoff & Rosenbauer 1984; and figure 4). The total water budget due to high temperature flow shall increase somewhat, but proportionally less so, if the integral in the equation above were replaced by a double integral taking the pressure dependence into account. By extrapolating from the results of Bischoff & Rosenbauer we find that at 350 °C the value of c_p decreases by ca. 15% at 450 bar relative to that at 250 bar, which is equivalent to the presumed depth of the reaction zone beneath a typical medium or fast-spreading ridge crest. The effects on the total water budget attributed to high temperature axial flow, integrated over the path length of that flow, will be commensurately smaller than this adjustment. We may therefore discount pressure dependence as falling well within the accumulated errors in these calculations and elect to consider only the single integral form of (2.19).

The major impact of pressure on calculations of heat flux stems from the potential for phase separation of the fluids. By increasing pressure at a given temperature such that the P–T regime falls below the critical point of phase separation, c_p decreases dramatically. The heat content of a given volume of hydrothermal fluid would drop significantly relative to fluids of the same T but somewhat lower P which fall on the other side of the critical point.

Pressure dependence of c_p would have essentially no effect on lower-temperature axial diffuse flow and off-axis flow as c_p achieves a nearly constant (pressure-independent) value for temperatures below ca. 200 °C.

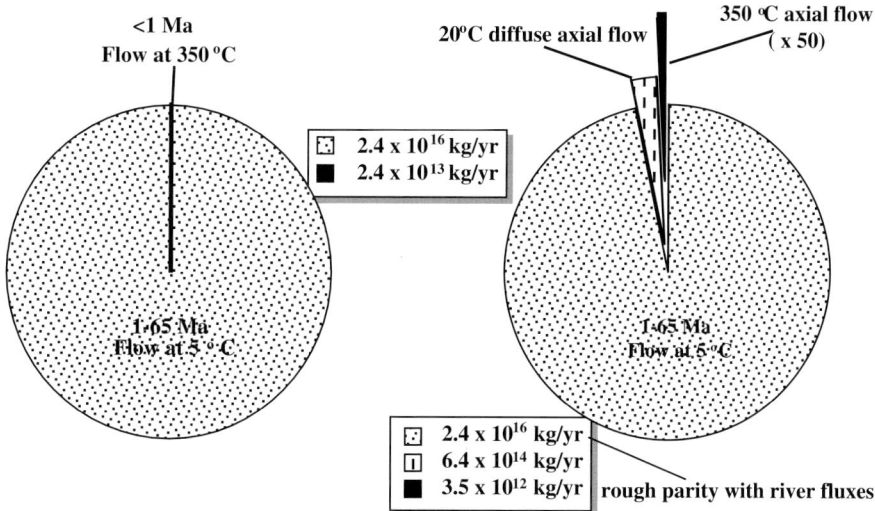

Figure 6. Total budget of hydrothermal fluids circulating through the oceanic lithosphere. The heat flux which drives the circulation (left). The mass of water necessary to cool the lithosphere, divided into axial (0–1 Ma) and off-axis (1–65 Ma) components (right). The axial component is calculated by presuming that 10% of the flow takes place at 350 °C, and 90% at 20 °C.

If we take the conservative view in terms of total volume of water transported that temperatures are raised 5 °C above ambient, we see that on average the equivalent of about 1 Sv (Sverdrup, or 10^6 m^3 s^{-1}) of seawater is transported through the lithosphere by hydrothermal circulation. This is approximately 1/30th of the total transport of seawater through the Gulf Stream, and in rough parity with the total flux of river water entering the world oceans (Elderfield & Schultz 1996).

Given the three-orders-of-magnitude difference in apparent water budgets associated with axial and off-axis regimes, the axial component of flow appears of little consequence to the total global water budget. This is in stark contrast to the impact of the high temperature axial component of seafloor hydrothermal flow on ocean chemistry. Although flank flow systems are quantified only poorly, the current evidence is that high temperature flow is by far the dominant hydrothermal contributor of both sources and sinks to global ocean chemistry (Elderfield & Schultz 1996). Furthermore, if one considers the contributions of lower temperature diffuse axial flow to the heat budgets, it becomes apparent that the associated water budgets increase dramatically.

We take the heat flux values presented in figure 3, accommodating axial diffuse heat flow in the ratio of 10:1 with axial high temperature flow, and convert these into water fluxes (equation (2.19)). This is seen in figure 6. The total axial flow due to diffuse modes, assuming these take place uniformly at a temperature elevated 20 °C above ambient, is 6.4×10^{14} kg yr^{-1}, or roughly 3% of that associated with low temperature off-axis flow. The corresponding water budget due to high temperature 350 °C flow drops to 3.5×10^{12} kg yr^{-1}. While still a small percentage of the total water circulating through the oceanic lithosphere, axial diffuse flow takes place at higher temperatures, and as seen later, is involved in considerable chemical exchange with the oceanic lithosphere and oceans.

3. Diffuse flow is an intrinsic feature of high temperature flow

A theoretical and observational framework has been provided showing that hydrothermal circulation persists throughout the oceanic lithosphere to ages as great as 65 ± 10 Ma. We consider in this section the nature of hydrothermal circulation in the youngest, hottest part of that system, the axial region (0–1 Ma). On axis, hydrothermal circulation may be modelled as that due to flow within a highly cracked permeable medium. Cracks within the sheeted dyke complex and extending up into the pillow basalt layer likely provide the predominate pathways by which fluids heated at the reaction zone are transported rapidly up to the seafloor.

Within the permeable subdomains in this region fluid flow is governed by Darcy's Law which describes the relationship between the pressure gradient driving the flow with a given velocity and the opposing viscous forces,

$$\bar{u} = -\frac{k}{\mu}(\nabla P + \Delta \rho g \hat{z}), \tag{3.1}$$

where $\bar{u} = (\hat{x}u + \hat{z}w)$ is the transport velocity in Cartesian coordinates $(\hat{x}, \hat{z})$, k is the permeability of the medium, μ is the dynamic viscosity of fluids within the medium, P is the pressure, $\Delta \rho$ is the density difference determined by the temperature distribution, and g is gravitational acceleration (Phillips 1991). For Darcy flow in a two-dimensional medium with isotropic permeability the resulting equations are,

$$\left.\begin{aligned} \frac{\partial^2 p}{\partial x^2} + \frac{\partial^2 p}{\partial z^2} &= \rho_0 \alpha g \frac{\partial T}{\partial z}, \\ \frac{\partial^2 \Psi}{\partial x^2} + \frac{\partial^2 \Psi}{\partial z^2} &= -\frac{\rho_0 k \alpha g}{\mu} \frac{\partial T}{\partial x}, \\ \frac{\partial \Psi}{\partial z}\frac{\partial T}{\partial x} - \frac{\partial \Psi}{\partial x}\frac{\partial T}{\partial z} &= \kappa \left[\frac{\partial^2 T}{\partial x^2} + \frac{\partial^2 T}{\partial z^2}\right], \end{aligned}\right\} \tag{3.2}$$

where Ψ is the stream function field and α is the coefficient of thermal expansion.

While appropriate for defining the flow and temperature fields within any given permeable region, this formulation is inappropriate to describe the flow within a crack bounding such a region. We consider flow within a crack to be represented by turbulent pipe flow which obeys the relationship

$$\frac{\partial p}{\partial z} = \frac{\rho_i f v_f^2}{4 a_p}, \tag{3.3}$$

where ρ_i is the density of the fluid in the pipe, f is an empirical friction factor, v_f is the turbulent fluid velocity, and a_p is the radius of the pipe (Turner & Campbell 1987).

(a) A model of flow within a cracked permeable domain

We take a model (figure 7) of a permeable medium surrounding a pipelike crack to represent a high temperature fluid source, originating at depth within the seafloor and communicating with the reaction zone deep within the hydrothermal system. The crack is embedded within the permeable seafloor and acts to channel high temperature fluids from the source and up into the water column. On a finer scale, such a model might also serve to represent the flow within an individual hydrothermal edifice.

In order to solve equations (3.1) and (3.2) for the given geometry we set the following boundary conditions. The temperature at the base of the structure ($z = L/2$) is set to a constant 360 °C at the origin, and extending along the bottom a distance $L/2$ from that point. The temperature then decreases exponentially from that point such that $T = 0$ at $x = L$. The origin is the centreline of the vertical pipe, and the model is two dimensional and symmetric about that axis.

The temperature of the pipe wall is set to 360 °C isothermally, which we take to be the temperature of the 'black smoker' source fluids flowing upward through the pipe. The pipe wall is made permeable such that hydrothermal source fluids may seep into the surrounding region and mix there with seawater advected into the structure. The advection takes place in response to the lateral pressure gradients set up within the permeable medium in response to the vertical flow within the pipe.

Variable boundary conditions are set at the top of the permeable subdomain. In locations where the solution requires effluent to flow outward from the structure, we require the vertical gradient of temperature at the seafloor–ocean interface (i.e. the heat flux due to thermal conduction) to be zero. In locations where the solution requires advection of cool seawater into the top of the permeable region, we require the temperature (rather than its vertical gradient) at that boundary to be zero. Thus, we alternate between Dirichlet and Neumann conditions depending upon the local behaviour of the solution.

For the model shown in figure 7, the geometry represents a permeable block with sides elevated above that of the surrounding seafloor, following Schultz et al. (1992). In this case, the boundary conditions along the side at $x = L$ are equivalent to those at the top, i.e. zero horizontal temperature gradient for zones where effluent flows outward from the sides of the structure, and zero temperature for zones where cool seawater is advected into the structure. We permit no vertical component to the effluent velocity at this boundary, thus $\Psi_x = 0$.

The top panel of figure 7 is diagnostic. The stream function field reveals the pattern of seepage into the structure of hydrothermal source fluids from within the pipe. There are strong horizontal gradients in Ψ extending from the origin to nearly $x = L/2$, at which point the horizontal gradient drops precipitously. The vertical velocity of the effluent flowing within the permeable medium is derived from the horizontal gradient of the stream function, thus near $x = L/2$ the upward flow of effluent out of the structure ends, and a broader zone of much slower flow of seawater into the top and sides of the structure begins. Within the upflow zone in the part of the structure near the origin ($x < L/2$), the strong lateral gradient in Ψ indicates there is a rapid drop-off in vertical effluent velocity as one moves away from the origin (i.e. the 'pipe' or plume feeder source).

The part of the mound centred around $x = L/2$ is a zone of intense mixing between in-flowing seawater and out-flowing effluent. There is a notable vertical gradient in Ψ, associated with a significant horizontal component to flow of effluent within the structure seen in the bottom right-hand quadrant of the top panel of figure 7. This indicates there is a preferential band of in-flow along the thermal boundary layer that has formed (this is seen clearly in the bottom panel, i.e. the temperature field).

The relationship between the permeability of the structure and the temperature and velocity fields observed at the top boundary is shown in figure 9. Here three different sets of solutions to (3.1) and (3.2) are presented for the geometry shown previously in figure 7. Each solution corresponds to the identical geometry but a different isotropic permeability, and thus a different Rayleigh number. In the first instance, for the geometry shown and for permeabilities no greater than 10^{-12} m^2,

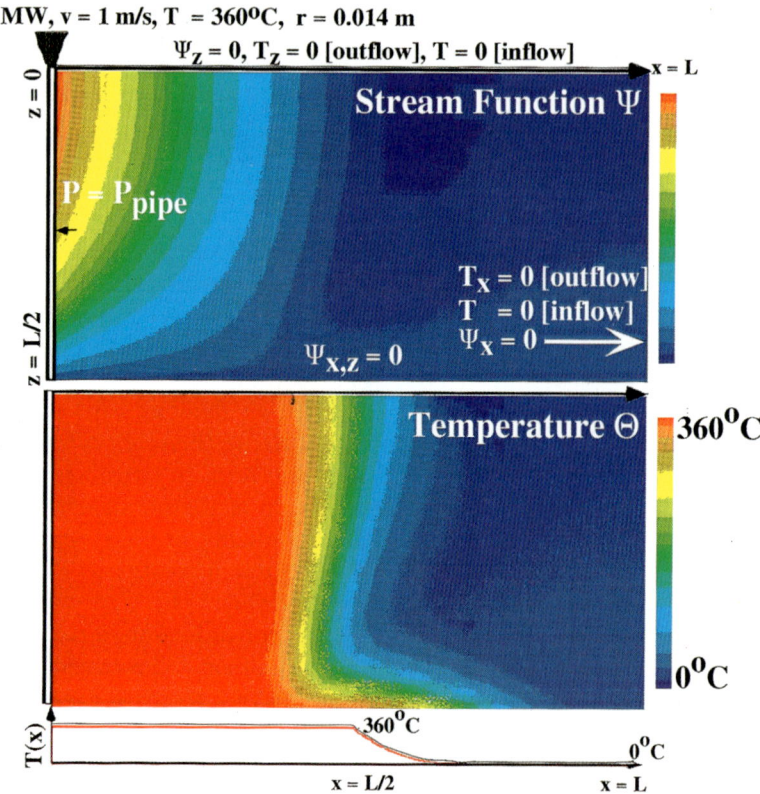

Figure 7. Solution for temperature and stream function fields within a two-dimensional permeable block with embedded crack. The flow within the crack, located at the axis of symmetry of this model, is taken to obey laminar pipe flow law. Solution by P. Dickson using a sparse iterative steady-state finite difference solver modified from Dickson et al. (1995).

the solution for Darcy flow simplifies to that of a diffusion equation, i.e. advection terms become unimportant and the Rayleigh number becomes small. At the highest permeability shown, $k = 10^{-11}$ m^2, the Rayleigh number increases and the effluent vertical exit velocity at the top boundary rises sharply (right panel of figure 9). The isotherms move inward toward the origin (left panel) as the effect of the cool advected seawater on the periphery of the structure is observed. For structures of different geometry the permeability at which advection will dominate over diffusion will differ from that found for this particular case, but there will exist a range of parameters for which solutions of this sort hold.

It is possible to obtain a stable numerical solution for Darcy flow for the present geometry and using the sparse iterative finite difference solution employed here (Dickson et al. 1995) only for low permeability. At higher permeabilities, the Darcy assumption of a balance between pressure gradients and viscous drag begins to reach a limiting case and it becomes impossible to drive fluid more quickly through the permeable region. In the true seafloor system it is likely that flow is instead dominated by the presence of small-scale cracks which provide highly permeable pathways in which higher velocity pipe flow, sheet flow, or other modes prevail, and in which Darcy flow is not appropriate. In the area immediately surrounding each of these small permeable pathways however, Darcy flow would hold locally, and the effects seen in figure 9 would, under the appropriate range of conditions, appear.

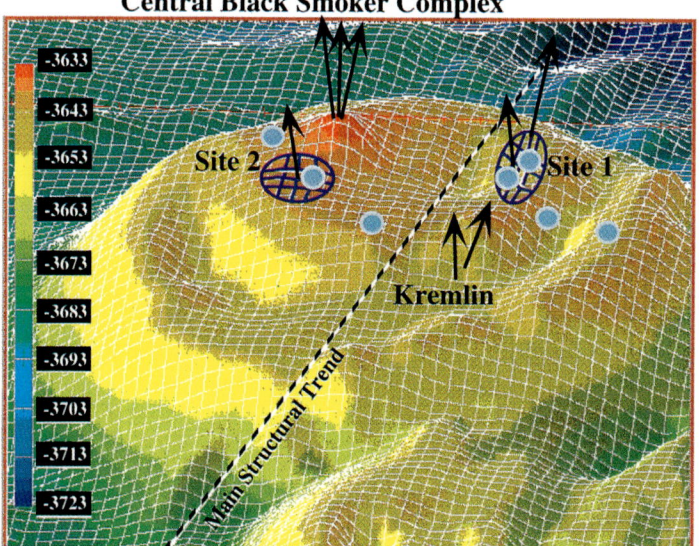

Figure 8. Bathymetric map of TAG mound. Flow emanating from central black smoker complex and from individual high temperature venting sites likely results from fluids channelled upward into the mound via highly permeable cracks of the sort represented in figure 7. The colour scale indicates depth in metres, and the horizontal scale of the mound is *ca.* 200 m across. Bathymetric data are from AMS-120 sidescan sonar survey, courtesy of P. Shaw, M. Kleinrock and S. Humphris.

In conclusion, figures 7 and 9 demonstrate the principle that in a crack-dominated system analogous to high temperature hydrothermal venting, there will exist a 'halo' of lower velocity broad-scale diffuse hydrothermal efflux which arises as a passive and intrinsic feature of the faster source flow. Horizontal pressure gradients are set up within a permeable medium surrounding a crack through which fluids flow, and in response seawater is entrained into such a body, mixes with the hydrothermal effluent entering the body from the walls of the source conduit, and then exits the top of the structure in a broad zone. We shall consider in the following section aspects of the chemical interactions that take place between the rocks making up the permeable region and the mixture of hydrothermal effluent and seawater flowing within it.

4. Combined chemical and physical measurements of diffuse effluent

One of the most important sets of measurements on submarine hydrothermal systems is those that characterize the fluxes of heat, water and chemical constituents and determine how they vary spatially and temporally. Because of the coupling of heat and chemical fluxes, these estimates are essential to determine the impact of hydrothermal systems on ocean chemistry (Elderfield & Schultz 1996). In this, we must recognize that high-temperature ('black smoker') flow and lower-temperature ('white smoker') and diffuse flow all must be sampled. Most work has emphasized the more spectacular high-temperature parts of hydrothermal systems. However, experiments to quantify the balance between diffuse (low temperature) and discrete (high temperature) sources on the Juan de Fuca Ridge (Schultz *et al.* 1992) have suggested that discrete flow only accounts for 1/5th to 1/10th of the axial heat flow. Thus, determination of the relative importance of these two flow paths is vital.

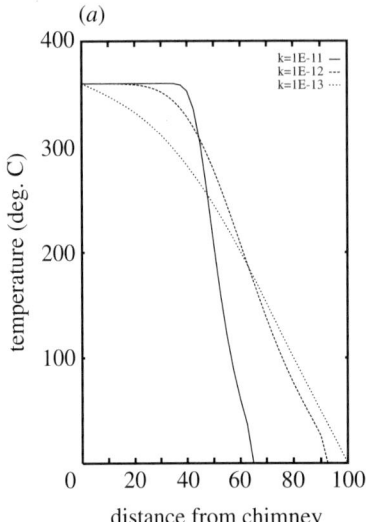 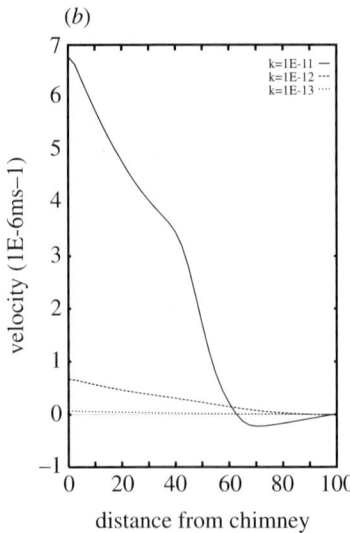

Figure 9. (a) Temperature in °C at the top surface (seafloor) of the two-dimensional permeable block vs horizontal distance in metres from the embedded crack for the model shown in the figure 7. (b) As (a), but vertical exit velocity of the hydrothermal effluent. Both (a) and (b) contain three sets of curves representing the flow solutions for permeabilities of $k = 10^{-11}$ m^2 (solid), 10^{-12} m^2 (long-dashed) and 10^{-13} m^2 (short-dashed) respectively.

Rudnicki & Elderfield (1992) calculated a total thermal flux of 500–900 MW for the TAG vent field from plume data, a figure which compares with 120 MW for the main black smoker venting complex at the summit of the TAG mound (Rona et al. 1990). This indicates that possibly 3/4 to 6/7th of the total TAG hydrothermal mound heat flux derives from low temperature systems which are entrained into the plume. This does not take into account diffuse flow that is not entrained in plumes, a discussion of which appears later.

(a) Diffuse effluent chemistry

The documentation of the chemistry of the on-axis diffuse flow is extremely poor. It relies principally on inferences based on the composition of hot spring deposits (Alt et al. 1987) and efforts to sample diffuse fluids directly by our group and by a few others (Butterfield & Massoth 1994). Recently, Mills et al. (1993) have provided some evidence of diffuse fluid chemistry from analyses of pore waters from metalliferous sediments at TAG but the only lower temperature fluids which had previously been sampled directly at TAG were 'white smokers', from a small cluster of sphalerite-rich dome structures to the southeast of the mound (the 'Kremlin' area ca. 50 m S of 'Site 1' in figure 8), and comprise ca. 15% seawater and ca. 85% high-temperature hydrothermal fluid with exit temperatures of ca. 300 °C (Edmond et al. 1995). However, diffuse flow of low-temperature fluids is ubiquitous over the top and sides of the mound, and is potentially of greater importance for the formation of ore deposits.

The development of the 'Medusa' system for sampling diffuse hydrothermal effluent and measuring its physical properties in situ (Schultz et al. 1996) enabled in 1994 and 1995 the first direct sampling of diffuse effluent exiting the TAG mound. The time series records and chemical samples discussed here were obtained from a 'Medusa' system deployed at 'Site 1' for a period of six months in 1994–1995. Spot

Table 2. *TAG fluid compositions*

(Composition of black smoker, white smoker and diffuse flow fluids from TAG, extrapolated to Mg = 0 (data from Edmond *et al.* (1995) and James (1995)).)

element	black smoker	white smoker	diffuse flow
Li (µM)	411	383	366
Rb (µM)	9.1	9.4	9
Ca (mmol kg^{-1})	0.8	27	22
Sr (µmol kg^{-1})	103	91	71
^{87}Sr/^{86}Sr	0.7038	0.7046	0.70304
pH	3.35	3	—
H$_2$S (mM)	2.5–3.5	0.5	< 0.04
Si (mM)	20.75	19.1	17
Cl (mM)	636	—	640
Fe (µM)	5590	3830	3260
Mn (µM)	680	750	635
Zn (µM)	46	300–400	62
Cu (µM)	120–150	3	< 3
La (pmol kg^{-1})	3710–4610	3590	2810
Ce (pmol kg^{-1})	8820–10200	4360	6970
Nd (pmol kg^{-1})	5250–6990	1590	4460
Sm (pmol kg^{-1})	1040–1450	250	920
Eu (pmol kg^{-1})	3390–3690	13800	9710
Gd (pmol kg^{-1})	895–1330	160	838
Dy (pmol kg^{-1})	635–907	110	528
Er (pmol kg^{-1})	281–336	46	203
Yb (pmol kg^{-1})	169–249	46	182
Lu (pmol kg^{-1})	21.4–30.6	4	24

measurements of diffuse effluent velocity and temperature were obtained from broad areas of the mound from sites representative of the diffuse flow regime found in the given areas. Concurrent samples of diffuse flow chemistry were obtained from seven sites on the mound (marked in figure 8 by filled cyan-coloured circles).

There appear to be two main venting centres on the mound. The first is the central black smoker complex, the trend of which continues southward to 'Site 2' (figure 8). The second is a feature extending from 'Site 1' along the trend of a grabenlike feature orientated NE–SW, and then south to the 'Kremlin' area. Areas of extensive black smoker flow are indicated on the figure by black arrows.

For comparative purposes with TAG black smoker (mainly from the central black smoker complex) and white smoker fluids (from the 'Kremlin' area), the concentration data for the diffuse flow (from the points marked on figure 8) have been extrapolated to zero Mg in table 1.

Data for several species (e.g. Li, Mn, Cl and Rb) in the diffuse flow fluids is best described as a mixture of high-temperature ('black smoker') fluid and seawater. This

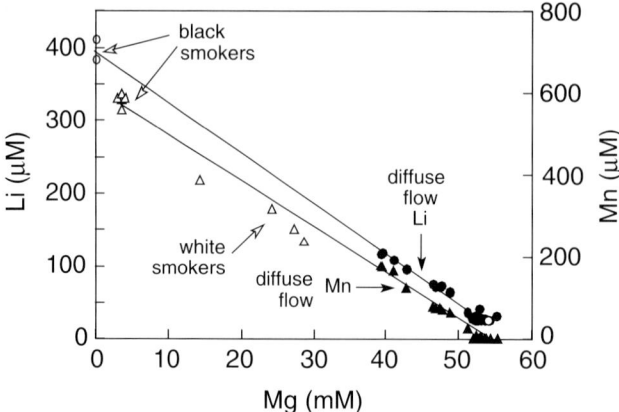

Figure 10. Li and Mn versus Mg in diffuse hydrothermal fluids at TAG. Also shown are compositions of black and white smoker fluids.

is shown in figure 10 using Mg as an index of seawater and implies that these species behave conservatively throughout the subsurface mixing process.

However, concentrations of other elements in the diffuse flow fluids are not conservative (figure 11). Fe in the diffuse flow endmember is ca. 44% depleted relative to that in black smoker fluids and H_2S is below the limits of detection. Si is also depressed by 20%, and Ca and Sr by 30%, relative to black smoker fluids. This is due to the precipitation of Fe-sulphides, silica and anhydrite as the high-temperature fluids mix with seawater within the mound. ODP drilling has shown that these three phases are commonly found within the interior of the TAG mound (Humphris et al. 1996). Thus, the fluid chemistry demonstrates the active precipitation of the three principal mineral types found at TAG as a consequence of mixing of entrained seawater with high temperature black smoker fluids, as expected by the flow simulation seen previously in figure 7.

Another group of elements, including Cu, Zn, REEs and U, show complex behaviours as a result of their migration within the mound by the process known as 'zone refining' (Franklin et al. 1981; Edmond et al. 1995) and this is described in detail by James (1995) and James & Elderfield (1996).

Because the chemical anomalies in the diffuse fluids attributable to precipitation of sulphide, anhydrite and silica within the hydrothermal deposit are known (figure 11), it is possible to estimate the rates of deposition of these minerals as a function of heat flux (figure 12). The formation rate (R) is given simply by

$$R = \frac{\Delta M Q_a}{c_p T_p}, \qquad (4.1)$$

where ΔM is the concentration difference between the diffuse flow endmember and the black smoker endmember at zero Mg, Q_a is the (advective) heat flux, and T_p is the temperature of mineral precipitation.

In this case, we define 'diffuse endmember' as the product of high temperature hydrothermal effluent that has mixed internally with seawater entrained into the interior of the mound, and which has reacted within the interior with the minerals making up the fabric of the mound. We do not refer here to high temperature effluent that has been diluted with seawater within the near-bottom water column. The

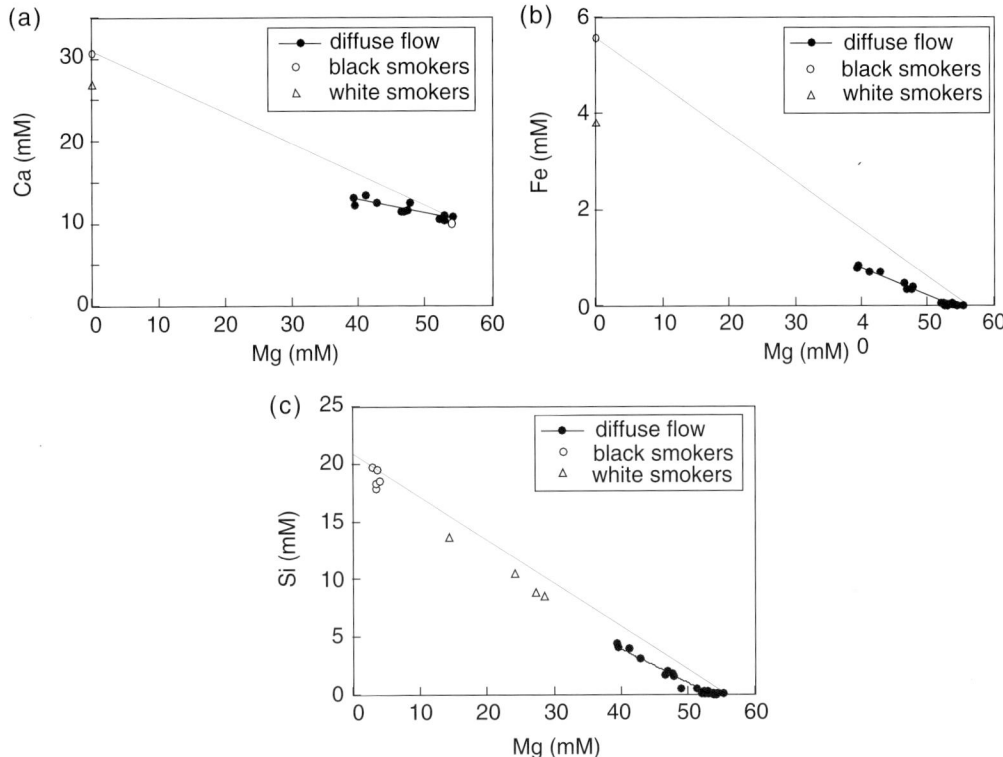

Figure 11. (a) Ca versus Mg, (b) Fe versus Mg, and (c) Si versus Mg in diffuse hydrothermal fluids at TAG. Also shown are compositions of black and white smoker fluids.

'Medusa' sampling system has been designed specifically to inhibit such dilution and thus to maximize the probability of obtaining uncontaminated diffuse endmember fluids. We consider this a prerequisite for accurate analysis of the chemical signature of both high and low temperature water–rock interactions.

From these estimates of R, the time taken (τ) for the mound to have attained its current inventory (A) of minerals (Tivey et al. 1996) can be estimated, thus providing an index of preservation (figure 13b):

$$\tau = A/R. \tag{4.2}$$

(i) *Areal measurement of heat flux density and the longevity of the TAG system*

A series of 'Medusa' spot measurements of effluent temperature and velocity were taken over the surface of the TAG mound (e.g. figures 8 and 12).

Each cluster of points in figure 12 represents a single sampling episode where the manned submersible used to position the instrument has removed the instrument's sensor head (Schultz et al. 1996) from the submersible's instrument basket and placed it directly atop diffuse flow emanating from the seafloor (in this case a surface of sulphide rubble atop the TAG mound). A single sampling episode lasts approximately 5–10 min, during which time there are continuous measurements made of effluent and ambient seawater temperature and vertical effluent velocity. (Transmissometry has recently been added to the capabilities of the instruments.) The gaps between clusters

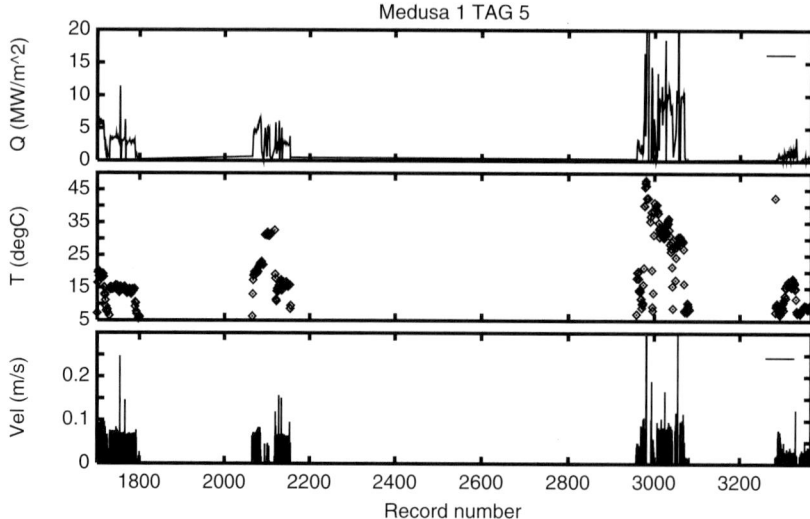

Figure 12. Typical spot measurements of diffuse effluent velocity, temperature and heat flux density taken at four closely spaced locations on the TAG mound (these sample locations are marked by filled circles in figure 8) using the 'Medusa' sampling system.

of points represent time periods where the submersible has replaced the sensor head in the instrument basket, and is traversing the surface of the mound en route to the next sampling location. At each sampling location it is possible to trigger any of a set of six 170 ml titanium syringe sample bottles which draw effluent from the interior of the sensor head, thus providing a means of concurrent sampling of the physical parameters and chemistry discussed within this section.

From the suite of spot measurements, the heat flux density over areas of the mound associated with visible diffuse efflux has been estimated as 2.5–8 MW m^{-2}. We observe diffuse flow of this magnitude over at least 1% of the total surface area of the mound. Over large areas of the remainder of the mound from which there is no visual indication of diffuse efflux, conductive heat flow measurements reveal vertical temperature profiles that are consistent with conductive rather than advective cooling (Becker & Von Herzen 1996). Over the remainder of the mound, diffuse advective discharge may exist, but below the limits of visual detectability, and also below the sensitivity threshold of the present configuration of our measurement system (i.e. very much less than 1.0 mm s^{-1}).

A lower bound on the total heat flux due to diffuse flow is therefore 780–2513 MW, based only on areas over which there is visible efflux, which suggests that the diffuse heat flux is roughly (and conservatively) an order of magnitude greater than the high-temperature heat flux of 120 MW. The mean and upper limit of this bound exceeds the upper limit of the Rudnicki & Elderfield (1992) estimate of 500–900 MW for the TAG vent field from plume data. This may be indicative of a part of the diffuse flow that is not entrained into the buoyant plume originating in the high temperature vent systems within TAG, and thus unaccounted for by measurements taken exclusively in the water column at the presumed height of neutral buoyancy of the main plume (i.e. ca. 200–300 m).

Anhydrite precipitates up to an order of magnitude more rapidly than sulphides and silica. Taking a figure of 2000 MW as that due to diffuse flow, mineral precip-

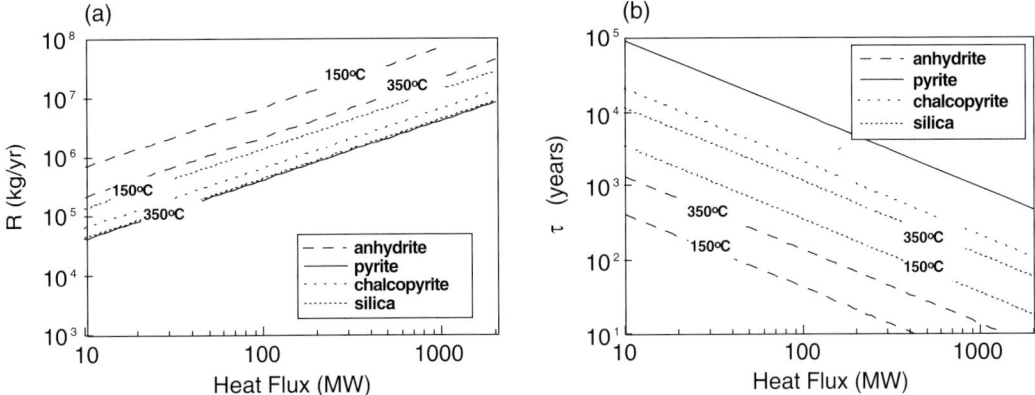

Figure 13. (a) Precipitation rate of various minerals versus heat flux attributed to diffuse flow, and (b) time taken for TAG mound to attain its current inventory of minerals. Based on James (1995), James & Elderfield (1996) and Tivey et al. (1995).

itation rates are of the order of 10^6–10^7 kg yr^{-1} of sulphide and silica and about 5–15×10^7 kg yr^{-1} of anhydrite.

According to Lalou et al. (1990), the TAG mound has grown episodically over the last 20 000 years and active hydrothermalism resumed about 50 years ago. On the basis of figure 13, it would seem that the mound anhydrite is contemporary, irrespective of choice of heat flux, and may represent as little as a few years of growth whereas mound sulphides and silica are clearly not contemporary, and represent 300–3000 years of growth. This is less than the age of the oldest hydrothermal deposits (20 000 years) and reflects, in part, periods where no mineral precipitation occurs and, in part, is indicative of reworking and oxidation of the sulphides.

5. Time series measurements at TAG

The spatial sampling of the physical and chemical properties of diffuse effluent at TAG has illustrated the range of contemporary mineralization processes taking place within this active hydrothermal system and has shed light on the episodic nature of hydrothermalism. Episodicity over tens to thousands of years is one extreme on the continuum of temporal variations in the hydrology of the system. The predominant modes of variability in the physical properties of diffuse effluent that we have observed over a six month instrument deployment period at TAG are those linked to tidal periodicities.

(a) Previous observations of tidal variability

Variations with tidal periods in both high temperature and low temperature flow have been observed for nearly a decade. Little et al. (1988) observed a strong semi-diurnal variation (peak-to-peak) of ca. 2 °C in temperatures measured over a twelve-day period a few centimetres above the seafloor in a diffuse flow area of the Guaymas Basin. This variation was taken to represent the modulation of the height of the near-bottom thermal boundary layer by tidal ocean currents (the modulation of tidal height at nearby Guaymas, Mexico is predominantly diurnal, thus the tidal (horizontal) speed variations are half this period, or semi-diurnal).

Schultz et al. (1992) analysed six weeks of continuous diffuse effluent and ambient temperature and diffuse effluent velocity data obtained in 1988 from the Endeavour Segment, Juan de Fuca Ridge. Here too, strong variations in temperature (several degrees peak-to-peak) with predominantly semi-diurnal periodicity were observed. In addition, a fixed array of ten thermocouples suspended directly above a high temperature smoker vent orifice detected strong semi-diurnal variability in black smoker plume temperatures (Monfort & Schultz 1988). The high temperature variability was interpreted as resulting from the lateral translation of the plume across the array of fixed temperature sensors as the plume was wafted back-and-forth by ocean tidal flow.

Another possible interpretation of the diffuse flow data was contamination of fluids entering the diffuse flow instrument (which in this case consisted of a vertical tube joined at the bottom to a funnel which was emplaced directly atop a percolating sulphide structure, and in which effluent velocity was detected by electromagnetic induction methods). In this instrument, the base of the flow-concentrator funnel was not sealed directly to the seafloor, making it possible for seawater to be advected into the base of the instrument where it would have the possibility to mix with diffuse endmember fluids. This advection could be modulated by the flow of tidal waters. Finally, it was considered that there may be pathways of sufficient permeability within the sulphide structure that broader tidal motions of the water column may be channelled into the interior of the structure and thereby modulate effluent temperatures.

(b) Tidal variability at TAG

Time series of diffuse effluent velocity and temperature, and the temperature of ambient seawater obtained during a $2\frac{3}{4}$ day period in September 1994 from 'Site 1' on the TAG mound (see figure 8) are shown in figure 14. The thicker curves represent low-pass filtered calibrated data (Schultz et al. 1996), using a robust moving average filter with 2 h corner period (Schultz et al. 1992). The thin curves represent the best fitting diurnal and semi-diurnal equilibrium tidal components (specifically the K_1, P_1, O_1, K_2, S_2, M_2 and N_2 lines). The tidal fits to these time series were obtained by calculating the discrete Fourier transform at these specific diurnal and semi-diurnal frequencies (Hendershott 1981), although the short duration of the time series section relative to the tidal periods makes it impossible to distinguish between the individual diurnal and semi-diurnal lines, rather only broad spectral peaks may be resolved.

Spectral resolution is illustrated in figure 15. Here the power spectrum for the diffuse effluent temperature was calculated by using the prolate taper function method of Thomson (1982) which was applied to a time series section spanning approximately two months in October–December 1994. The spectrum is plotted within an envelope representing ±1 standard error. A number of distinct peaks are revealed showing that the effluent temperature is modulated primarily by semi-diurnal variations and harmonics, although there is significant power in the diurnal band, as well as even longer period variations extending to two days as well.

The undisturbed time series section is too short to show quantitatively that variations with periods greater than about two days exist. There was an episode of substantial disturbance to flow that is attributed to ODP drilling activities at TAG. The relationship between these activities, the effluent temperatures, and the penetration of specific anhydrite breccia petrological horizons is discussed in Schultz et al. (1996). These effects have limited our ability to resolve longer period natural variability, although direct visual examination of the time series, which extend from

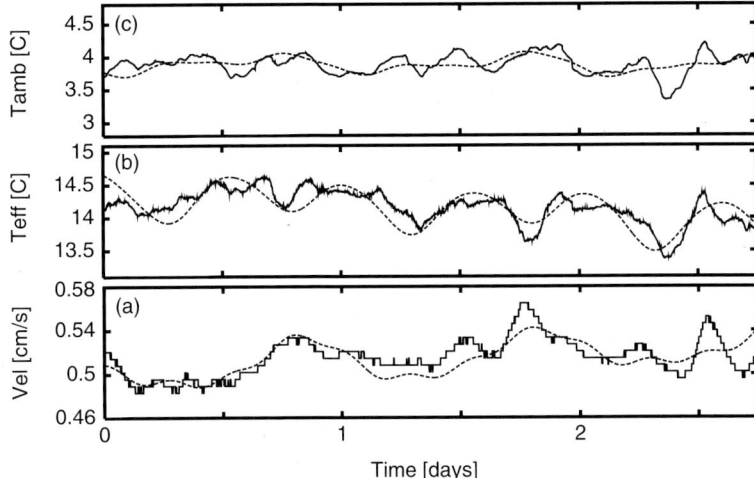

Figure 14. Time series of diffuse effluent (a) vertical exit velocity, (b) temperature, and (c) ambient seawater temperature measured by Medusa monitoring system 1 at TAG Site 1 for a three day period in 1994 before initiation of ODP drilling at TAG. The thin smooth curves are the time series obtained from the best-fitting diurnal and semi-diurnal tidal variations while the rougher curves are the actual measurements.

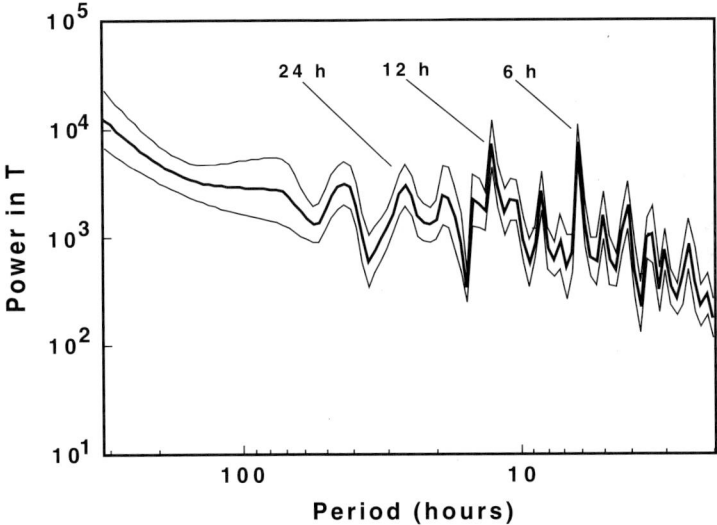

Figure 15. Power spectrum for diffuse hydrothermal effluent temperature variations as recorded as TAG Site 1 by Medusa 1 during the period 19 October to 25 December 1994. The spectrum shown is a robust estimate based on taking a series of 792-point 30% overlapped sections, each of which is tapered in the time domain by applying a time-bandwidth product 4 prolate spheroidal taper function. The time series was prewhitened by applying a 48-point autoregressive filter to a lowpass filtered version of the original time series. Well-defined peaks in the spectrum are seen at the tidal periods and their harmonics. The spectrum is shown within an envelope representing one standard error.

August 1994 to the beginning of January 1995 do suggest that there is fortnightly variability as well. There is also evidence of a secular shift in the locus of heating of the TAG mound during the six month measurement period (Schultz et al. 1996).

Spectral analysis of diffuse effluent velocity (rather than temperature) time series reveal that these variations are predominantly diurnal. The power in the diurnal peak is approximately fives times that of the semi-diurnal peak for effluent velocity. The different spectral forms for the effluent temperature and velocity may be diagnostic of the nature of the internal hydrology of TAG. This is the subject of current investigation. Mechanisms that may modulate the response of a hydrothermal system to tidally modified boundary conditions are proposed below.

(c) Mechanisms for tidal influence on hydrothermal flow and measurement

Ocean tides modulate the seafloor pressure field, changing the value of hydrostatic pressure within the hydrothermal system. The variation in height of the water column loads the seafloor, which has finite rigidity, and which deforms in response to this tidal loading. The solid Earth also responds to astronomical tidal forces and acts to deform the oceanic lithosphere from below thus changing the state of stress at the seafloor. Changes in the horizontal velocity of the water column have an effect on the thermal boundary conditions on the seafloor. There are measurement issues related to the Bernoulli effect due to lateral tidal motions on open crack-dominated hydrothermal flow. We shall consider each of these cases in turn.

(i) Temporal variability in seafloor thermal boundary conditions

In §3 we considered the effect of in-flow of seawater and out-flow of effluent on a model of flow within a cracked permeable structure. In regions of out-flow we imposed a boundary condition of zero thermal gradient, and in regions of in-flow a condition of zero temperature.

Tidal motions of the water column will alter the seafloor thermal boundary conditions similarly, forcing approximate Dirichlet and Neumann conditions to hold, depending upon the horizontal motions. At periods of peak tidal exchange, the strong motion of the tides will act to transport heat advectively and efficiently away from the outer surface of the cracked permeable hydrothermal system. At such times we may reasonably impose an approximate zero temperature condition at the boundary between the sea and the structure. Given peak tidal velocities within Mid-Ocean Ridge hydrothermal systems of ca. 20 cm s^{-1}, (Lukashin, personal communication; Rudnicki 1990), very efficient lateral advection of heat into the water column is to be expected.

At periods of slack tidal exchange, a thin thermal boundary layer of the sort proposed by Little et al. (1988) is expected. In such a layer, vertical gradients in temperature are minimized.

The major effect of alternating between the peak-tidal-exhange and slack-tidal-exchange thermal boundary conditions is to perturb the temperature field inside the permeable medium, but only very close to its outer surface. There is little effect on the pattern of flow within the structure. While seemingly of little consequence, it is at the seawater–structure interface that conductive and advective heat flow measurements are made. The tidal variation in thermal boundary conditions may therefore have some effect on measurements made within the thickness of the thermal boundary layer.

(ii) Bernoulli effects

The lateral motion of the water column past an open channel either penetrating the interior of the permeable structure (e.g. a vent chimney), or any measurement system situated directly above the structure, would experience a vertical pressure gradient imposed by the lateral motion of fluids past the top of the chamber.

Bernoulli's equation relates the variation of pressure along a given streamline to the variation in speed of flow,

$$\tfrac{1}{2}\rho|\bar{u}|^2 + p = C_S, \tag{5.1}$$

where $\bar{u}$ is the velocity of the flow, ρ is the density of the fluid, $\tfrac{1}{2}\rho|\bar{u}|^2$ is the kinetic energy per unit volume of the fluid, p is the pressure, and C_S is a constant along a given streamline S. One sees immediately that at points along a given streamline where velocity is high, pressure is low.

The practical consequence of Bernoulli's equation is that lateral flow of seawater around an obstacle such as a hydrothermal vent structure will tend to generate a pressure drop immediately above the opening of a vent or instrument orifice. This is the basic principle behind both the venturi tube and the pitot tube, devices which make use of the Bernoulli effect to measure the velocity of moving fluids.

It is conceivable that Bernoulli effects may have an impact on the velocity field immediately within and above vent orifices and sulphide structures, or in some configurations within instruments deployed to measure the temperature or velocity of effluent in situ. These effects may be larger in the presence of slip, i.e. when there are strong vertical gradients in the lateral flow of seawater past a structure. The presence of a viscous boundary layer on the outer surface of the hydrothermal mound or vent may provide such an environment: This might have an impact on time series measurements of the temperature of black smoker fluids made by inserting sensors directly into the vent orifice.

(iii) *Response of the seafloor to ocean tides and solid Earth deformations*

Seafloor hydrostatic pressure is modulated directly by ocean tides. Peak tidal amplitudes in the open ocean (generally semi-diurnal peaks are predominant) are no more than *ca.* 1 m (Schwiderski 1980). This presents a hydrostatic pressure variation of roughly one part in 2.5–4.0×10^3 at the surface of a typical seafloor hydrothermal system. The direct effect of this on the governing equations of flow, e.g. (3.1), (3.2) and (3.3), is small and depends on the confined or lithostatic pressure of the system, although a time-dependent element is introduced, and some small-scale effects may arise. This may be contrasted with the effects of changes in barometric pressure on continental aquifers and wells. Total barometric pressure on the Earth's surface may change by approximately five parts in 1×10^2 during the transition from low-to-high pressure fronts, an effect one-to-two orders of magnitude greater than that experienced by seafloor hydrothermal systems by tidal variations in hydrostatic pressure.

In continental systems changes in barometric pressure may depress or elevate the level of water in open artesian wells according to the relation (Bredehoeft 1967)

$$\frac{dp_b}{dp} = -B, \tag{5.2}$$

where p_b is the barometric pressure. The constant of proportionality B, the 'barometric efficiency' a function of the porosity of the system, is seen to vary over the full range of 0 to 1 in continental wells. Thus water levels in continental artesian wells may be influenced strongly by changes in barometric pressure. While this is not expected to be the case in seafloor systems, it is imprudent to dismiss completely the possible influence of time variations in hydrostatic pressure on the level of submarine aquifers.

The tidal dilation of continental aquifers also acts to displace the water level within the aquifer. Bredeheoft (1967) calculate the changes in aquifer water level associated with tidal potentials and find these range between 10^{-3} and 10^{-2} m. This is quite similar to water level fluctuations reported as due to Earth tides in slightly fractured crystalline rock (ca. 10^{-2} m) and in sediments (ca. 10^{-3} m) in the southeastern United States (Marine 1975). It is reasonable to expect a similar process to operate within submarine aquifers. The change in water levels in such a submarine system, if sufficiently large, might influence the advective flow of heat to the surface which might manifest itself as a signal with tidal periodicities. Further analysis is required before the plausibility of this mechanism can be evaluated properly.

Davis & Becker (1994) report on an anticorrelation between temperature and pressure within ODP Hole 858G, a drill hole penetrating deep into a sedimented hydrothermal system on the Juan de Fuca Ridge. Fang et al. (1993) also note a similar effect within pore pressure measured within marine sediments, and D. Orange (personal communication, 1996) has observed related effects on the sedimented seafloor of Monterey Canyon in California. Wang & Davis (1996) note that a pressure variation at the seafloor induces both an essentially instantaneous change in pressures at all depths within the seafloor (i.e. an elastic response) as well as an additional component of the pressure signal that diffuses downward with a depth-dependent attenuation (see also Crawford et al. 1991). The phase shift and amplitude attenuation of the tidally induced pressure change is frequency dependent and a function primarily of hydraulic diffusivity. The skin depth of the diffusion of the seafloor pressure disturbance is proportional to the reciprocal square-root of the product of diffusivity and period. For typical marine sediments, this is only a few metres. Wang & Davis (1996) suggest that internal contrasts in elastic properties (e.g. the presence of free gas) can give rise to large instantaneous pressure changes across layer boundaries, thus amplifying the otherwise small effect.

(iv) *Seafloor tidal deformation and modulation of permeability structure*

We consider a final consequence of the loading of the seafloor by ocean tides and the deformation of the solid Earth. Under the appropriate conditions it may be plausible to consider the sensitivity of the bulk permeability structure of the lithospheric hydrothermal system to seafloor deformations.

The compliance of the seafloor is the complex frequency domain function (i.e. transfer function) relating the deformation of the seafloor to the vertical normal stress due to the pressure imposed upon the seafloor by the water column. The vertical component of compliance may be written (Crawford et al. 1991)

$$\zeta(\omega) = \frac{u_z}{\tau_{zz}} = u_z \bigg/ \left[\lambda \left(\frac{\partial u_x}{\partial x} + \frac{\partial u_z}{\partial z} \right) + 2\mu \frac{\partial u_z}{\partial z} \right], \qquad (5.3)$$

where u_j is the particle displacement in the x_j direction, μ is the rigidity or shear modulus (the first Lamé parameter), λ is the second Lamé parameter, ρ is the density of the material, and τ_{zz} is the vertical stress imposed on the seafloor by the water column. The magnitude of vertical displacement of the seafloor is therefore related to the imposed stress and the material properties of the seafloor (i.e. the Lamé parameters).

The unknown and potentially considerable degree of spatial heterogeneity of the elastic properties of an active hydrothermal system make it difficult to apply these concepts for modelling finer-scale tidally induced deformations. In such a case the approach of Crawford et al. (1991) in measuring simultaneously the seafloor pressure and displacement fields, and then inverting estimates of the vertical compliance for these material properties is quite promising. This provides a means to recover, usually in concert with allied seismic observations, the elastic properties of the lithosphere.

On the global scale, one may refer to globally averaged models of the Earth's elastic parameters. We have explored such an approach by making use of a series of computer programs distributed by H. G. Wenzel of the University of Karlsruhe. These codes use the PREM radially symmetric seismic model of elastic parameters of the Earth's crust and mantle, and by referring to Dehant (1987), Wahr (1981) and Zschau & Wang (1987), make it possible to relate the tidal potential functions to solid Earth deformations. The Earth-tide induced deformations of the Earth's surface are given (Bredehoeft 1967)

$$u_r = \bar{h}\frac{W_2}{g}, \qquad u_\theta = \frac{\bar{l}}{g}\frac{\partial W_2}{\partial \theta}, \qquad u_\phi = \frac{\bar{l}}{g}\sin\theta\frac{\partial W_2}{\partial \phi}, \qquad (5.4)$$

where $\bar{h}$ and $\bar{l}$ are 'Love numbers' (taken to be constant at the Earth's surface) which relate the true displacements to those predicted by equilibrium tide theory, g is gravitational acceleration, W_2 is the lunar/solar disturbing potential, r is the radial direction, θ is the colatitude, and ϕ is the longitude, positive eastward. The Love numbers may be determined from the elastic properties and the density of the material in question (Takeuchi 1950).

The solid Earth tidal displacements of the Earth's surface relative to the geoid may be calculated using the approach above. Displacements of nearly 1 m peak are calculated. Of course, this value is based on an underlying radially symmetric elastic model which may not reflect accurately the local structure within the lithospheric hydrothermal system.

Tsuruoka et al. (1995) have used a similar approach to study the correlation between tides and triggering of earthquakes. This work includes the contributions to the seafloor stress field of both solid Earth tides and ocean tidal loading forces and thus is particularly appropriate for the present discussion. The maximum incremental tidal shear stress reported by Tsuruoka et al. is ca. 4000 Pa, peak-to-peak with a predominantly semi-diurnal periodicity.

In an isotropic medium it is possible to relate the normal stresses to the normal strains through the material constants λ and μ (Bredeheoft 1967)

$$\sigma_{rr} = \lambda \Delta + 2\mu\epsilon_{rr}, \qquad \sigma_{\theta\theta} = \lambda\Delta + 2\mu\epsilon_{\theta\theta}, \qquad \sigma_{\phi\phi} = \lambda\Delta + 2\mu\epsilon_{\phi\phi}, \qquad (5.5)$$

where the dilatation $\Delta = \epsilon_{rr} + \epsilon_{\theta\theta} + \epsilon_{\phi\phi}$.

By substituting the radial normal component of strain the dilatation is written

$$\Delta = \frac{1-2\nu}{1-\nu}(\epsilon_{\theta\theta} + \epsilon_{\phi\phi}), \qquad (5.6)$$

where ν is Poisson's ratio. In a spherical radially symmetric Earth the horizontal components of the strain may be solved for by calculating the displacements and their horizontal derivatives,

$$\epsilon_{\theta\theta} = \frac{1}{r}\left(\frac{\partial u_\theta}{\partial \theta} + u_r\right), \qquad \epsilon_{\phi\phi} = \frac{1}{r}\left(\frac{1}{\sin\theta}\frac{\partial u_\phi}{\partial \phi} + u_\theta\frac{\cos\theta}{\sin\theta} + u_r\right). \qquad (5.7)$$

Presuming that compressional stresses due to tidal effects are of similar order to the shear stresses calculated by Tsuruoka et al. (1995), i.e. ca. 4000 Pa peak-to-peak, and taking $\nu \approx \frac{1}{4}$ we find that $\Delta \sim 10^{-8}$, a figure in general agreement with that calculated for the lunar semi-diurnal component by Bredehoeft (1967), and by Wenzel's computer code.

(v) A model of lithospheric permeability

While Δ is only a small volumetric change, the effects of this change on the effective permeability of a cracked permeable system may be significant. We examine this by representing the permeability of an idealized medium as that due to a network of fine interconnected capillary tubes. Within such a tube, 'Hagen–Poiseuille' or pipe flow will hold where

$$u(r_{\mathrm{p}}) = -\frac{\partial p}{\partial x}\frac{(a_{\mathrm{p}}^2 - r_{\mathrm{p}}^2)}{4\mu}, \tag{5.8}$$

where x is the direction of the axis of the pipe, a_{p} is the radius of the pipe, r_{p} is the radial distance from the centre of the pipe, μ is the viscosity of the fluid, and $u(r_{\mathrm{p}})$ is the velocity of the fluid in the x direction. The Reynolds number is defined $Re = \rho \hat{u} 2a_{\mathrm{p}}/\mu$, where $\hat{u}$ is the average velocity in the pipe, given below. For a given pressure, the average velocity of fluid flowing through the capillary tube is proportional to the square of the tube's radius.

The mass flux of fluid passing through the pipe is given by,

$$\int_0^{a_{\mathrm{p}}} \rho 2\pi r_{\mathrm{p}} u(r_{\mathrm{p}}) \, \mathrm{d}r_{\mathrm{p}} = -\frac{\pi \rho}{8\mu}\frac{\partial p}{\partial x} a_{\mathrm{p}}^4, \tag{5.9}$$

thus the mass flux is proportional to the fourth power of the pipe radius and the average velocity $\hat{u}$ is the mass flux divided by $\pi \rho a_{\mathrm{p}}^2$.

Hagen–Poiseuille flow holds for pipe geometries of small Reynolds number. For a given set of conditions there exists a critical pipe radius (i.e. Reynolds number) where the flow begins to transition from laminar to turbulent. In turbulent flow the flow rate through the pipe will be less for a given pressure gradient than for laminar flow (Tritton 1988).

The permeability of a medium consisting of n parallel-aligned pipes of equal radius aligned at an angle θ to the pressure gradient, embedded within an impermeable matrix is (Phillips 1991)

$$\boldsymbol{k} = \frac{\phi a_{\mathrm{p}}^2}{32}\boldsymbol{A}, \tag{5.10}$$

where the porosity $\phi = n\pi a_{\mathrm{p}}^2$, and where $\boldsymbol{A}$ is a 3×3 matrix of sines and cosines. For capillary tubes orientated randomly relative to the pressure gradient,

$$k = \frac{\phi a_{\mathrm{p}}^2}{96}. \tag{5.11}$$

This discussion does not depend critically on the assumption that permeability is regulated entirely by systems of capillary tubes (obviously a gross simplification). Phillips (1991) notes that if tubes are replaced by an isotropic system of plane fractures of crack width δ_{w}, the permeability is

$$k = \frac{\phi \delta_{\mathrm{w}}^2}{36}. \tag{5.12}$$

Section 3 makes it clear that providing a sufficient supply of effluent to the surface of a hydrothermal system requires that the permeable system must have cracks embedded within it. Throughout much of the volume of the structure flow will obey Darcy's Law, but purely Darcy flow cannot explain the large velocities and fluxes observed in typical diffuse flow fields. Therefore in such systems bulk transport of fluids is accelerated by the presence of cracks.

The permeability of typical oceanic basalts is constrained poorly, but values in the range 10^{-19}–10^{-13} m^2 have been measured, e.g. in DSDP Hole 504B, Costa Rica Rift (Williams et al. 1986).

If the bulk permeability is crack-dominated, then to achieve the small observed permeabilities of oceanic basalt implies that the crack widths δ_w of (5.12), or pipe radii a_p of (5.11) are on average likely to be exceedingly small. The dilatation Δ is likely to be taken up largely through constriction of these small voids, thus the commensurate change in the size of the cracks is likely to be large. This is an area of current research, and we are investigating coupling boundary element solutions for the deformation of cracks and tubes to models of lithospheric strain due to ocean tidal loading and solid Earth deformations.

In the event that the $O(10^{-8})$ typical dilatation does lead to significant changes in mean pipe radius, particularly if the mean pipe flow path is represented by convoluted and tortuous pipes and cracks, the small dilatation may result in nearly unbounded changes in bulk permeability. If the constriction forces the Reynolds number beyond the critical range, flow will transition from laminar to turbulent, amplifying these effects further. Such modulations of effective permeability will lead to changes in the flow velocity field with tidal periodicities.

It has already been noted that the mass flux scales with the fourth power of pipe radius. The advective heat flux through a pipe will therefore also scale with the fourth power, which would then couple into the permeable part of the system, in a complex way, through (3.2). It may be expected qualitatively that tidal effects on the temperature of the fluids exiting the hydrothermal system may be more pronounced that those seen in the velocity of those fluids.

The differences between the spectral characteristics of the temperature and the velocity measurements of diffuse flow on the TAG hydrothermal mound have been pointed out previously. The frequency-dependence of seafloor compliance may provide a means of modulating the spectrum of the effluent temperature and velocity in different ways. The stresses imposed on the lithosphere by diurnal ocean tides will tend to penetrate to greater depths than those due to semi-diurnal tides. This too is an area of current research.

(vi) Continental analogues

Rinehart (1980) discusses the relationship between continental solid Earth tidal deformations and the average interval between geyser eruptions in geothermal systems. In addition to the familiar diurnal and semi-diurnal modulations, he considers further the fortnightly, semi-annual and even longer period astronomical tidal cycles and corresponding deformations. Rinehart concurs that the elastic effect on competent rocks results in dilatations of order 10^{-7}–10^{-8}. He interprets the inelastic deformation to arise from the compression and dilatation of the pores and voids within such geothermal systems, and that this may be governing the tidal response.

In continental systems, the diurnal and semi-diurnal tides have little appreciable effect on geyser activity, despite these being associated with the largest elastic dilatation of any of the solid Earth tides. There is, however, a striking correlation between the fortnightly tidal peak, as well as even longer period tidal variations extending to months and years, and the mean interval between geyser eruptions at the Riverside, Old Faithful, Grand and Steamboat geysers in North America (Rinehart 1980).

The possibility that very long period, or secular variations in the permeability structure of the oceanic lithosphere may be linked to tidal loading and deformation may lead to very considerable uncertainty in efforts to quantify hydrothermal fluxes made from repeated spot measurement episodes over sequential field seasons. Rather, combining hydrothermal measurements with geodetic observations, and doing so in the form of quasi-permanent seafloor observatories may prove the most effective means of studying the interaction between the oceanic lithosphere and the hydrosphere.

This paper was prepared under the partial support of grants received from the Natural Environment Research Council/BRIDGE Programme, and the European Union MAST III Programme. The authors thank P. Dickson for her calculations of flow within a cracked permeable medium, and R. James and M. Greaves for their chemical analysis of the Medusa fluid samples. We acknowledge the dedication of our workshop staff in designing and building the instruments used to obtain the samples (particularly M. Walker, S. Riches, and P. Smith), and also the crews of the RV Keldysh and Atlantis II, and submersibles MIR1, MIR2 and Alvin. A.S. is indebted to the late Clive Lister, to whom this paper is dedicated, for many valuable conversations over the years, for assistance in construction of critical parts of his earlier instruments, and for being a voice of reason. Contribution number 4828, Department of Earth Sciences and Institute of Theoretical Geophysics, University of Cambridge.

References

Alt, J. C., Lonsdale, P., Haymon, R. & Muehlenbachs, K. 1987 Hydrothermal sulfide and oxide deposits on seamounts near 21° N East Pacific Rise. *Geol. Soc. Am. Bull.* **98**, 157–168.

Baker, P. A., Stout, P. M., Kastner, M. & Elderfield, H. 1991 Large-scale lateral advection of seawater through oceanic crust in the central equatorial Pacific. *Earth Planet. Sci. Lett.* **105**, 522–533.

Baker, E. T., Massoth, G. J., Walker, S. L. & Embley, R. W. 1993 A method for quantitatively estimating diffuse and discrete hydrothermal discharge. *Earth Planet. Sci. Lett.* **118**, 235–249.

Baker, E. T., German, C. R. & Elderfield, H. 1995 Hydrothermal plumes over spreading-center axes: global distributions and geological inferences. In *Seafloor hydrothermal systems* (ed. S. E. Humphris, R. A. Zierenberg, L. S. Mullineaux & R. E. Thomson), pp. 47–71. (Geophysical Monograph 91.) Washington, DC: AGU.

Becker, K. & Von Herzen, R. P. 1996 Pre-drilling observations of conductive heat flow at the TAG active mound using Alvin. *Proc. ODP Initial Rep.* **158**, 23–29.

Bischoff, J. L. & Rosenbauer, R. J. 1985 An empirical equation of state for hydrothermal seawater (3.2 percent NaCl). *Am. Jl Sci.* **285**, 725–763.

Bischoff, J. L. & Rosenbauer, R. J. 1989 Salinity variations in submarine hydrothermal systems by layered double-diffusive convection. *J. Geol.* **97**, 613–623.

Bredehoeft, J. D. 1967 Response of well-aquifer systems to Earth tides. *J. Geophys. Res.* **72**, 3075–3087.

Butterfield, D. A. & Massoth, G. J. 1994 Geochemistry of North Cleft Segment vent fluids – temporal changes in chlorinity and their possible relation to recent volcanism. *J. Geophys. Res.* B **99**, 4951–4968.

Butterfield, D. A., McDuff, R. E., Mottl, M. J., Lilley, M. D., Lupton, J. E. & Massoth, G. J. 1994 Gradients in the composition of hydrothermal fluids from the Endeavour Segment vent field – phase-separation and brine loss. *J. Geophys. Res.* B **99**, 9561–9583.

Carlson, R. L. & Johnson, H. P. 1994 On modeling the thermal evolution of the oceanic upper mantle: an assessment of the cooling plate model. *J. Geophys. Res.* B **99**, 3201–3214.

Chen, C.-T. A. & Marshall, W. L. 1982 Amorphous silica solubilities. IV. Behavior in pure water and aqueous sodium chloride, sodium sulfate, magnesium chloride, and magnesium sulfate solutions up to 350 °C. *Geochim. Cosmochim. Acta* **46**, 279–287.

Corliss, J. B., Dymond, J., Gordon, L. I., Edmond, J. M., Von Herzen, R. P., Ballard, R. D., Green, K., Williams, D., Bainsbridge, A., Crane, K. & Van Andel, T. H. 1977 Submarine thermal springs on the Galapagos Rift. *Science* **203**, 1073–1083.

Crawford, W. C., Webb, S. C. & Hildebrand, J. A. 1991 Seafloor compliance observed by long-period pressure and displacement measurements. *J. Geophys. Res.* B **96**, 16151–16160.

Davis, E. E. & Lister, C. R. B. 1974 Fundamentals of ridge crest topography. *Earth Planet. Sci. Lett.* **21**, 405–413.

Davis, E. E. *et al.* 1992 Flank flux: an experiment to study the nature of hydrothermal circulation in young oceanic crust. *Can. Jl Earth Sci.* **29**, 925–952.

Dehant, V. 1987 Tidal parameters for an inelastic Earth. *Phys. Earth Planet. Int.* **49**, 97–116.

Delaney, J. R., Mogk, D. W. & Mottl, M. J. 1987 Quartz-cemented breccias from the Mid-Atlantic Ridge: sampled of a high-salinity hydrothermal upflow zone. *J. Geophys. Res.* B **92**, 9175–9192.

Dickson, P., Schultz, A. & Woods, A. 1995 Preliminary modelling of hydrothermal circulation within mid-ocean ridge sulphide structures. In *Hydrothermal vents and processes* (ed. L. M. Parson, C. L. Walker & D. R. Dixon), pp. 145–158. Geological Society Special Publication no. 87.

Edmond, J. M., Campbell, A. C., Palmer, M. R., Klinkhammer, G. P., German, C. R., Edmonds, H. N., Elderfield, H., Thompson, G. & Rona, P. 1995 Time series studies of vent fluids from the TAG and MARK sites (1986, 1990) Mid-Atlantic Ridge: a new solution chemistry model and a mechanism for Cu/Zn zonation in massive sulphide orebodies. In *Hydrothermal vents and processes* (ed. L. M. Parson, C. L. Walker & D. R. Dixon), pp. 77–86. Geological Society Special Publication no. 87.

Edmonds, H. N. & Edmond, J. N. 1995 A three-component mixing model for ridge-crest hydrothermal fluids. *Earth Planet. Sci. Lett.* **134**, 53–67.

Elder, J. W. 1965 Physical processes in geothermal areas. In *Terrestrial heat flow* (ed. W. H. K. Lee), pp. 211–239. (Geophysical Monograph 8.) Washington, DC: AGU.

Elderfield, H. & Schultz, A. 1996 Mid-ocean ridge hydrothermal fluxes and the chemical composition of the ocean. *A. Rev. Earth Planet. Sci.* **24**, 191–224.

Fang, W. W., Langseth, M. G. & Schultheiss, P. J. 1993 Analysis and application of *in situ* pore pressure measurements in marine sediments. *J. Geophys. Res.* **98**, 7921–7938.

Fehn, U., Green, K. E., Von Herzen, R. P. & Cathles, L. M. 1983 Numerical models for the hydrothermal field at the Galapagos spreading center. *J. Geophys. Res.* B **88**, 1033–1048.

Franklin, J. M., Sangster, D. M. & Lydon, J. W. 1981 Volcanic-associated massive sulphide-deposits (ed. B. J. Skinner). *Econ. Geol.* **75**, pp. 485–627.

Gallahan, W. E. & Duncan, R. A. 1994 Spatial and temporal variability in crystallization of celadonites within the Troodos ophiolite, Cyprus: implications for low-temperature alteration of the oceanic crust. *J. Geophys. Res.* B **99**, 3147–3161.

Hartline, B. K. & Lister, C. R. B. 1981 Topographic forcing of supercritical convection in a porous medium such as the oceanic crust. *Earth Planet. Sci. Lett.* **55**, 75–86.

Hendershott, M. C. 1981 Long waves and ocean tides. In *Evolution of physical oceanography* (ed. B. A. Warren & C. Wunsch), pp. 292–341. Cambridge, MA: MIT Press.

Honnorez, J. 1981 The aging of the oceanic crust at low-temperatures. In *The sea*, vol. 7: *The oceanic lithosphere* (ed. E. Emiliani), pp. 525–588. New York: Wiley.

Humphris, S. E., Herzig, P. M., Miller, D. J. & ODP leg 158 Shipboard Scientific Party 1995 The internal structure of an active sea-floor massive sulphide deposit. *Nature* **377**, 713–716.

James, R. H. 1995 Chemical processes in submarine hydrothermal systems at the Mid-Atlantic Ridge. Ph.D. thesis, University of Cambridge.

James, R. H. & Elderfield, H. 1996 Chemistry of ore-forming fluids and mineral formation rates in an active hydrothermal sulphide deposit on the Mid-Atlantic Ridge. *Geology* **24**, 1147–1150.

Johnson, H. P., Becker, K. & Von Herzen, R. 1993 Near-axis heat flow measurements on the Northern Juan de Fuca Ridge: implications for fluid circulation in oceanic crust. *Geophys. Res. Lett.* **20**, 1875–1878.

Kadko, D., Baker, E., Alt, J. & Baross, J. 1994 Global impact of submarine hydrothermal processes. Report of the RIDGE/VENTS Workshop, US Ridge Inter-Disciplinary Global Experiments, Boulder, Colorado.

Kennedy, G. C. 1950 A portion of the system silica-water. *Econ. Geol.* **45**, 629–653.

Lalou, C., Thompson, G., Arnold, M., Brichet, E., Druffel, E. & Rona, P. 1990 Geochronology of TAG and Snakepit hydrothermal fields, Mid Atlantic Ridge: witness to a long and complex hydrothermal history. *Earth Planet. Sci. Lett.* **97**, 113–128.

Le Pichon, X. & Langseth Jr, M. G. 1969 Heat flow from mid-ocean ridges and sea-floor spreading. *Tectonophys.* **8**, 319–344.

Lister, C. R. B. 1972 On the thermal balance of a mid-ocean ridge. *Geophys. Jl R. Astr. Soc.* **26**, 515–535.

Little, S. A., Stolzenbach, K. D. & Grassle, F. J. 1988 Tidal current effects on temperature in diffuse hydrothermal flow: Guaymas Basin. *Geophys. Res. Lett.* **15**, 1491–1494.

Marine, I. W. 1975 Water level fluctuations due to Earth tides in a well pumping from a slightly fractured crystalline rock. *Water Resources Res.* **11**, 165–173.

McKenzie, D. P. 1967 Some remarks on heat flow and gravity anomalies. *J. Geophys. Res.* **72**, 6261–6273.

Melchior, P. 1956 Sur l'effet des marées terrestes dans les variations de nivaeu observée dan les puits, en particular au sondage de Turnhout (Belgium). *Commun. Obs. R. Belgique* **108**, 7–28.

Mills, R. A., Thomson, J., Elderfield, H. & Rona, P. A. 1993 Pore water chemistry of metalliferous sediments from the Mid-Atlantic Ridge: diagenesis and low-temperature fluxes. *Eos* **74**, 10.

Morton, J. L. 1984 Oceanic spreading centers: axial magma chambers, thermal structure, and small scale ridge jumps. Ph.D. thesis, Stanford University, California.

Morton, J. L. & Sleep, N. H. 1985 A mid-ocean ridge thermal model: constraints on the volume of axial hydrothermal heat flux. *J. Geophys. Res.* B **90**(13), 11 345–11 353.

Morton, J. L., Sleep, N. H., Normark, W. R. & Tompkins, D. H. 1987 Structure of the Southern Juan de Fuca Ridge from seismic reflection records. *J. Geophys. Res.* B **92**, 11 315–11 326.

Monfort, M. & Schultz, A. 1988 Timeseries measurements of hydrothermal vent temperature and diffuse percolation velocity: results from an ALVIN submersible program, Endeavour Segment, Juan de Fuca Ridge. *Eos* **69**, 1484.

Parsons, B. & Sclater, J. G. 1977 An analysis of the variation of ocean floor bathymetry and heat flow with age. *J. Geophys. Res.* **82**, 803–827.

Phillips, O. M. 1991 *Flow and reactions in permeable rocks*. New York: Cambridge University Press.

Phipps Morgan, J. & Chen, Y. J. 1993 The genesis of oceanic crust: magma injection, hydrothermal circulation, and crustal flow. *J. Geophys. Res.* B **98**, 6283–6297.

Rinehart, J. S. 1980 *Geysers and geothermal energy*. New York: Springer.

Rona, P. A. & Trivett, D. A. 1992 Discrete and diffuse heat transfer at ASHES vent field, Axial Volcano, Juan de Fuca Ridge. *Earth Planet. Sci. Lett.* **109**, 57–71.

Rona, P., Hannington, M. D., Raman, C. V., Thompson, G., Tivey, M. K., Humphris, S. E., Lalou, C. & Peterson, S. 1990 Active and relict sea-floor hydrothermal mineralization at the TAG hydrothermal field. *Econ. Geol.* **88**, 1989–2017.

Rosenberg, N. D., Lupton, J. E., Kadko, D., Collier, R., Lilley, M. D. & Pak, H. 1988 Estimation of heat and chemical fluxes from a seafloor hydrothermal vent field using radon measurements. *Nature* **334**, 604–607.

Rudnicki, M. D. 1990 Hydrothermal plumes at the Mid-Atlantic Ridge. Ph.D. thesis, University of Cambridge.

Rudnicki, M. D. & Elderfield, H. 1992 Theory applied to the Mid-Atlantic Ridge hydrothermal plumes: the finite difference approach. *J. Volcanology Geotherm. Res.* **50**, 163–174.

Schultz, A., Delaney, J. R. & McDuff, R. E. 1992 On the partitioning of heat flux between diffuse and point source seafloor venting. *J. Geophys. Res.* **97**, 12299–12314.

Schultz, A., Dickson, P. & Elderfield, H. 1996 Temporal variations in diffuse hydrothermal flow at TAG. *Geophys. Res. Lett.* **23**, 3471–3474.

Schwiderski, E. W. 1980 On charting global ocean tides. *Rev. Geophys. Space Phys.* **18**, 243–268.

Sclater, J. G., Jaupart, C. & Galson, D. 1980 The heat flow through oceanic and continental crust and the heat loss of the Earth. *Rev. Geophys. Space Phys.* **18**, 2269–2311.

Sleep, N. H. 1975 Formation of oceanic crust: some thermal constraints. *J. Geophys. Res.* **80**, 4037–4042.

Stein, C. A. & Stein, S. 1992 A model for the global variation in oceanic depth and heat flow with lithospheric age. *Nature* **359**, 123–129.

Stein, C. A. & Stein, S. 1994 Constraints on hydrothermal heat flux through the oceanic lithosphere from global heat flow. *J. Geophys. Res.* B **99**, 3081–3095.

Stein, C. A., Stein, S. & Pelayo, A. M. 1995 Heat flow and hydrothermal circulation. In *Seafloor hydrothermal systems* (ed. S. E. Humphris, R. A. Zierenberg, L. S. Mullineaux & R. E. Thomson), pp. 425–445. (Geophysical Monograph 91.) Washington, DC: AGU.

Takeuchi, H. 1950 On the Earth tide of the compressible Earth of variable density and elasticity. *Eos* **31**, 651–689.

Thomson, D. J. 1982 Spectrum estimation and harmonic analysis. *Proc. IEEE* **70**, 1055–1096.

Tivey, M. K., Humphris, S. E., Thompson, G., Hannington, M. D. & Rona, P. A. 1995 Deducing patterns of fluid-flow and mixing within the TAG active hydrothermal mound using mineralogical and geochemical data. *J. Geophys. Res.* B **100**, 12527–12555.

Tritton, D. J. 1988 *Physical fluid dynamics*, 2nd edn. Oxford: Oxford Science Publications, Clarendon Press.

Tsuruoka, H., Ohtake, M. & Sato, H. 1995 Statistical test of the tidal triggering of earthquakes: contribution of the ocean tide loading effect. *Geophys. Jl Int.* **122**, 183–194.

Turner, J. S. & Gustafson, L. B. 1978 The flow of hot saline solutions from vents in the sea floor – some implications for exhalative massive sulfide and other ore deposits. *Econ. Geol.* **73**, 1082–1100.

Turner, J. S. & Campbell, I. H. 1987 A laboratory and theoretical study of the growth of 'black smoker' chimneys. *Earth Planet. Sci. Lett.* **82**, 36–48.

Vera, E. E. & Diebold, J. B. 1994 Seismic imaging of oceanic layer 2A between $9°30'$ and $10°$ N on the East Pacific Rise from two-ship wide-aperature profiles. *J. Geophys. Res.* B **99**, 3031–3041.

Von Damm, K. L., Edmond, J. M., Grant, B., Measures, C. I., Walden, B. & Weiss, R. F. 1985 Chemistry of submarine hydrothermal solutions at $21°$ N, East Pacific Rise. *Geochim. Cosmochim. Acta* **49**, 2197–2220.

Wahr, J. M. 1981 Body tides on an elliptical, rotating, elastic and oceanless Earth. *Geophys. Jl R. Astr. Soc.* **64**, 677–703.

Wang, K. & Davis, E. E. 1996 Theory for the propagation of tidally induced pore pressure variations in layered subseafloor formations. *J. Geophys. Res.* B **101**, 11483–11495.

Wheat, C. G. & Mottl, M. J. 1994 Hydrothermal circulation, Juan de Fuca Ridge eastern flank: factors controlling basement water composition. *J. Geophys. Res.* B **99**, 3067–3080.

Williams, C. F., Narasimhan, T. N., Anderson, R. N., Zoback, M. D. & Becker, K. 1986 Convection in the oceanic crust: simulation of observations from Deep Sea Drilling Project Hole 504B, Costa Rica Rift. *J. Geophys. Res.* B **91**, 4877–4889.

Zschau, J. & Wang, R. 1987 Imperfect elasticity in the Earth's mantle: implications for Earth tides and long period deformations. In *Proc. 9th. Int. Symp. on Earth Tides* (ed. J. T. Kuo), pp. 605–629. Stuttgart: Schweizerbartsche Verlagsbuchhandlung.

Where are the large hydrothermal sulphide deposits in the oceans?

By Y. Fouquet

IFREMER, Centre de Brest, BP70, 29280 Plouzané, France

Large sulphide deposits have been identified on slow and fast spreading ridges and back-arc basins. Their formation is controlled by a combination of several conditions, each of which alone is often only compatible with the formation of small and unstable deposits. The geological control of deposits has to be considered both at the regional and local scales. The convective system is dependent on the morphology of the heat source (magma chamber) and the magma supply. Major sites are controlled by regional topographic highs that are the locus of the highest magma and heat supply along the ridge. On slow spreading ridges the flow of hydrothermal fluids can also be controlled by major regional rift valley faults. The discharge within a field is controlled by the local near surface permeability related to faulting or permeability of rocks. Recent discoveries considerably enlarge the potential locations of hydrothermal activity. On slow spreading ridges we have now to consider the base and top of the rift valley walls and the non-transform offsets, in addition to the relatively well documented control by volcanic topographic highs. Known sites also demonstrate that slow spreading ridges are more favourable for the formation of extensive mineralization. On fast spreading ridges, deposits are numerous and very small because the upflow zone is relatively narrow and subject to perturbation by frequent tectonic and volcanic activity. However, near fast spreading ridges, first order sulphide deposits can be formed on off-axial seamounts. Geological and physical conditions are key parameters controlling the morphology and potential size of deposits. Among these parameters, boiling, mixing within the crust, or precipitation under an impermeable cap rock, can enhance the formation of extensive subsurface mineralization within the oceanic crust. However, the knowledge of these deposits requires further investigation in the vertical dimension.

1. Introduction

Since the discovery of the first black smoker, more than 100 hydrothermal fields have been found in the oceans. About 15 of these deposits are large enough to be considered as an ore deposit if they were located on land. Their size and grade are similar to those of fossil sulphide deposits now mined on land. The 146 known sulphide sites occur in four different tectonic settings (figure 1). The first sulphide mineralization on the seafloor was discovered in 1978 (Cyamex et al. 1979; Francheteau et al. 1979; Spiess et al. 1980) at 21° N on the East Pacific Rise and the Galapagos ridge (Corliss et al. 1979). These discoveries were the proof that hydrothermal activity was a major process associated with the formation of young oceanic crust. Numerous cruises have now confirmed that hydrothermal processes are responsible for the formation of large sulphide deposits (Rona 1988; Rona & Scott 1993; Scott 1987). Exploration conducted during the past 15 years has shown a wide variety of hydrothermal activities and deposits in the oceans. This variability is first due to the various geodynamic settings and nature of source rocks. Hydrothermal deposits are now known on fast spreading ridges (Embley et al. 1988; Francheteau et al. 1979; Fouquet et al. 1988), super fast spreading ridges (Backer et al. 1985; Renard et al. 1985; Fouquet et al. 1994), slow spreading ridges (Rona et al. 1986a; Thompson et al. 1988; Honnorez et al. 1990; Langmuir et al. 1996), sediment covered ridges (Peter & Scott, 1988; Goodfellow & Franklin, 1993; Koski et al. 1988), young and mature back-arc basins (Fouquet et al. 1991; Halbach et al. 1989, Fouquet et al. 1993), island arcs (Urabe et al. 1987; Herzig et al. 1994) and fracture zones (Bonatti et al. 1976).

The median size for significant volcanogenic massive sulphides in Canada is 1.3×10^6 tonnes and for every 'significant deposit' there are probably 100 or more small deposits (Scott 1985). Several modern seafloor deposits have size and grades comparable to these ancient deposits. This paper presents the variability of geological controls on the location of hydrothermal fields, with a specific emphasis on major sulphide deposits. Large hydrothermal deposits are listed in table 1. They are located in three main volcanic settings: slow and fast spreading ridges and back-arc basins. To better understand the different mechanisms controlling their formation an overview of the knowledge of the location of hydrothermal fields in the ocean is presented. Figure 2 shows that the majority (42%) of hydrothermal fields are located on fast spreading ridges. However most of these sites are not compatible with the formation of large sulphide deposits. The compositions of deposits and the special case of the Red Sea are not discussed in this short paper.

2. Geological control on the location of major sulphide deposits in the oceans

Considering our knowledge of the location of major hydrothermal fields the geological control of hydrothermal activity has to be considered both at the regional and local scales (figure 3). The flow of hydrothermal fluids is controlled by major structural and volcanic elements (neovolcanic ridges, rift valley faults,...) but the discharge within a field is controlled by the local near surface permeability.

(a) Slow spreading

On slow spreading ridges a typical morphology is a 60 km long, 20 km wide and 1 km deep segment with fractures zones at both ends (figure 3a). The recent volcanic activity is concentrated on a narrow neovolcanic ridge generally located at the centre of the segment. The morphology of the neovolcanic ridge typically has a topographic high at its centre, indicating a higher magmatic budget in this area. On slow spreading ridges the volcanic ridge is very often an alignment of localized volcanic centres rather than a continuous ridge as seen on a typical fast spreading ridge. Volcanic ridges are punctuated by hundreds of discrete axial and off-axis volcanoes (Smith & Cann 1990). Less frequent tectonic events may promote long lived and more stable structures for hydrothermal upflow. Two types of *regional* controls are identified for major deposits: the topographic high (figure 3a), and the base (figure 3b) and top (figure 3c) of the rift valley walls. Major fields, such as TAG, controlled by graben walls faults, are also located at the latitute of the topographic high indicating a preferential location near the hot domain of the segment where the magmatic budget is

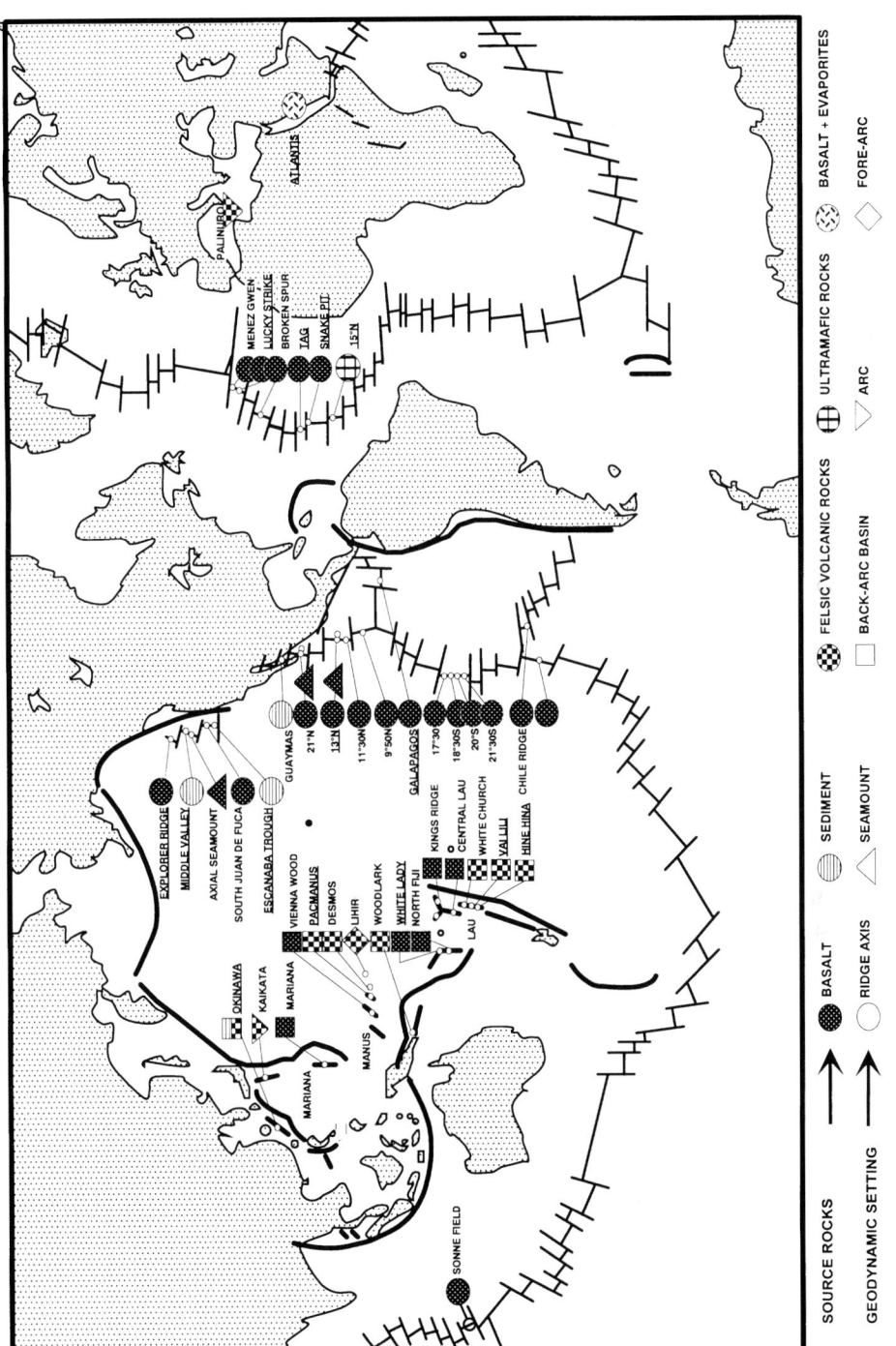

Figure 1. Location of hydrothermal sulphide deposits in the ocean, with indication of geodynamic setting and source rocks. Names underlined correspond to large deposits.

Table 1. Principal characteristics of major sulphide deposits in the oceans

site	geological setting	geological control	depth	cm yr^{-1}	source rock	type/size
Lucky Strike	slow spreading	topographic high + N–S faults + caldera	1650 m	2.2	enriched MORB	1 km × 1 km
TAG	slow spreading	valley wall + crossing faults + volc. centre	3650 m	2.6	MORB	250 × 45 m high + 2 mounds
Snake Pit	slow spreading	topographic high + volc. centre + graben	3465 m	2.6	MORB	3 mounds, 250 m × 55 m
14° 45′ N	slow spreading	valley wall + axial high + tranverse faults	3000 m	2.6	harzburgites	300 m × 125 m × 80 m
13° N off-axis	fast spreading	Seamount + pit crater + lava flows lid	2650 m	12	MORB	200 m × 70 m high, 800 × 200 m
Explorer ridge	intermediate spreading	neovolcanic ridge + crossing faults	1800 m	6	enriched MORB	200 × 25 high, + other mounds
Galapagos	intermediate spreading	faults + double rift + central rift faults	2850 m	6	MORB	3 mounds, 1000 × 150 × 35 m
Vai Lili	immature back-arc	topographic high, volcanic ridge	1710 m	6	basalt–andesite–dacite	400 × 100 m
Pere Lachaise	mature back-arc	triple junction + volcanic high	1950 m	7	MORB	1 km × 1 km
Pacmanus	immature back-arc	topographic high + pull-apart ridge	1675 m	?	Basalt-Dacite-Rhyolite	mounds discontinuous 3 km
Jade	immature back-arc	pull apart	1610 m	7.3	basalt–rhyolite/sediments	mounds and chimneys
Middle Valley	sedimented ridge	control by faults + sill + sediment lid	2500 m	6	basalt–sediment	75 × 35 × 95 m high
NESCA	sedimented ridge	volcanic centre + sediment lid	3300 m	2.2	sediment–basalt	several mounds (100 m long)
Atlantis 2	immature ocean	axial basin + Brines as a lid	2000 m	2	sediments	100 MT

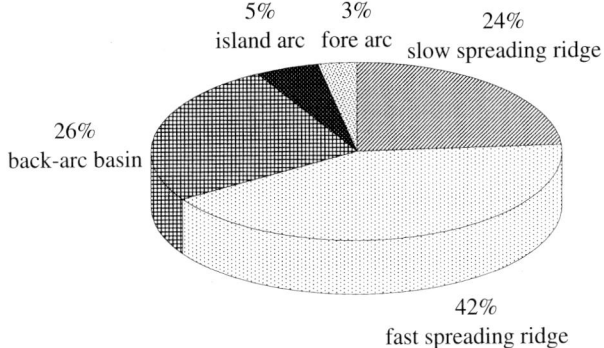

Figure 2. Relative abundance of identified hydrothermal fields in major geodynamic settings.

high. Typical locations at the topographic high include the Snake Pit (Fouquet et al. 1993b), Lucky Strike and Menez Gwen fields (Langmuir et al. 1996; Fouquet et al. 1996). The best example of a location at the base of the graben wall is the TAG field (Rona et al. 1993). Recent investigations at 14° 45′ N have demonstrated that the top of graben wall is also a potential site for high temperature venting (Krasnov et al. 1995). Recent investigations demonstrate that the Azores triple junctiun domain is hydrothermally more active than the rest of the ridge. At a *local scale* the control is an axial summit lenticular graben at Snake Pit (figure 3e) or a caldera (figure 3f) for Lucky Strike and possibly a discrete volcanic centre for TAG. Thus the local volcanic controls tends to be the opposite of the regional controls: for a regional volcanic control, the local control is tectonic and for a regional tectonic control the local control tends to be volcanic. Recent cruises in the Azores domain gave significant information on a possible third type of setting for hydrothermal activity on slow spreading ridges. Side scan sonar and plume particles indicate that non-transform offsets play a role in focusing hydrothermal flow (German et al. 1995) (figure 3d). Further submersible investigations are necessary in this environment, to determine if it is compatible with the formation of large deposits. The last type of setting is represented by stockwork like mineralization occurring within fracture zones (Bonatti et al. 1976).

(b) Fast spreading

On fast spreading ridges the regional control tends to be the topographic high between two major fracture zones (figure 3l) where the hydrothermal activity is more developed. This model was proposed by Francheteau & Ballard (1983) and demonstrated by Bougault et al. (1993) for the 13° N area on the EPR. However each segment between the major fractures has to be considered as independent for volcanic, tectonic and hydrothermal activity. At a local scale the style and location of activity depends on which stage the segment is at (figure 3h − −j). During the volcanic stage, vents are controlled by axial summit caldera and lava lakes while at the tectonic stage vents tend to be controlled by graben faults (figure 3j). In most cases the instability of these two-dimensional convective systems does not favour the formation of large sulphide deposits at the axis. However, off-axis seamounts (figure 3k) are more stable systems compatible with long lived three-dimensional convective cells that are more efficient for the formation of large sulphide deposits (Fouquet et al. 1996).

216 Y. Fouquet

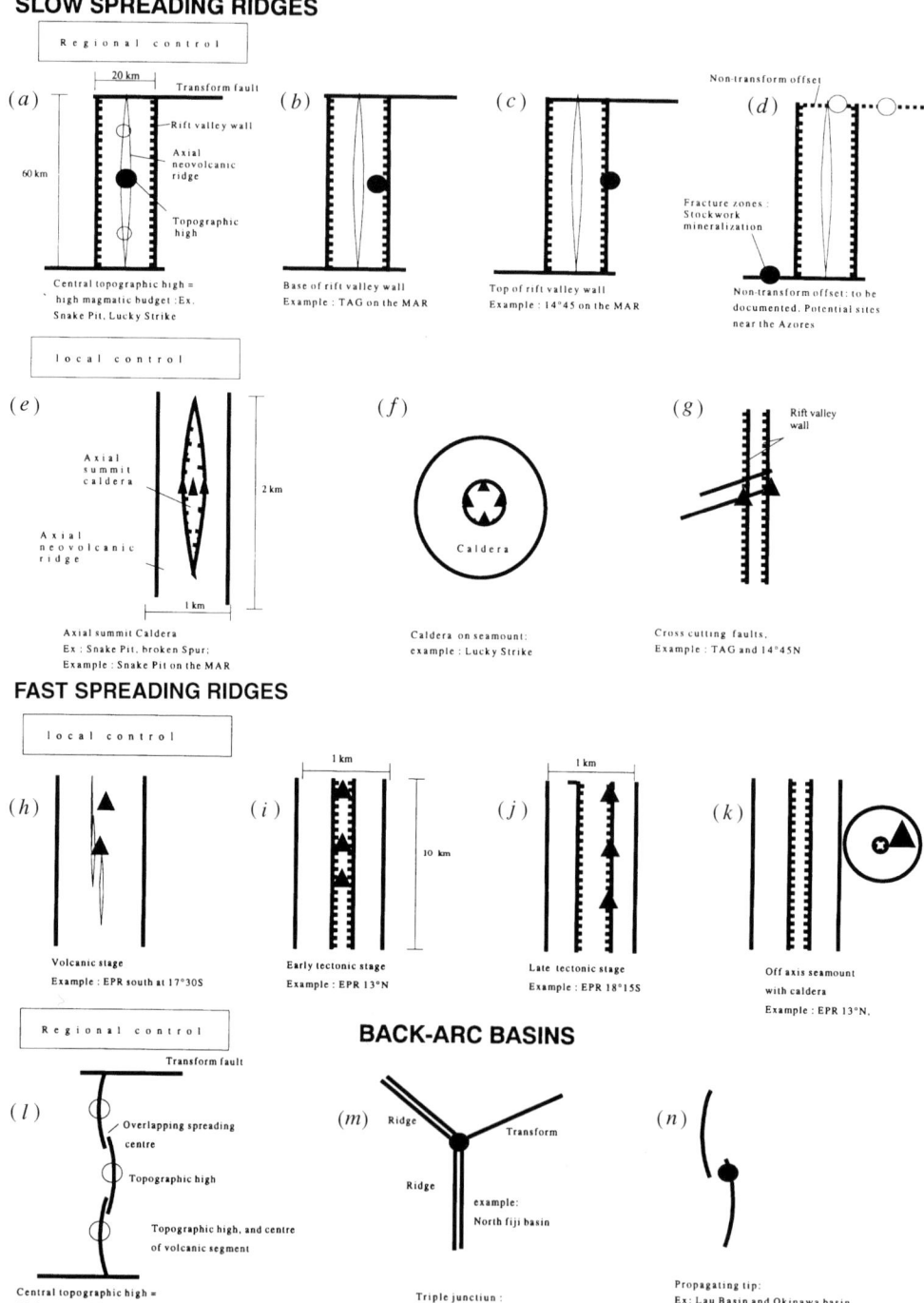

Figure 3. Geological controls of hydrothermal fields on spreading ridges.

(c) *Back-arc*

In back-arc environments hydrothermal processes are similar to that on mid-ocean ridges. However due to their instability and/or degree of maturity, some specific controls have to be considered. Typical locations on topographic highs are documented at several sites: North Fiji basin (Bendel *et al.* 1993), Lau Basin (Fouquet *et al.* 1993a), and Manus basin (Binns & Scott 1993). Seamounts, for example the Franklin seamount in the Woodlark intracontinental rifting (Binns *et al.* 1993b) are also good targets for hydrothermal deposits. The northern North Fiji basin is the best example of a large deposit controlled by a triple junction system where cross cutting faults play a major role in focusing the vents (figure 3m) (Bendel *et al.* 1993). In the Lau basin the most active and largest known sulphide deposit is close to an overlapping system along the spreading ridge (figure 3n). The Kuroko-type Okinawa deposit is controlled by a collapse system interpreted as a pull apart depression in the back-arc system (Halbach *et al.* 1993).

3. Geological, physical and chemical factors controlling morphology and size of deposits

Investigations on fossil deposits have shown that several factors can effect the morphology and size of the deposits (Large 1992). The same factors have to be considered in the modern ocean. In addition to the geological control of hydrothermal discharge, the formation of large sulphide deposits requires an efficient mechanism for precipitating and trapping the sulphide minerals from the fluid. In a typical black smoker chimney, 97% of the total amount of metals is dispersed in the ambient sea water due to mixing and thus rapid dilution of the fluid in the open ocean (Converse *et al.* 1984). Thus the formation of a large sulphide deposit needs to involve specific conditions resulting in a lowering of this percentage.

(a) Mixing

Restricted mixing within the mound or mixing of cold seawater with the ascending hydrothermal fluid will result in rapid precipitation of metallic sulphides and calcium and barium sulphates (sulphate being derived from ambient seawater) to produce the black smoker plume. Restricted mixing conditions within the chimneys are also necessary for the concentration of some elements such as gold (Herzig *et al.* 1993). At the scale of the mound, the presence of impermeable cap rocks are key factors to prevent rapid mixing and dilution of the fluid in ambient seawater. These cap rocks also play the role of a geochemical barrier. Hydrothermal solutions are conductively cooled and precipitate sulphides before their emission on the seafloor. Examples of this type are described at Lucky Strike and Lau Basin. Mixing of seawater and ascending hydrothermal fluid within the crust appears to be a process that can potentially form large deposits. This can happen in highly permeable rocks such as faulted grabens (figure 4Ib) or porous volcaniclastic rocks (figure 4Ic). However, further studies are necessary to better document the vertical extent of these types of deposits.

(b) Permeability

The type of permeability existing on the seafloor plays a major role on focusing the hydrothermal discharge which in turn is important to produce large deposits. In impermeable volcanic sequences, such as massive lava flows, significant fluid flow can only be achieved along major faults (figure 4Ia). A common situation is at axial ridges where water is focused along cracks. This situation has a high potential for production of large mound shaped deposits on slow spreading ridges where convective

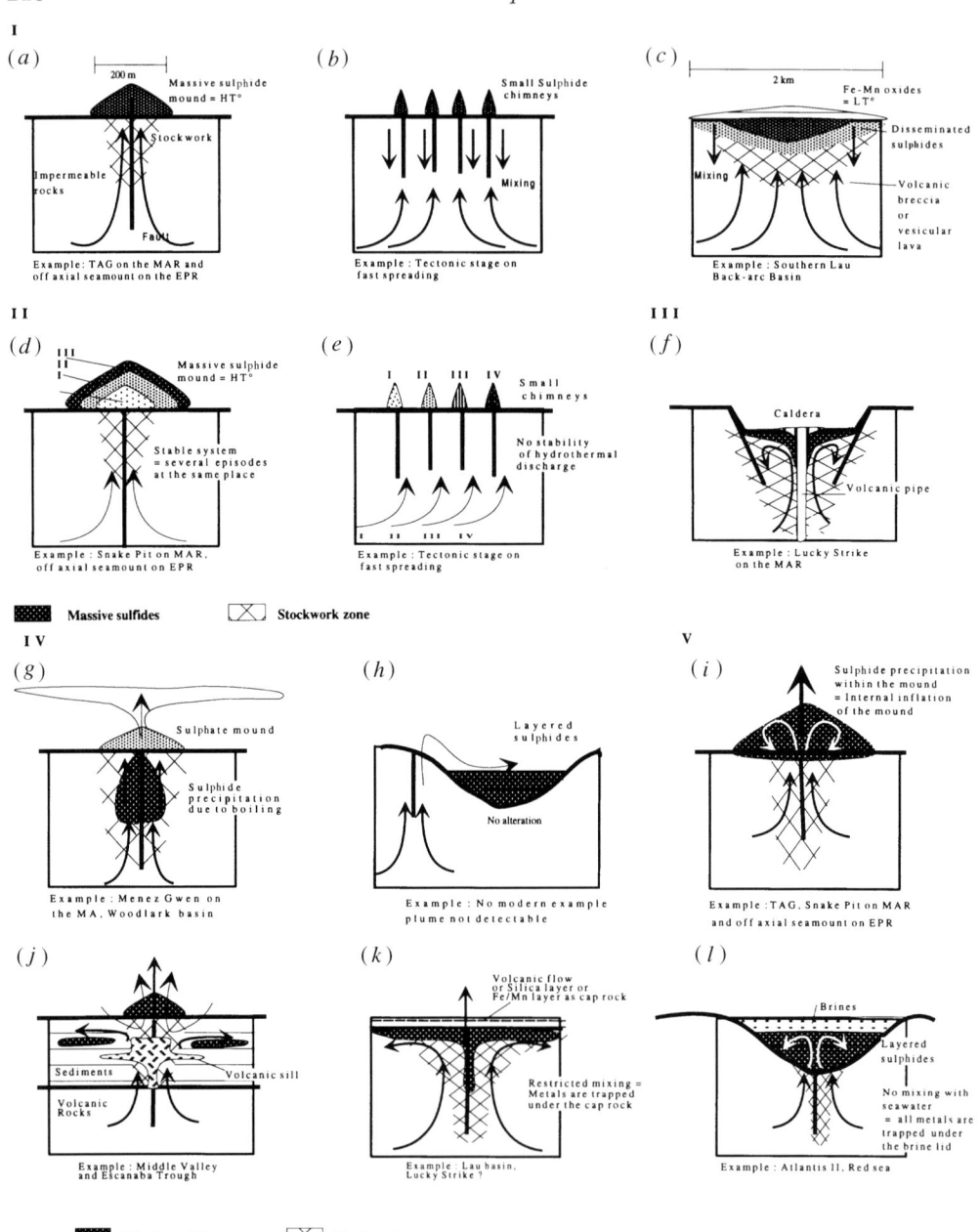

Figure 4. Factors controlling size and morphology of submarine hydrothermal sulphide deposits. (a), focused discharge; (b), diffuse discharge in tectonised lava; (c), diffuse discharge in permeable rocks; (d), stable system; (e), unstable system; (f), caldera; (g), venting of vapour phase = low salinity fluid; h, venting of brine = high salinity fluids; (i), trap is the mount itself; (j), trap is sediment cover; (k), trap is impermeable rock; (l), trap is a brine pool. I, permeability; II, stability of venting system; III, geometry of system; IV, water depth/phase separation; V, geological trap.

cells are stable. The permeability at the upper part of the convective system can increase for two reasons: (1) faulting and brecciation of lava and (2) permeability of volcanic rocks. At the end of the tectonic stage on fast spreading ridges, the crust is highly fissured (figure 4Ib) and highly permeable. This configuration gives numerous pathways both for hot ascending hydrothermal fluid and cold descending seawater. The mixing front is located within the crust and at the surface both hot fluids and diffuse low temperature discharge are observed. Deposits are numerous and very small because the upflow zone is relatively narrow and subject to perturbation by frequent tectonic and volcanic activity. This situation is well documented on the East Pacific Rise at 13° N and on the southern EPR. On slow spreading the system is less permeable and hydrothermal discharge better focused.

In permeable volcanic sequences (figure 4Ic), such as volcaniclastic material or highly vesicular volcanic rocks, hydrothermal flow is less focused. This situation is common in felsic volcanic environments where volcanic rocks are highly vesicular and brecciated. Few faults occur at the surface of the ridge and hydrothermal fluids are mixing with cold seawater within the permeable and porous volcanic rocks. The result is the formation of a low temperature Fe/Mn or Si crust at the seafloor. This crust can act as a lid on the system and allow the formation of massive sulphide by replacement of the pervasively altered volcanic rocks within the oceanic crust (Fouquet *et al.* 1993*a*). The morphology of the deposit is not a mound but occurs both as massive sulphide and disseminated mineralization within the crust.

(*c*) *Stability of hydrothermal system*

The depth and size of heat source and the stability of structure have a profound effect on the longevity of a vent area and therefore on the size of a deposit. The construction of a large mound implies the circulation of a large amount of hydrothermal fluid at the same location. On fast spreading ridges it has been observed that within a period of a few years the hydrothermal activity moves along the axis (figure 4IIe). This is not favourable for the generation of large sulphide deposits. The formation of first order sulphide mounds requires that the convective cells be in the same place for several successive hydrothermal episodes (figure 4IId). For example at TAG several hydrothermal episodes have been documented at the same place for a period of a more than 26 000 years during which at least five hydrothermal episodes are documented (Lalou *et al.* 1993). This is also the case on the Snake Pit and probably the Lucky Strike sites. Thus we see that more stable convective systems at slow spreading are more favourable to the formation of large deposits than are the highly unstable fast spreading hydrothermal systems. However, it was recently demonstrated that a similar stable configuration can happen on off-axial seamount close to the ridge on fast spreading ridges (Fouquet *et al.* 1996). A particular situation is the border of a caldera (figure 4IIIf). The particular morphology of the system produces extensive deposits, with a massive part having the shape of a lens with a relatively flat or concave surface contrasting with the typical conical shape of a mound. Many seamounts have a summit caldera, a structural feature common to the environment of formation of some ancient massive sulphide deposits (Ohmoto 1978). Calderas are areas of high heat flow and intense fracturing, two important requirements for the formation of large deposits.

(*d*) *Boiling: water depth*

A typical vent fluid at 350 °C at a water depth of 3000 m (figure 4IVa) is well below the boiling point for this pressure and will precipitate sulphide as a cooling product when the fluid reaches the seafloor. In shallow water boiling may occur. Boiling and separation of a steam phase leaves a residual liquid cooler, more saline and enriched in metals (due to partitioning of NaCl into the liquid phase) and depleted in H_2S (due to its partitioning into the vapour phase). This process may result in the formation of stockwork-dominated mineralization (figure 4IVa) as a network of veins within the crust, whereas at the surface only low temperature and metal depleted mineralization is formed. In deep-water environments metal deposition is concentrated at the seawater interface with limited stockwork development. After phase separation, the venting fluids may result in two types of smokers the first will be venting fluids of low salinity and high gas content. A typical example is seen at the shallow (800 m) Menez Gwen site on the MAR. The fluid has a low salinity, is enriched in gas and both fluid and mineral precipitates are depleted in metal content (Fouquet et al. 1996). The surface precipitates are dominated by barite and anhydrite. This means that metals are probably trapped deeper in the system. The discharge of the more dense fluid may occur as bottom-seeking fluids. For the moment there is no way to identify this type of discharge from conventional equipment used to detect the hydrothermal plumes, however several distal fossil deposits are considered to have been formed through this process (Scott 1985). On fast spreading ridges recent investigations showed that the fluids could have higher or lower salinity than seawater at 2600 m water depth. On tectonized ridges the highly saline fluids are enriched in metals, and may correspond to the venting of dense brines generated during an earlier phase separation along the segment. Conversely the low salinity fluids are related to recent lavas and thus to early phase separation enhanced during basaltic eruptions (Charlou et al. 1996). A similar example is also known in the Woodlark back-arc basin (Binns et al. 1993a) where extensive stockwork mineralization is inferred from the fluid composition.

(e) Geological trap/cap rocks

Different types of traps may enhance the accumulation of metals and therefore the efficiency of the system. The first trap is the mound itself (figure 4Vi). Old systems are sealed and enhance the focused discharge of deep fluids through discrete vents. As the mound is growing, mixing does not occur only in the open ocean as is the case for a typical smoker, but some restricted mixing occurs within the mound and allows a higher amount of metal to be precipitated. In other words, the mound acts as a cap on the hydrothermal system. This process was recently well documented when drilling through the TAG mound (Humphris et al. 1995).

Probably a more efficient system is the sediment cover on a ridge (figure 4Vj). It has been demonstrated for the Guaymas Basin (Bowers et al. 1985) that compared to a typical black smoker fluid resulting from the interaction of seawater with basalt, the metal contents are highly depleted in fluids from sedimented ridges. One possible explanation is that a significant amount of metal was lost during interaction of the end-member hydrothermal fluid with the sediment during the ascent of the fluid. This again allows metal precipitation in environments where mixing and thus rapid dilution by seawater is restricted during cooling of the fluid. The morphology of these deposits can be a mound at the surface but also sill like replacement levels within the sediments. However, the knowledge of the associated deposit requires further investigation in the vertical dimension. ODP drilling on a sulphide mound at Middle Valley showed that sulphide bodies in these environments are particularly thick (at least 90 m) (Mottl et al. 1994).

Another potential trap is an impermeable layer acting as a physical cap and chemical barrier on the hydrothermal system. This layer can be a silica, carbonate or sulphate layer, or a series of lava flows. Few examples are known in the modern ocean. At Lucky Strike on the MAR, a layer of SiO_2 clearly acts as a barrier to the ascending fluid and may enhance the formation of extensive subsurface sulphide precipitate (Fouquet et al. 1996). The occurrence of an impermeable cap rock (silica) on highly permeable rocks (volcanic and tectonic breccia) is a favourable configuration to form large deposits within the crust. A similar situation was seen in the Lau back-arc basin where massive sulphide deposits are actively forming under a Fe/Mn crust, which acts as a cap to the highly permeable volcaniclastic breccia on the ridge (Fouquet et al. 1993a). Again here, further investiagtions are needed to document processes occurring along the vertical section of the hydrothermal system. In all these cases the morphology of the deposits will not be a mound but a lenticular body within the crust. In addition these deposits have a high potential to be preserved because they are protected from rapid oxidation by direct contact with seawater.

4. Conclusions and new perspectives for hydrothermal exploration

After 15 years of exploration, various hydrothermal systems are now well known in the ocean. The formation of a large deposit is controlled by the combination of several conditions, each of which alone is often only compatible with the formation of small and unstable hydrothermal systems.

Recent discoveries considerably enlarge the potential locations of hydrothermal activity. On slow spreading ridges we have now to consider the base and top of graben wall and the non-transform offsets in addition to the relatively well documented control of the volcanic topographic high. On fast spreading ridges, off-axial seamounts are potential sites that have to be explored to determine their preferential control for the formation of large deposits. Back-arc basins have more complex tectonic histories that may enhance the combination of the several factors necessary for the formation of large deposits. They are likely to represent the closest equivalent to major massive sulphide deposits on land. According to these results future strategy to explore hydrothermal systems on mid-oceanic ridges needs to be revised.

In addition to the typical formation of a mound shaped deposit, there is now clear evidence that major sulphides deposits can be formed within the crust as stockwork mineralization related to boiling or precipitation of sulphides due to restricted mixing under a geological lid. A great deal of work is still to be done to document these deposits because drilling operations are necessary to describe or even to discover them.

Finally, a number of deposits cannot be identified using the conventional equipment used to detect hydrothermal fields. Most active fields are identified through their buoyant plume using vertical or dynamic hydrocasts. These techniques will not be efficient at detecting venting of dense brines that cannot rise to form a plume. These brines will accumulate in depressions with a minimum of mixing, preserving most of the metallogenic potential of the fluid.

References

Backer, H., Lange, J. & Marching, V. 1985 Hydrothermal activity and sulphide formation in axial valleys of the East Pacific Rise crest between 18° and 22° S. *Earth Planet Sci. Lett.* **72**, 9–22.

Bendel, V., Fouquet Y., Auzende, J. M., Lagabrielle, Y., Grimaud, D. & Urabe T. 1993 The white lady active hydrothermal field, North Fidji Back-Arc Basin, SW Pacific. *Econ. Geol.* **88**, 2037–2049,

Binns, R. A. & Scott, S. D. 1993 Actively-forming polymetallic sulphide deposits associated with felsic volcanic rocks in the eastern Manus back-arc basin, Papua New Guinea. *Econ. Geol.* **88**, 2226–2236.

Binns, R. A., Scott, S. D., Bogdanov, A., Listzin, A. P., Gordeev, V. V., Gurvich, E. G., Finlayson, E. J., Boyd, T., Dotter, L. E., Wheller, G. E. & Muravyev, G. 1993a Hydrothermal oxide and gold-rich sulphate deposits of Franklin Seamount, western Woodlark Basin, Papua, New Guinea. *Econ. Geol.* **88**, 2122–2153.

Binns, R. A., Scott, S. D., Bogdanov, Y. A., Listzin, A. P., Gordeev, V. V., Gurvich, E. G., Finlayson, E. J., Boyd, T., Dotter, L. E., Wheller, G. E. & Muravyev, K. G. 1993b Hydrothermal oxide and gold-rich sulphate deposits of Franklin seamounts, western Woodlark basin, Papua New Guinea. *Econ. Geol.* **88**, 2122–2153.

Bonatti, E., Guerstein-Honnorez, B. M. & Honnorez, J. 1976 Copper–iron sulphide mineralizations from the equatorial Mid-Atlantic Ridge. *Econ. Geol.* **71**, 1515–1525.

Bougault, H., Charlou, J. L., Fouquet, Y., Needham, H. D., Vaslet, N., Appriou, P., Jean Baptiste, P., Rona, P. A., Dmitriev, L. & Silantiev, S. 1993 Fast and low spreading ridges: structure and hydrothermal activity, ultramafic topographic highs, and CH_4 output. *Geophys. Res.* **98**, 9643–9651.

Bowers, T. S., Von Damm, K. L. & Edmond, J. M. 1985 Chemical evolution of mid-ocean ridge hot springs. *Geochim. Cosmochim. Acta* **49**, 2239–2252.

Charlou, J. L., Fouquet, Y., Donval, J. P., Auzende, J. M., Jean-Baptiste, P., Stievenard, M. & Michel, S. 1996 Mineral and gas chemistry of hydrothermal fluids on an ultra fast spreading ridge: east Pacific Rise, 17° to 19° S (Naudur Cruise 1993). Phase separation controlled by volcanic and tectonic activity. *J. Geophys. Res.* **101**, 15899–15919.

Converse, D. R., Holland, H. D. & Edmond, J. M. 1984 Flow rates in the axial hot springs of the East Pacific Rise (21° N): implications for the heat budget and the formation of massive sulphide deposits. *Earth Planet. Sci. Lett.* **69**, 159–175.

Corliss, J., Dymond, J., Gordon, L. I., Edmond, J., von Herzen, R. P., Richard, P., Ballard, R. D., Green, K., Williams, D., Bainbridge, A., Crane, K. & van Andel, T. H. 1979 Submarine thermal springs on the *Galapagos Rift. Science* **203**, 1073–1083.

Cyamex, , Francheteau, J., Needham, H. D., Choukroune, P., Juteau, T., Seguret, M., Ballard, R. D., Fox, P. J., Normark, W. R., Carranza, A., Cordoba, D., Guerrero, J. & Rangin, C. 1979 Massive deep-sea sulfide ore deposits discovered on the East Pacific Rise. *Nature* **277**, 523–528.

Embley, R. W., Jonasson, I. R., Perfit, M. R., Franklin, J. M., Tivey, M. A., Malahoff, A., Smith, M. F. & Francis, T. J. G. 1988 Submersible investigation of an extinct hydrothermal system on the Galapagos ridge: sulphide mounds, stockwork zone and differential lavas. *Canadian Mineralogist* **26**, 517–539.

Fouquet, Y., Auclair, G., Cambon, P. & Etoubleau, J. 1988 Geological setting and mineralogical and geochemical investigations on sulphide deposits near 13° N on the east pacific rise. *Marine Geology* **84**, 145–178.

Fouquet, Y., Auzende, J. M., Ballu, V., Batiza, R., Bideau, D., Cormier, M. H., Geistdoerfer, P., Lagabrielle, Y., Sinton, J. & Spadea, P. 1994 Variabilité des manifestations hydrothermales actuelles le long d'une dorsale ultra-rapide: exemple de la dorsale Est Pacifique entre 17°S et 19° S (campagne NAUDUR). *C.R. Acad. Sci., Paris* **319**, 1399–1406.

Fouquet, Y., Knott, R., Cambon, P., Fallick, A., Rickard, D. & Desbruyères, D. 1995 Formation of large sulphide mineralizations along fast spreading ridges. Example from off axial deposits at 12° 43′ N on the East Pacific Rise. *Earth Planet. Sci. Lett.* **144**, 147–162.

Fouquet, Y., Von Stackelberg, U., Charlou, J. L., Erzinger, J., Herzig, P. M., Mühe, R. & Wiedicke, M. 1993a Metallogenesis in back-arc environments—The Lau basin example. *Econ. Geol.* **88**, 2150–2177.

Fouquet, Y., Von Stakelberg, U., Charlou, J. L., Donval, J. P., Erzinger, J., Foucher, J. P., Herzig, P., Mühe, R., Soakai, S. M. W. & Whitechurch, H. 1991 Hydrothermal activity and metallogenesis in the Lau Basin. *Nature* **349**, 778–781.

Fouquet, Y., Wafik, A., Mevel, C., Cambon, P., Meyer, G. & Gente, P. 1993b Tectonic setting, mineralogical and geochemical zonation in the Snake Pit sulphide deposit. *Econ. Geol.* **88**, 2018–2036.

Fouquet, Y. *et al.* 1997 Hydrothermal processes on shallow volcanic segments: MAR near the Azores triple junction. *J. Geophys. Res.* Submitted.

Francheteau, J. & Ballard, R.D., 1983 The East Pacific Rise near 21° N, 13° N and 20° S: inferences for along strike variability of axial processes of the Mid-Ocean Ridge. Earth Planet. Sci. Lett. **64**, 93–116.

Francheteau, J., Needham, H. D., Choukroune, P., Juteau, T., Seguret, M., Normark, W., Ballard, R., Fox, P., Carranza, A., Cordoba, D., Guerrero, J., Rangin, C., Bougault, H., Cambon, P. & Hékinian, R. 1979 Massive deep-sea sulphide ore deposits discovered on the East Pacific Rise. *Nature* **277**, 523–528.

German, C. R. *et al.* 1995 Hydrothermal exploration at the Azores Triple Junction: tectonic control of venting at slow-spreading ridges? EPSL **138**, 93–104.

Goodfellow, W. D. & Franklin, J. M., 1993 Geology, mineralogy, and geochemistry of sediment-hosted clastic massive sulphides in shallow cores, Middle Valley, northern Juan de Fuca Ridge. *Econ. Geol.* **88**, 2033-2064.

Halbach, P. *et al.* 1989 Probable modern analogue of Kuroko-type massive sulphide deposits in the Okinawa back-arc basin. *Nature* **338**, 496–499.

Halbach, P., Pracejus, B. & Märten, A. 1993 Geology and mineralogy of massive sulphide ores from the Central Okinawa trough Japan. *Econ. Geol.* **88**, 2206–2221.

Herzig, P., Hannington, M., McInnes, B., Stoffers, P., Villinger, H., Seifert, R., Binns, R. & Liebe, T., 1994. Submarine volcanism and hydrothermal venting studied in Papua, New Guinea. Eos, Wash **75**, 513–516.

Herzig, P. M. D. H., Fouquet, Y., Stackleberg, U. & Petersen, S. 1993 Gold-rich polymetallic sulphides from the Lau Back-Arc and implications for the geochemistry of gold in seafloor hydrothermal systems of the SW Pacific. *Econ. Geol.* **88**, 2182–2209.

Honnorez, J., Mével, C. & Honnorez-Guerstein, B. M. 1990 Mineralogy and chemistry of sulphide deposits drilled from hydrothermal mound of the Snake Pit active field. *MAR, Ocean Drilling Program, Scientific Results*, 145–162.

Humpris, S. E., Herzig, P. M., Miller, D. J., Alt, J. C., Becker, K., Brown, D., Brügman, G., Chiba, H., Fouquet, Y., Gemmel, B., Guerin, G., Hannington, M., Holm, N. G., Honnorez, J., Iturrino, G. J., Knott, R., Ludwig, R., Nakamura, K., Peterszen, S., Reysenbach, A.L., Rona, P. A., Smith, S., Struz, A. A., Tivey, M. K. & Zhao, X. 1995. The internal structure of an active seafloor massive sulphide deposit. *Nature* **377**, 713–716.

Krasnov, S. G. *et al.* 1995. Detailed geological studies of hydrothermal fields in the North Atlantic. *Geological Society Special publication, Hydrothermal vents and processes* **87**, 43–64.

Koski, R. A., Shanks, W. C., Bohrson, W. A. & Oscarson, R. L. 1988 The composition of massive sulphide deposits from the sediment-covered floor of Escanaba Trough, Gorda Ridge: implications for depositional processes. *Can. Min.* **26**, 655–673.

Lalou, C., Reyss, J. L., Brichet, E., Arnold, M., Thompson, G., Fouquet, Y. & Rona, P. A. 1993 New age data for Mid-Atlantic Ridge hydrothermal sites: TAG and Snakepit chronology revisited. *J. Geophys. Res.* B **98**, 9705–9713.

Langmuir, C. H. *et al.* 1997 Description and significance of hydrothermal vents near a mantle Hot Spot: the Lucky Strike vent field at 37° N on the Mid Atlantic Ridge. *J. Geophys. Res.* (In the press.)

Large, R. L. 1992 Australian volcanic-hosted massive sulphide deposits: features, styles and genetic models. *Econ. Geol.* **87**, 471–510.

Mottl, J. M., Wheat, C. G. & Boulègue, J. 1994 Timing of ore deposition and sill intrusion at site 856: evidence from stratigraphy, alteration, and sediment pore water composition. *Proc. Ocean Drilling Program, Scientific Results* **139**, 679–693.

Ohmoto, H. 1978 Submarine calderas: a key to the formation of volcanogenic massive sulphide deposits? *Min. Geol.* **28**, 219–231.

Peter, J. M. & Scott, S. D. 1988 Mineralogy, composition and fluid-inclusion microthermometry of seafloor hydrothermal deposits in the southern trough of Guaymas basin, Gulf of California. *Can. Min.* **26**, 567–587.

Renard, V., Hekinian, R., Francheteau, J., Ballard, R. D. & Backer, H. 1985 Submersible observations at the axis of the ultra-fast-spreading East Pacific Rise (17°30′ to 21°30′ S). *Earth Planet. Sci. Lett.* **75**, 339–353.

Rona, P. & Scott, S. D. 1993 A special issue on seafloor hydrothermal mineralization: new perspectives. *Preface. Economic Geology* **88**, 1935–1975.

Rona, P. A. 1988 Hydrothermal mineralization at oceanic ridges. *Can. Min.* **26**, 431–465.

Rona, P. A., Hannington, M. D., Raman, C. V., Thompson, G., Tivey, M. K., Humpris, S. E., Lalou, C. & Petersen, S. 1993 Active and relict sea-floor hydrothermal mineralization at the TAG hydrothermal field, Mid-Atlantic Ridge. *Econ. Geol.* **88**, 1989–2017.

Rona, P. A., Klinhkammer, G., Nelsen, T. A., Trefry, J. H. & Elderfield, H. 1986 Black smokers, massive sulphides and vent biota at the Mid-Atlantic Ridge. *Nature* **321**, 33–37.

Scott, S. D. 1985 Seafloor polymetallic sulphide deposits: modern and ancient. *Mar. Mining* **2**, 191–212.

Scott, S. D. 1987 Seafloor polymetallic sulphides: scientific curiosities or mines of the future? *Proceedings, NATO advanced research workshop on marine mineral resources assessment strategies* **194**, 277–300.

Scott, S. D. 1992 Polymetallic sulphide riches from the deep: fact or fallacy? In *Use and misuse of the seafloor* (ed. K. J. Hsü & J. Thiede), pp. 87–115. New York: Wiley-Interscience.

Spiess, F. N. *et al.* 1980 East Pacific Rise: hot springs and geophysical experiments. *Science* **207**, 1421–1433.

Thompson, G., Humphris, S. E., Schroeder, B., Sulanowska, M. & Rona, P. A. 1988 Active vents and massive sulphides at 26° N (TAG) and 23° N (Snake Pit) on the Mid-Atlantic Ridge. *Can. Min.* **26**, 697–711.

Urabe, T., Yuasa, M., Nakao, S. & co workers 1987 Hydrothermal sulphides from a submarine caldera in the Shichito-Iwojima Ridge, northwestern Pacific. *Marine Geology* **74**, 295–299.

Sea water entrainment and fluid evolution within the TAG hydrothermal mound: evidence from analyses of anhydrite

By R. A. Mills[1] and M. K. Tivey[2]

[1] Department of Oceanography, Southampton Oceanography Centre, Southampton, SO14 3ZH, UK
[2] Woods Hole Oceanographic Institution, Woods Hole, MA 02543, USA

Entrainment of sea water into the active deposit has now been recognized as an important control on the structure and composition of the Trans-Atlantic Geotraverse (TAG) active hydrothermal mound. One key manifestation of sea water–fluid interactions is the precipitation of anhydrite from heated mixtures of sea water and hydrothermal fluids. Submersible and drilling recovery of anhydrite from the TAG active hydrothermal mound has generated a wide range of samples from depths of up to 120 m below seafloor (mbsf). Information on trapping temperatures and salinities from analyses of fluid inclusions in anhydrite crystals and the Sr, Ca, Mg and Sr isotopic compositions of anhydrite samples are used here as tracers of entrainment, fluid mixing and evolution during mound circulation. Trapping temperatures in samples from the central and southeast quadrant of the active mound sites are all high, with many higher than the current measured temperature of fluids venting from black and white smokers. There is also a trend of increasing temperature with depth in both of these areas, with cooler temperatures near the mound surface close to the highest measured temperatures of fluids venting from the mound. Salinities of fluids in all inclusions analysed are consistent with phase separation at depth below the seafloor. Mg partitioning into anhydrite is minimal due to mismatch in cationic radius, though high apparent partition coefficients for Mg are observed in surface anhydrite samples which are inferred to be caused by the presence of a fine-grained Mg bearing phase (e.g. talc). The inferred Sr partition coefficient for anhydrite precipitation within the TAG mound is < 1, which results in fluid evolution to high Sr/Ca ratios with ongoing circulation and precipitation and explains the observed variability in Sr/Ca ratios of individual anhydrite crystals. By considering rates of fluid flow and amounts of subsurface precipitation, it is estimated that the existing $\sim 2 \times 10^4$ m^3 of anhydrite present within the TAG active mound could have been deposited in 80 to 800 yr. The convective process of entraining and conductively heating sea water may be responsible for cooling black smoker fluids from > 380 °C to the temperature of 366 °C currently measured in orifices of chimneys that comprise the black smoker complex. The occurrence and geochemistry of anhydrite delineates areas of high permeability and extensive mixing in the upper mound, with entrainment and heating of sea water with limited mixing with rising fluids occurring at depth in the mound.

1. Introduction

It has long been speculated that sea water may be being entrained locally at active seafloor vent sites, and that such entrainment may play a significant role in the formation of massive sulphide deposits (e.g. Janecky & Shanks, 1988; Schultz et al. 1992). Results of recent submersible and drilling studies at the Trans-Atlantic Geotraverse (TAG) active hydrothermal mound on the Mid-Atlantic Ridge (MAR), 26° N, have provided dramatic evidence for this process. Blocks of massive anhydrite have been recovered from the surface of the mound (Thompson et al. 1988; Tivey et al. 1995), and significant amounts of anhydrite were recovered from within the mound during drilling on Ocean Drilling Program (ODP) Leg 158 (Humphris et al. 1995; Shipboard Scientific Party, 1996). In addition, fluids exiting the surface of the mound have been sampled and analysed (Campbell et al. 1988; Edmond et al. 1995; Charlou et al. 1996; Edmonds et al. 1996; Gamo et al. 1996; James & Elderfield, 1996; Mills et al. 1996). Differences in compositions of two distinct high temperature fluids exiting the mound (black and white smoker fluids) can be explained by mixing of hydrothermal fluid with sea water, conductive cooling and precipitation within the mound (Edmond et al. 1995; Tivey et al. 1995). Likewise, comparison of black smoker fluid compositions with compositions of lower temperature diffuse fluids provides evidence for deposition of anhydrite, pyrite and silica within the mound, that is, concentrations of Ca, Si, Fe and H_2S are less than would be predicted from conservative mixtures of sea water and hydrothermal fluid (James & Elderfield, 1996). Pore fluids collected from the mound periphery also reflect this internal mixing and fluid evolution (Mills et al. 1996). The presence at TAG of significant anhydrite reflects the importance of sea water entrainment (Humphris et al. 1995), and the variability in composition of fluids currently venting from the mound demonstrates that complex processes of precipitation and remobilization are occurring as a result of this entrainment.

The TAG active hydrothermal mound is located at a depth of 3670 metres on 100 000 year old crust (based on spreading rates) near the foot of the eastern wall of the median valley within the larger TAG hydrothermal field (Rona et al. 1993a). The geological setting is described in detail by Rona et al. (1993a) and Kleinrock & Humphris (1996). The active mound measures roughly 150–200 meters in diameter, and rises ~ 50 metres above the surrounding seafloor. The mound surface is covered entirely by hydrothermal precipitates. Active venting from the mound includes strongly focused black smoker activity localized northwest of the apex of the mound at the Black Smoker Complex (BSC), and extensive diffuse flow of lower temperature fluids over patchy areas of the top and sides of the mound, and through the apron surrounding the mound (Thompson et al. 1988; Tivey et al. 1995; BRAVEX, 1995). From 1986 to 1990, focused lower temperature white smoker (273–301 °C) fluids emanated from many small (1–2 m) bulbous chimneys in the southeastern quadrant of the mound known as the 'Kremlin' area (Thompson et al. 1988; Edmond et al. 1995; Tivey et al. 1995). Dive observations from 1993 to 1996 suggest that this white smoker venting activity has ceased, and black smoker fluids have been observed venting from bulbous, anhydrite-dominated chimneys in the northeast quadrant of the mound (BRAVEX, 1995). Samples recovered from the surface of the mound include Cu–Fe sulphide-rich black smoker chimneys, Cu–Fe and Fe-sulphide rich 'crusts', and blocks of massive intermixed pyrite and anhydrite from the BSC, Zn-rich white smoker chimneys from the Kremlin area, and Zn- and/or Fe-rich massive sulphides and amorphous Fe-oxyhydroxides from the rest of the upper surface of the mound;

blocks of Fe-rich massive sulphide and ochreous material are exposed on the slopes of the mound (Thompson et al. 1988; Tivey et al. 1995). The range of sample types and their distribution and chemistry have been used to deduce patterns of fluid flow and mixing within the mound (Tivey et al. 1995). The proposed model involves entrainment of sea water into the mound at two levels: into the sides of the BSC above ~ 3650 m below sea level (mbsl), and into the main portion of the mound below ~ 3650 mbsl resulting in mixing of sea water and hydrothermal fluid, precipitation of minerals, generation of a more acidic fluid and subsequent metal remobilization (Tivey et al. 1995).

Geochemical modelling calculations that considered formation of white smoker fluids from black smoker fluids corroborated these interpretations, and predicted deposition from mixing within the mound of ~ 19 parts anhydrite to 8 parts pyrite to 1 part chalcopyrite (Tivey et al. 1995). Measurements of conductive heat flow over the mound surface provided further support for this proposed model. At the BSC the conductive heat flow values are extremely variable, very high heat flow values are observed on the south and southeast sides of the mound, and in sediments on the seafloor surrounding the mound, and a band of very low heat flow 20 to 50 m west of the BSC suggests an area of shallow recharge (Becker & Von Herzen, 1996; Becker et al. 1996).

In 1994 the Ocean Drilling Program drilled 17 holes into the TAG hydrothermal mound to reveal its internal structure. Material recovered from all holes provides evidence that the deposit is dominantly a breccia pile, with current hydrothermal activity forming precipitates both at the surface and within the mound (Humphris et al. 1995). Not surprisingly, one of the significant findings was that there is abundant anhydrite present within the mound. Six major zones were identified (figure 1): (I) an upper zone (down to 10 to 20 mbsf) of clast-supported sulphide breccias, (II) an anhydrite-rich zone composed of matrix-supported pyrite–anhydrite breccias (down to ~ 30 mbsf) underlain by (III) pyrite–silica–anhydrite breccias (down to ~ 45 mbsf) with anhydrite veining common (composite veins are up to 45 cm thick), (IV) a zone of silicified and brecciated wallrock (down to ~ 100 mbsf) composed of pyrite–silica breccias underlain by (V) silicified wallrock breccias and (VI) a zone of chloritized basalt breccia (at depths > 100 mbsf) composed of altered basalt fragments cemented by quartz and pyrite and cross-cut by veins of pyrite and/or quartz. In the TAG-1 area, located ~ 20 m southeast of the BSC, the TAG-2 area, located on the southeast edge of the mound near the Kremlin area, and the TAG-5 area, located 20 to 30 m north–northeast of the BSC, anhydrite was present in core recovered from all depths down to the base of the deepest holes (125 mbsf at TAG-1, 54.3 mbsf at TAG-2 and 59.4 mbsf at TAG-5; Shipboard Scientific Party, 1996). The anhydrite grains are present in vugs, in veins up to 45 cm thick, as matrix in sulphide–silicate–sulphate breccias and on broken surfaces of pyrite–silica, siliceous wallrock and chloritized basalt breccias. Anhydrite was not recovered from either the TAG-3 area, located ~ 55 m south of the BSC, or from the TAG-4 area, located ~ 5 to 25 m west of the BSC.

The importance of the abundance and distribution of anhydrite within the TAG mound, and within seafloor hydrothermal systems in general, stems from its retrograde solubility (it becomes less soluble with increasing temperature, and in sea water is undersaturated at temperatures less than ~ 150 °C). Sulphate, present in sea water, is virtually absent in hydrothermal fluids (Von Damm 1990 and references therein). The δ^{34}S compositions of anhydrite from a number of different vent sites

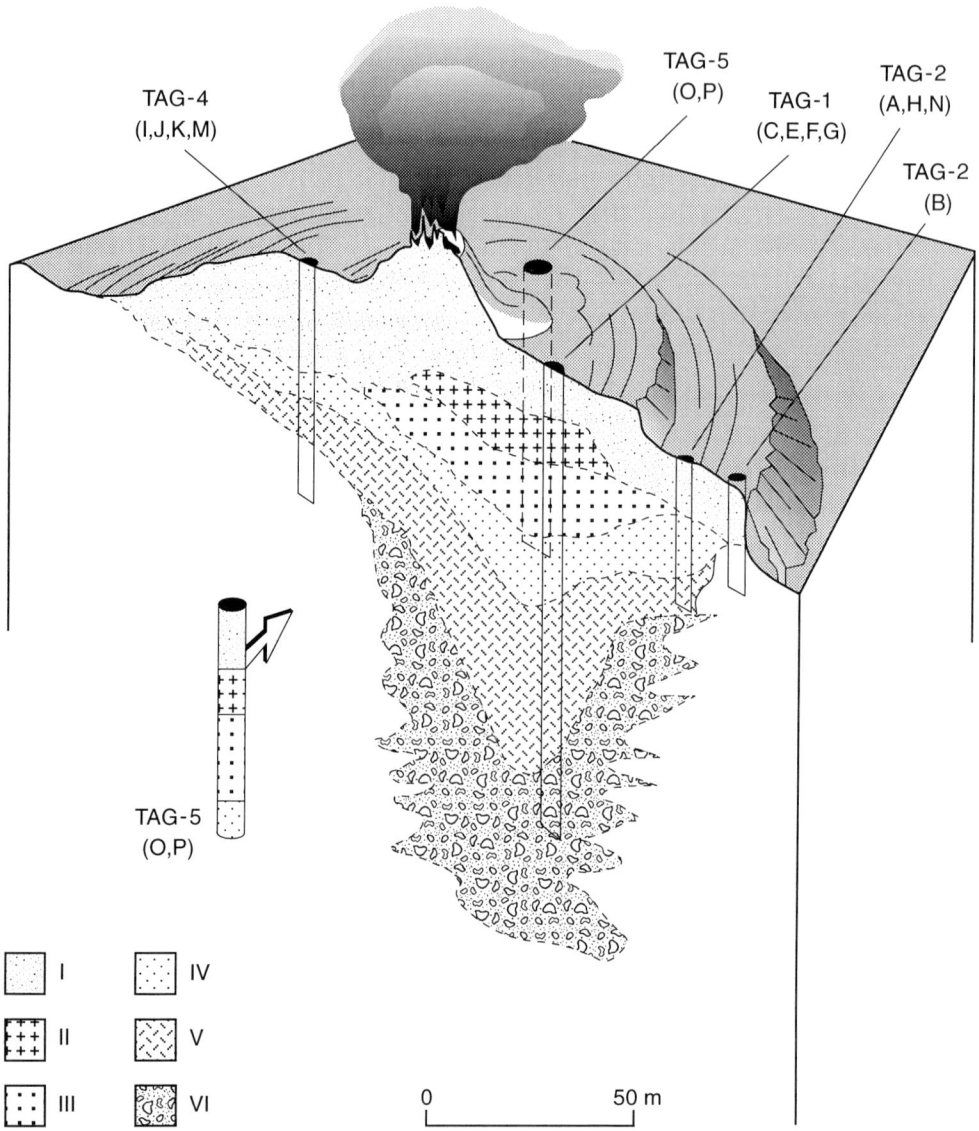

Figure 1. Lithology of the TAG active mound showing the distribution of black smoker venting, a simplified internal structure, and the drilling areas TAG-1, TAG-2 and TAG-5 host anhydrite sampled in this study. Six brecciated lithological units have been identified and are (I) massive pyrite, (II) pyrite–anhydrite, (III) pyrite–silica–anhydrite, (IV) pyrite–silica, (V) silicified wall-rock and (VI) chloritized basalt. Units I, II and III comprise the pyrite-dominated mound unit, IV, V and VI comprise the stockwork. Adapted from Humphris et al. (1995).

are all close to that of sea water suggesting that SO_4^{2-} in anhydrite is solely derived from sea water, and not from oxidation of reduced sulphur present in hydrothermal fluids (e.g. Styrt et al. 1981; Kusakabe et al. 1982; Kerridge et al. 1983; Chiba et al. 1998). In contrast, Ca and Sr are present in both hydrothermal fluids (30.8 mmol/kg and 103 µmol/kg respectively at TAG) and sea water (10.3 mmol/kg and 88 µmol/kg respectively; Elderfield et al. 1993; Edmond et al. 1995). Analyses of Sr isotopic compositions of anhydrite grains have been used to demonstrate that the Ca and Sr compositions of vent deposit anhydrite reflect mixing between hydrothermal fluids and sea water (e.g. Albarede et al. 1981). If there is no fractionation of Sr isotopes during mineralization, the Sr isotopic composition of any fluid or mineral will record the extent of mixing of high temperature non-radiogenic hydrothermal fluid with radiogenic sea water (Mills & Elderfield, 1995; James & Elderfield, 1996; Mills et al. 1996).

The occurrence and composition of anhydrite within hydrothermal systems can thus be used to trace processes associated with sea water entrainment and mixing of hydrothermal fluids and sea water. In addition, fluid inclusions in individual anhydrite grains can be examined to yield information on the temperature and salinity of the fluid mixtures that formed the anhydrite crystals. We present here results of a comprehensive study of anhydrite samples recovered both from within and from the surface of the TAG active hydrothermal mound to examine the processes and consequences of sea water entrainment, mixing, precipitation and fluid evolution within the mound. Microthermometric analyses of fluid inclusions in anhydrite provide information on the temperatures and salinities of fluids that formed the anhydrite, and Sr isotope analyses of the same samples allow determination of the proportions of hydrothermal fluid and sea water comprising these fluids. These analyses, coupled with data on the relative proportions of Sr, Ca and Mg in the anhydrite grains, allow reconstruction of fluid evolution history within the TAG mound. Precipitation and dissolution of anhydrite through time has the potential to extensively modify the structure of the mound, the composition of the solid phases and the evolved hydrothermal fluids present within the mound and exiting the surface of the mound. Such behaviour is not only important within modern seafloor deposits, but has implications for on-land analogues such as Cyprus-type ore bodies (Constantinou & Govett, 1973).

2. Sampling and methodology

A total of 31 samples of anhydrite, from the surface and from a range of depths within the mound, were analysed. Surface samples were recovered using the Alvin and Mir submersibles in 1986, 1990 and 1991 (Thompson et al. 1988; Rona et al. 1993b; Tivey et al. 1995). All subsurface material was recovered from drilling during ODP Leg 158. Individual 0.5 to 3 mm long tabular to acicular anhydrite crystals were picked from massive surface samples and from broken surfaces, matrix, vugs and veins in drillcore samples (tables 1 and 2).

Fluid inclusion analyses on individual anhydrite crystals were carried out using a Fluid Inc. adapted US Geological Survey gas-flow heating and freezing stage, following the procedures of Roedder (1984). Synthetic fluid inclusion samples were used to standardize the stage by measuring phase transitions at -56.6, 0.0 and $+374.1\,°C$. Progressive heating and cycling were used to avoid and test for leaking of inclusions (e.g. Kelley et al. 1993). Temperature measurements during heating and freezing were reproducible to $\pm 0.5\,°C$ and $\pm 0.1\,°C$, respectively. All measurements were made along single arrays or on clusters of primary inclusions in which all inclusions displayed similar phase ratios, and homogenization temperatures were within 5–10 °C of adjacent inclusions. Freezing point depression temperatures were used to calculate salinities (expressed as wt% NaCl equivalent) using data and equations from Bodnar (1993). Trapping temperatures were calculated from homogenization temperatures using the equation of Zhang & Frantz (1987) to correct for effects of pressure, assuming hydrostatic conditions.

Sr, Mg and Ca concentrations in solutions made from dissolved anhydrite crystals were determined by inductively coupled plasma atomic emission spectrometry and have a reproducibility of better than 1%. Sample preparation included picking ~ 20 mg of anhydrite, crushing crystals in agate with 1 ml of quartz distilled (QD) water and diluting the resultant solution to 2–3 ml with QD H_2O. Dissolution was brought to completion by adding a drop of 6N HCl and leaving the mixture on a hot plate in a capped tetrafluoroethylene Savillex beaker.

Sr separation for isotopic determination was carried out at the University of Cambridge by cation exchange. Samples were loaded in acidified QD water and eluted with 1.75M HCl, procedural blanks were < 200 pg for Sr. Sr isotopic compositions ($^{87}Sr/^{86}Sr$) were determined by thermal ionization mass spectrometry at the Southampton Oceanography Centre by loading eluted samples on single Ta filaments with TaF_5 in phosphoric acid and were analysed using a Fisons Sector 54 multicollector by multidynamic peak switching. External precision for these measurements is better than 2.5×10^{-5} (2σ standard error (SE)).

Geochemical calculations of mixing and heating of hydrothermal fluid and sea water were carried out using the computer code MINEQL (Westall et al. 1976), modified for use at high temperature (Tivey & McDuff, 1990), using the SUPCRT92 database (Johnson et al. 1992). Database modifications included addition of data for HCl from Sverjensky et al. (1991), CuCl from Ding & Seyfried (1992a), and $FeCl_2$ from Ding & Seyfried (1992b).

3. Results

(a) Fluid inclusions

Fluid inclusions in anhydrite crystals, rectangular to circular in shape, are two-phase, liquid–vapour inclusions, ranging in size from 3 to 30 µm. Primary inclusions occur in abundance parallel to crystal faces. Thin, tubular inclusions were avoided owing to concerns about leakage of fluid inclusions and exchange. Secondary inclusions are less common, range in size from 4 to 25 µm, occur along healed microfractures, and may be pseudosecondary inclusions, formed during fracturing and concomitant growth of the crystal (Roedder, 1984). In crystals from the TAG-1 and TAG-2 areas, inclusions are abundant. In contrast, in anhydrite crystals from the TAG-5 area there are fewer intact fluid inclusions, and a much greater abundance of what appear to be decrepitated, empty inclusions (Tivey et al. 1998).

Fluid inclusion homogenization temperatures (T_h) in all samples range from 168 to 361 °C, corresponding to trapping temperatures (T_t) of 187 to 388 °C (table 1). For the same sample, homogenization and trapping temperatures in secondary inclusions are similar to those for primary inclusions, and the range in T_t is generally less than ~ 20 °C. In samples from the TAG-1 and TAG-2 areas (except at 9.1 mbsf in the TAG-2 area), the range in trapping temperatures is small (338 to 388 °C). Trapping temperatures in inclusions from the shallowest hole in the TAG-2 area are lower, and range from 280 to 313 °C, within 12 °C of measured white smoker fluid temperatures (273 to 301 °C; Edmond et al. 1995). Similarly, maximum trapping temperatures in samples from near the top of the mound in the TAG-1 area are about 10 °C less than the highest measured temperature of black smoker fluid venting from the mound (366 °C; Edmond et al. 1995). Similarities between vent fluid and

Table 1. Sample descriptions and fluid inclusion data for TAG anhydrite samples

sample	depth (mbsf)	sample description	total number of measurements	T_h (°C range)	$-T_m$ (°C range)	T_t (°C range)[a]	NaCl (wt% equiv)[b]
Central complex – TAG-1			103				
ALV 1677-2	0	Massive anhydrite	10	311–326	1.9–2.2	338–353	3.2–3.7
158-957C-7N-1	21	Vein in nodular siliceous pyrite–anhydrite breccia with py	14	311–329	1.7–2.6	338–356	2.9–4.3
158-957C-11N-1	32	Banded vein with some hm	1	336	2.1	363	3.5
158-957C-11N-3 Vein	35	Vein in pyrite–silica breccia with cp	10	331–340	0.9–2.8	358–368	1.6–4.6
158-957C-11N-3 Interstitial	35	Vein in pyrite–silica breccia with py	12	334–342	0.7–2.5	362–369	1.2–4.2
158-957C-15N-2	44	Vein in silicified wallrock breccia	18	348–352	2.4–2.9	375–379	4.0–4.8
158-957C-15N-4	47	Thin vein against surface of nodular pyrite–silica breccia with py	10	340–348	1.5–2.5	368–375	2.6–4.2
158-957E-12R-1	92	Broken surface of silicified wallrock breccia	14	349–359	1.9–2.8	376–386	3.2–4.6
158-957E-17R-1	116	Broken surface of chloritized basalt breccia	8	336–354	1.5–3.1	365–381	2.6–5.1
158-957E-18R-1	121	Vein in chloritized basalt breccia, with py	6	354–361	1.8–2.6	382–388	3.1–4.3
Kremlin area – TAG-2			46				
158-957H-1N-1	9.1	Vein and matrix of cp-rich porous massive pyrite	7	255–287	1.5–2.1	280–313	2.6–3.5
158-957H-3N-1	18	Matrix of porous nodular pyrite breccia	6	315–327	1.7–2.8	342–354	2.9–4.6
158-957H-5N-2	28	0.5 cm wide vein on side of py clast from silicifed wallrock breccia	10	320–328	2.2–2.7	347–355	3.7–4.5
158-957H-5N-2-79-84 matrix	28	Matrix in silicified wallrock breccia	5	321–331	0.8–2.6	349–358	1.4–4.3
158-957H-5N-2-79-84 vug	28	Large vug around silica clast in silicified wallrock breccia	8	326–332	1.5–2.5	353–359	2.6–4.2
158-957H-9X-1	45	Broken surface of massive pyrite breccia with cp and py	10	326–342	2.1–3.1	354–369	3.5–5.1
Northern area – TAG-5			27				
158-957O-4R-1	16	Broken surface of vein-related pyrite-anhydrite breccia	6	250–252	1.9–2.3	274–277	3.2–3.9
158-957P-6R-1	27	Vein-related pyrite–anhydrite breccia	16	293–309	1.9–2.2	319–337	3.2–3.7
158-957P-12R-4	58	Broken outer surface of massive granular pyrite	5	168–187	2.1–2.4	187–207	3.5–4.0

py = pyrite, cp = chalcopyrite, hm = haematite; T_h = homogenization temperature, T_m = final ice melting temperature, T_t = trapping temperature.
[a] Temperatures corrected for entrapment pressure using equations of Zhang & Frantz (1987).
[b] Salinities calculated using data and equations from Bodnar (1993).

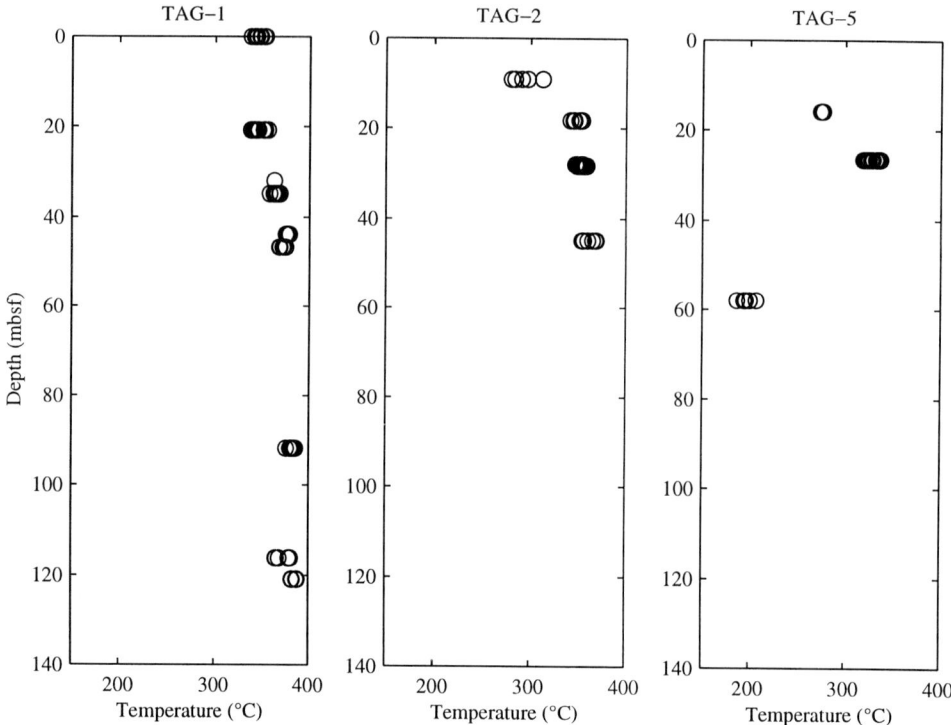

Figure 2. Trapping temperatures (pressure corrected) as a function of depth for fluid inclusions in anhydrite from the TAG-1 (103 samples), TAG-2 (46 samples) and TAG-5 (27 samples) areas.

fluid inclusion trapping temperatures have been observed in anhydrite samples from other active seafloor vent sites (LeBel & Oudin, 1982; Brett et al. 1987; Nehlig, 1991; Peter et al. 1994). While temperatures near the mound surface were close to those of venting fluids, there was a general increase in T_t with depth within the mound in the TAG-1 and TAG-2 areas to a maximum of 388 °C at 121 mbsf (figure 2). This consistent increase in temperature with depth, coupled with the similarity of trapping temperatures in the shallowest samples with measured temperatures of black and white smoker fluids, suggests that current temperatures deep within the mound are in excess of 380 to 390 °C (Tivey et al. 1998). In contrast to inclusions in samples from the TAG-1 and TAG-2 areas, fluid inclusions in anhydrite crystals from the TAG-5 area exhibit a much greater range of trapping temperatures, from 187 to 337 °C. A similar large range ($T_t = 267$–357 °C) was observed by Petersen et al. (1998), though in the samples they analysed the lower temperatures were measured in crystals from a depth of 22 mbsf. The observation of lower trapping temperatures, particularly at depth ($T_t = 187$ to 207 °C at 58 mbsf) indicates that temperatures of at least some fluids forming anhydrite at depth in the TAG-5 area were cool relative to those in the TAG-1 and TAG-2 areas, possibly representing paths of sea water influx into the mound 20–40 m north of the BSC; the observation, prior to heating and freezing measurements, of abundant decrepitated, empty inclusions in anhydrite crystals from the TAG-5 area suggests that this cooler fluid circulation may have been succeeded by a more recent high temperature event (Tivey et al. 1998).

Final ice melting temperatures for all fluid inclusions (T_m) range from -3.1 to -0.7 °C, and correspond to salinities of 5.1 to 1.2 wt% NaCl equivalent (table 1; figure 3). As discussed in detail by Tivey et al. (1998), it is possible for there to be some

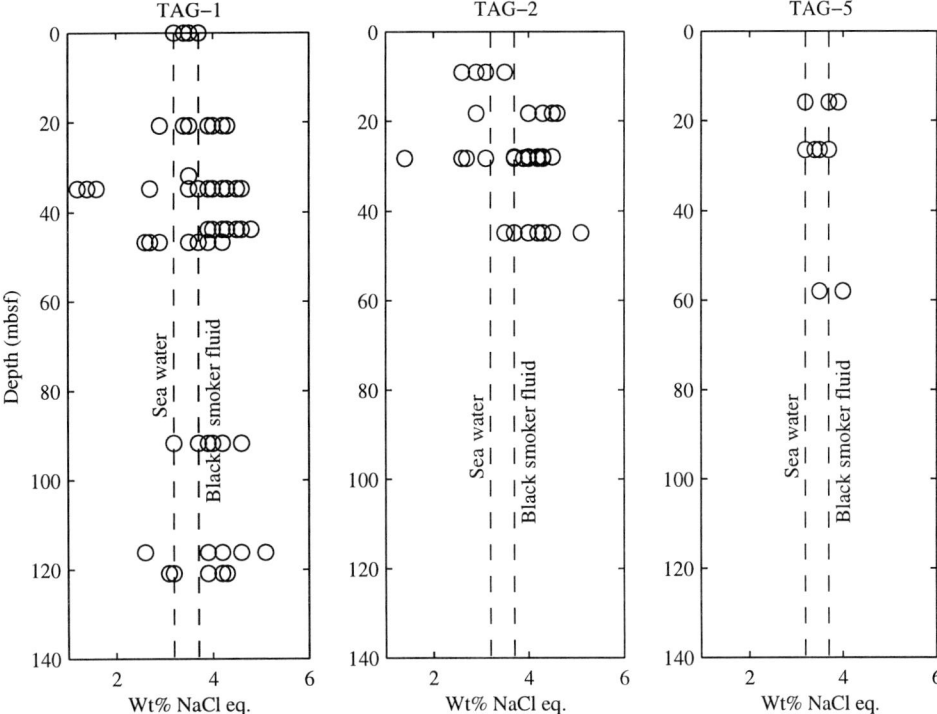

Figure 3. Salinities as a function of depth for fluid inclusions in anhydrite from the TAG-1 (103 samples), TAG-2 (46 samples) and TAG-5 (27 samples) areas. Dashed lines indicate salinities of sea water and TAG black smoker fluid.

increase in salinity in fluid inclusions hosted in anhydrite from anhydrite dissolution, but this is likely to be less than 0.66 wt%; the volume change accompanying such an increase in salinity is insignificant, of the order of 0.2 vol%, and thus homogenization temperatures would not be affected. Secondary inclusions in TAG samples exhibit similar ranges in salinity as primary inclusions for the same sample. There is no general trend of salinities with depth in the TAG mound, and all salinities fall well within the range measured in mid-ocean ridge hydrothermal fluids (~ 0.2 to ~ 7 wt% NaCl equivalent; Von Damm, 1990; Butterfield & Massoth, 1994; Von Damm et al. 1995; figure 3). TAG sample salinities are also similar to those reported from analyses of fluid inclusions in anhydrite crystals from vent deposits on the East Pacific Rise and southern Juan de Fuca ridge, and to discrete populations of low-salinity (~ 1 to 8 wt% NaCl equivalent), liquid-dominated fluid inclusions in modern and ancient igneous oceanic crust samples (LeBel & Oudin, 1982; Brett et al. 1987; Kelley & Robinson, 1990; Nehlig, 1991; Vanko et al. 1992; Kelley et al. 1993). The favoured explanation for generation of the range in salinities is phase separation (either boiling or condensation) of sea water-derived fluids and/or variable mixing of hydrothermal sea water with phase-separated brines and vapours (Kelley & Delaney, 1987; Von Damm, 1988; Kelley & Robinson, 1990; Nehlig, 1991; Vanko et al. 1992; Kelley et al. 1993; Saccocia & Gillis, 1995; Edmonds & Edmond, 1995; Kelley & Malpas, 1996).

(b) *Geochemistry*

Sr isotopic compositions of TAG anhydrite samples are shown in table 2, and are plotted as a function of depth in figure 4a. For comparison, sea water and hydrothermal

Table 2. Sample description, Sr/Ca, Mg/Ca and Sr isotope composition of anhydrite

sample	depth (mbsf)	sample description	Sr/Ca (mmol/mol)	Mg/Ca (mmol/mol)	$^{87}Sr/^{86}Sr$	% hydro-thermal	% Sea water	predicted Sr/Ca (mmol/mol)	predicted Mg/Ca (mmol/mol)	apparent D_{Sr}	apparent D_{Mg} (×10^{-3})	temp. (°C) of mixed fluid[b]
Central Complex – TAG-1												
ALV 1677-2	0	Massive anhydrite	13.7	11.7	0.707 214	31	69	5.59	2.30	2.5	5.1	210
MIR-1-74-2A	0	Massive anhydrite	10.9	28.2	0.707 363	28	72	5.74	2.45	1.9	12	201
MIR-1-74-2B	0	Massive anhydrite	11.4	313	0.707 586	24	76	5.99	2.69	1.9	122	172
ALV 2190-8-1B	0	Massive anhydrite	12.4	1.91	0.707 190	31	69	5.57	2.28	2.2	0.84	210
MIR-2-78-1	0	Massive anhydrite	12.4	19.5	0.706 861	36	64	5.26	1.98	2.4	9.9	210
158-957E-11R-1 (Piece 2)	87	Outer surface of massive granular pyrite sample	6.77	1.06	0.707 870	20	80	6.33	3.02	1.1	0.35	144
158-957E-15R-1 (Piece 2)	107	Outer surface of chloritized basalt breccia clast	6.78	3.37	0.707 690	23	77	6.11	2.80	1.1	1.2	165
158-957E-17R-1 (Piece 2)	116	Broken surface of chloritized basalt breccia	4.44	2.12	0.708 155	15	85	6.71	3.39	0.66	0.63	109
158-957E-18R-1 (Piece 4)	121	Vein in chloritized basalt breccia, with py	4.50	0.747	0.709 128	0.5	99.5	8.47	5.08	0.53	0.15	6.53
Kremlin area – TAG-2												
ALV 2187-1-2	0	White smoker chimney sample	14.5	16.2	0.705 308	64	36	4.18	0.911	3.5	18	324
158-957H-1N-1 (Piece 9)	9.1	Vein and matrix of cp-rich porous massive pyrite	6.24	1.72	0.707 251	30	70	5.63	2.34	1.1	0.74	215
158-957H-5N-1 (Piece 5B)	27	1 cm thick vein in nodular pyrite silica breccia	11.3	2.56	0.707 103	32	68	5.48	2.20	2.1	1.2	217
158-957H-6N-1 (Piece 5)	31	Vug in nodular pyrite–silica breccia	8.78	1.04	0.707 030	34	66	5.41	2.13	1.6	0.49	208
158-957H-9X-1 (Piece 4)	45	Broken surface of massive pyrite breccia with cp and py	7.74	2.76	0.706 866	36	64	5.27	1.98	1.5	1.4	210
Northern area – TAG-5												
158-957P-1R-1 (Piece 1)	0	Veins and vugs from massive pyrite-anhydrite breccia	12.4	11.6	0.707 383	28	72	5.76	2.47	2.2	4.7	201
158-957O-2R-1 (Piece 4)	8.0	Veins and outer surface from nodular py breccia	9.18	3.18	0.707 260	30	70	5.64	2.35	1.6	1.4	215
158-957P-11R-1 (Piece 7)	50	Broken outer surface of pyrite silica breccia	9.26	1.62	0.707 796	21	79	6.24	2.93	1.5	0.55	151
158-957P-12R-4 (Piece 11)	58	Broken outer surface of massive granular pyrite	6.72	1.10	0.708 847	4.7	95	7.87	4.51	0.85	0.24	35.3

py = pyrite, cp = chalcopyrite, hm = haematite.
[a] Calculated from the sea water and hydrothermal fluid end-member $^{87}Sr/^{86}Sr$ compositions (0.709 16 and 0.703 45 respectively) and contents (88 and 103 μmol/kg respectively) (Elderfield et al. 1993; Edmond et al. 1995) as described by Mills & Elderfield (1995).
[b] Calculated considering the heat capacities of 3.2 wt% NaCl solutions at 40 MPa at 2.7 °C and 366 °C (Bischoff & Rosenbauer 1985).

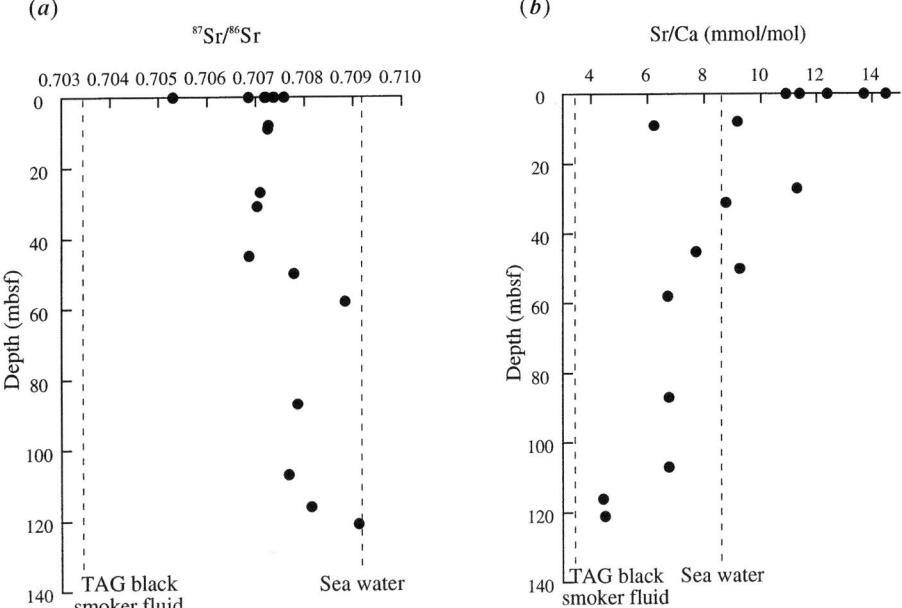

Figure 4. Downcore variation in (a) $^{87}Sr/^{86}Sr$ and (b) Sr/Ca ratios within the TAG mound. TAG black smoker fluid and sea water values are shown for comparison.

fluid Sr isotope compositions are also shown. End-member Sr isotope compositions are well established at 0.709 16 for sea water and 0.703 45 for mean TAG vent fluids from 1990 to 1993 (Elderfield et al. 1993; James & Elderfield, 1997). The $^{87}Sr/^{86}Sr$ ratio of TAG anhydrite varies from 0.705 308 to 0.709 128, suggesting circulation of sea water-dominated fluids throughout most of the active mound. Since end-member Sr concentrations and isotopic compositions are known, relative proportions of sea water and hydrothermal components in the fluid that formed the anhydrite crystals can be calculated from mass balance considerations assuming that the present-day black smoker fluid is identical to the black smoker fluid in the mound at the time the anhydrite precipitated (Mills & Elderfield, 1995; table 2). TAG anhydrite samples demonstrate a range of hydrothermal fluid–sea water mixtures all relatively sea water-dominated (36:64 to 0.5:99.5), except for anhydrite from the base of a white smoker chimney which exhibits an isotopic signature indicative of a hydrothermal-dominated fluid (64:36). Values close to sea water are observed within the stockwork at 58 mbsf in the TAG-5 area and 121 mbsf in the TAG-1 area, suggesting that these samples are from areas of sea water recharge and entrainment that are restricted from mixing with rising hydrothermal fluids.

Sr, Mg and Ca contents of TAG anhydrite samples are also shown in table 2. As with Sr isotopic compositions, there is a wide range in Sr/Ca and Mg/Ca ratios (4.4–14.5 mmol/mol and 0.7–313 mmol/mol for Sr/Ca and Mg/Ca ratios respectively). Sr/Ca ratios decrease systematically downcore and surface samples exhibit high values (figure 4b). Mg/Ca ratios are extremely high in some of the surface samples and consistently low throughout the rest of the mound. Using Sr isotope data as indicative of proportions of hydrothermal fluid and sea water, the Sr, Ca and Mg content of the parent fluid that precipitated the anhydrite within the mound can be estimated. Differences between the ratios inferred for the parent fluids and those observed in the anhydrites are due to element fractionation during precipitation. Predicted Sr/Ca and Mg/Ca ratios are shown in table 2, and can be compared to ratios observed in anhydrite samples. Neglecting the surface samples, observed Mg/Ca ratios are generally lower than predicted values for the parent fluid. This reflects the discrepancy between the ionic radius of Mg^{2+} (103 pm; Shannon, 1976) and that of Ca^{2+} (126 pm; Shannon, 1976) and therefore substitution into the anhydrite lattice is limited by this mismatch (Mills et al. 1998). The enhanced Mg/Ca ratios in the near-surface and surface samples is inferred to be due to coexistence of a Mg-rich phase (e.g. talc) rather than extensive lattice substitution of Mg into anhydrite. In contrast, the Sr/Ca ratios of TAG anhydrite samples are mostly similar to or higher than the ratios for the predicted parent fluids. This arises from the similarity in cationic radius ($Sr^{2+} = 140$ pm; $Ca^{2+} = 126$ pm; Shannon, 1976) and the charge balance between Sr and Ca allowing ready substitution of Sr into the $CaSO_4$ lattice during precipitation.

4. Discussion

The data presented above provide information on temperatures within the mound and on proportions of sea water and hydrothermal fluid mixing within the mound. Trapping temperatures from the TAG-1 and TAG-2 areas are all high, with many higher than the current measured temperatures of fluids venting from black smokers and temperatures increase with depth to a maximum of 388 °C at 121 mbsf. Sr isotope data indicate that fluids responsible for formation of anhydrite are sea water dominated. Combining information on mixing proportions and on temperatures within the mound with Sr/Ca ratios in anhydrite grains allows examination of processes of entrainment, mixing, conductive heating and fluid evolution within the mound.

(a) Fluid evolution

Sr/Ca ratios in anhydrite samples from within the mound are generally similar to or greater than Sr/Ca ratios calculated for corresponding parent fluids (table 2; figure 4b). This can possibly be explained by considering distribution coefficients. Doerner–Hoskins type models have been established to be an appropriate description of trace element partitioning into crystalline systems during unidirectional precipitation of minerals (Shikazono & Holland, 1983). This approach assumes that the partitioning of trace elements between solid and solution phase is controlled by a partition coefficient, D where

$$D_{Sr} = \frac{(Sr/Ca)_{solid}}{(Sr/Ca)_{fluid}}. \tag{1}$$

Large deviations from thermodynamic equilibrium will occur in systems such as TAG which undergo rapid fluctuations in fluid flow and large gradients in temperature and fluid composition. Apparent partition coefficients for TAG anhydrite samples can be established from the measured element/Ca ratio and the value calculated for the corresponding fluid (table 2), assuming that the Sr/Ca ratio of this parent fluid is not changing significantly as crystals precipitate, i.e. that Sr, Ca and SO_4 are being replenished as fluid flows through the vein or fracture. D_{Mg} values for surface samples are surprisingly high suggesting that some additional Mg removal is occurring at the mound–sea water interface, e.g. a talc phase as suggested above. No caminite (Mg-hydroxysulphate hydrate; MHSH) has been observed in TAG anhydrite samples, but fine-grained crystalline phases (possibly talc) have been observed optically in interstitial spaces between anhydrite crystals in surface samples. This material will be readily leached by the methodology used here leading to the high Mg/Ca ratios observed. The low D_{Mg} throughout the rest of the mound reflects the lattice mismatch of the Mg and Ca cations.

There is a wide range in apparent D_{Sr} in the anhydrite samples from TAG, from 0.5 to 3.5. Laboratory evaluations of the Sr partition coefficient for rectangular anhydrite fall in the range 0.27–0.35 (Shikazono & Holland, 1983). Acicular anhydrite, which was observed to precipitate from more extremely supersaturated solutions in laboratory experiments, was observed to precipitate with a higher D_{Sr} of 0.52–0.55. These values are all much lower than the calculated apparent partition coefficients. Factors possibly affecting partition coefficients include the extent of supersaturation and the rate of anhydrite precipitation (Shikazono & Holland, 1983), and temperature and fluid composition. A general inverse trend with increasing temperature is apparent in the literature (Kushnir, 1982; Shikazono & Holland, 1983) which extrapolates to a D_{Sr} value of < 0.1 at 350 °C. Again, this value is significantly less than calculated apparent partition coefficients. Attempts to quantify D_{Sr} for natural hydrothermal systems suggest a range from 0.63 to 1.1 (Shikazono & Holland, 1983; Mills et al. 1998; Teagle et al. 1998) all of which are higher than laboratory experimental values. While fluid inclusion trapping temperatures increase downcore and there is a concurrent decrease in apparent D_{Sr} within the mound anhydrite with depth (figure 5), there is no good correlation between measured fluid trapping temperature and partition coefficient. In general, temperatures are high throughout the mound, and are > 337 °C below 10 mbsf in the TAG-1 and TAG-2 areas. The high D_{Sr} values and the lack of correlation with temperature suggests that temperature plays only a minor role in controlling the apparent partition coefficient for Sr in TAG anhydrite and there is a far more important control on the anhydrite geochemistry.

A more probable explanation for differences between Sr/Ca ratios in anhydrite crystals and Sr/Ca ratios predicted for the corresponding fluids is that our assumptions for calculating the ratios of the fluids are inappropriate. The alternative we propose is that the Sr/Ca ratios of the fluids within the TAG mound evolve during sea water entrainment and repeated mixing with hydrothermal fluid. Two scenarios for explaining Sr/Ca ratios in TAG anhydrite samples firstly a replenished system and secondly a closed system, are discussed below.

(i) *Replenished system*

White smoker fluids have been previously utilized to infer subseafloor processes within the active TAG mound (e.g. Edmond et al. 1995) and geochemical modelling of major element and rare earth element (REE) compositions suggest that $\sim 0.25 \times 10^{-2}$ mol anhydrite/kg fluid precipitates within the mound prior to venting in the Kremlin area (Mills & Elderfield, 1995; Tivey et al. 1995). Sr and Ca measurements have also been made on the same fluid samples (Edmond et al. 1995) allowing estimation of the removal of Sr and Ca into anhydrite within the mound and hence the mean partition coefficient for anhydrite involved in this process. Changes in the Sr/Ca ratio of the fluid as anhydrite precipitation proceeds can be modelled as a Rayleigh-type distillation:

$$(\text{Sr/Ca})_{\text{final}} = (\text{Sr/Ca})_{\text{initial}} F^{(D_{Sr}-1)} \tag{2}$$

where F is the fraction of original Ca remaining in solution. Substituting values for black and white smoker fluids (Sr = 103 μmol/kg, Ca = 30.8 mmol/kg and

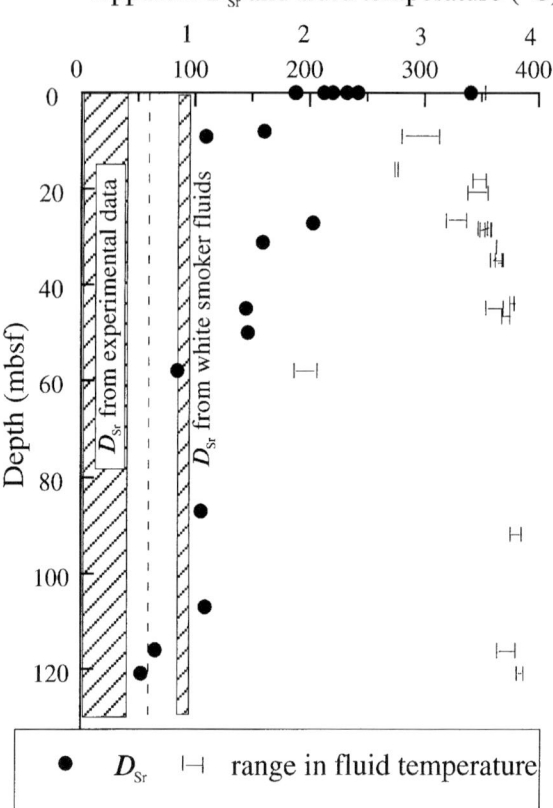

Figure 5. Downcore variation in apparent D_{Sr} values and trapping temperature (T_t) ranges observed in TAG anhydrite samples. Experimental data from Shikazono & Holland (1983) and D_{Sr} values inferred from white smoker fluid data are shown for comparison. The dashed line is for a D_{Sr} value of 0.6 which is used here for consideration of fluid evolution.

Sr = 91 µmol/kg, Ca = 27 mmol/kg in black and white smoker end-member fluids respectively; Edmond et al. 1995) gives a D_{Sr} value of 0.94 for fluids exiting from the Kremlin area. Precipitation of anhydrite within the mound with $D_{Sr} < 1$ will result in evolution of the fluid–sea water mixture to higher Sr/Ca ratios; as precipitation continues subsurface, so the Sr/Ca ratio will increase further.

The downcore decrease in apparent D_{Sr} is shown in figure 5 where values close to experimental data and the mean D_{Sr} inferred from white smoker fluid composition are observed at depth. This trend suggests that sea water entrainment to > 100 mbsf results in anhydrite precipitation after conductive heating and minor mixing with rising fluids with D_{Sr} values close to experimental conditions (Shikazono & Holland, 1983). Such anhydrite is present on broken surfaces of altered basaltic clasts in the stockwork zone. Whereas the sulphide-hosted anhydrite veins from the upper mound exhibit high apparent D_{Sr} values, suggesting that these samples precipitate from evolved fluid–sea water mixtures, with higher Sr/Ca ratios. As the fluids circulate within the mound, are heated and mix with upwelling hydrothermal fluids and precipitate anhydrite, the Sr/Ca ratio will evolve to higher values. Here we use a D_{Sr} estimate of 0.6 to quantify the effect on fluid evolution (figure 5).

The extent to which this process can affect Sr/Ca ratios can be demonstrated by

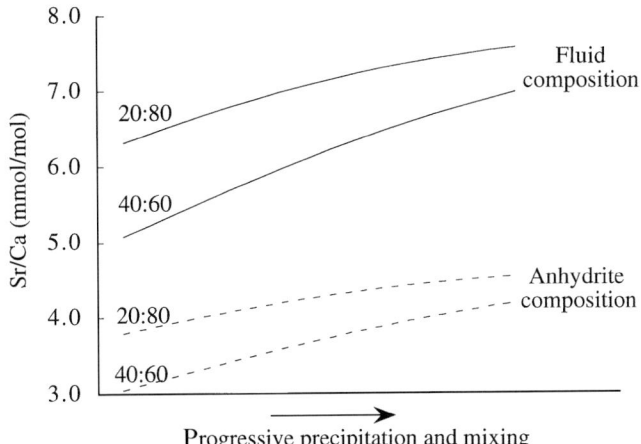

Figure 6. Sr/Ca compositional evolution for replenished system Rayleigh fractionation of two hydrothermal fluid–sea water mixtures. Fluid evolution is tracked using equation 2 assuming 5% Ca removal and subsequent mixing with the initial fluid composition, anhydrite composition is calculated from equation 1, $D_{Sr} = 0.6$.

considering progressive precipitation of anhydrite from a mixture of hydrothermal fluid and sea water, coupled with replenishment of Ca, Sr and SO_4 via continued entrainment of fluids into the permeable fluid flow system. Figure 6 depicts reaction paths for two mixtures of hydrothermal fluid and sea water. The initial conditions chosen are for hydrothermal fluid–sea water mixtures of 20:80 and 40:60 which, with the exception of four samples, covers the range inferred from Sr isotopic systematics in table 2. The evolution of the fluid composition is simulated by incrementally precipitating 5% of the Ca as anhydrite with a D_{Sr} of 0.6 and subsequently mixing the resultant evolved fluid with the initial fluid mixture in proportions of 95% evolved fluid to 5% initial fluid. The Sr/Ca ratio of the evolved fluid is calculated using equation (2) and the Sr/Ca ratio of anhydrite is calculated using equation (1). Concentrations of Ca and Sr in the initial mixtures are 14.3 mmol/kg and 90.8 μmol/kg respectively for a 20:80 mix, and 18.5 mmol/kg and 94 μmol/kg respectively for a 40:60 mix. The extent of fluid modification by this process is controlled by the D_{Sr} value and the composition of the newly entrained fluid and figure 6 demonstrates the potential for such processes to evolve the Sr/Ca ratio of both the fluid and the precipitated anhydrite within the TAG system.

(ii) *Closed system*

The discussion above considers fluid evolution in a replenished system where fluid is flowing relatively rapidly past crystals that are growing along boundaries of fractures. An alternative to consider is mixing and precipitation in a closed system. Here we consider as before two possibilities, one starting with an initial mixture of 20% black smoker fluid and 80% sea water, and the second with an initial mixture of 40% black smoker fluid and 60% sea water (i.e. the same initial conditions as for figure 6). Tivey *et al.* (1998), considering sea water and the composition of end-member black smoker fluid (Edmond *et al.* 1995), calculated that a mix of 80% sea water and 20% black smoker fluid at 350 °C would be in equilibrium with anhydrite (80% of total precipitate), MHSH (16% of total precipitate) and trace amounts of talc and haematite. If precipitation of talc, chrysotile and MHSH are suppressed since they have not been identified in drillcore samples (Shipboard Scientific Party, 1996), then anhydrite (14.3 mmol), chalcopyrite (0.0266 mmol) and haematite (0.37 mmol) are calculated to precipitate per kg of fluid. Only 0.06 mmol/kg Ca, or 0.4% of the initial Ca, remains in solution. The concentration of Sr remaining in solution can be calculated for this closed system by considering the fraction, F, of Ca remaining (0.004), the concentration of Sr in the 20:80 mixture (90.8 µmol/kg), and the partition coefficient, D_{Sr}:

$$[Sr]_{Solution} = \frac{F[Sr]_{Mixture}}{D_{Sr} - (FD_{Sr}) + F}. \tag{3}$$

For $D_{Sr} = 0.6$, the Sr concentration in the resultant fluid is 0.6 µmol/kg, the Sr/Ca ratio of the resultant solution is 10.5 mmol/mol, the Sr concentration in anhydrite is 90.2 µmol/kg and the Sr/Ca of the precipitated anhydrite is 6.3 mmol/mol. These amounts of Sr and Ca in the resultant evolved fluid are too small to greatly affect the signature of anhydrite precipitating from a mix of this fluid with either black smoker fluid or sea water.

We can also consider fluid evolving from an initial mix of 40% black smoker fluid and 60% sea water. With precipitation of talc, chrysotile and MHSH suppressed, such a mixture at 350 °C is in equilibrium with 16.1 mmol anhydrite and trace haematite (0.35 mmol) per kg fluid. Initial concentrations of Ca, SO_4 and Sr are 18.5 mmol, 16.8 mmol and 94 µmol per kg fluid. Ca remaining in solution at equilibrium is 2.36 mmol/kg, or 12.8% (i.e. $F = 0.128$). Using equation (3) above and a D_{Sr} of 0.6, Sr remaining in solution is 18.5 µmol/kg, and thus Sr in anhydrite is 75.5 µmol/kg, Sr/Ca in solution is 7.8 mmol/mol and Sr/Ca in anhydrite is 4.7 mmol/mol. The resulting 'evolved fluid' still has a Sr isotopic ratio of 0.706 66 since the isotope systematics are unaffected by anhydrite precipitation, indicating the original mix of 40% black smoker fluid and 60% sea water. A mixture of 84% of this fluid and 16% sea water results in a solution with a Sr isotope ratio of 0.707 87, identical to a mixture of 20% black smoker fluid and 80% sea water. However, this new solution contains 29.6 µmol/kg Sr, 3.6 mmol/kg Ca and 5.2 mmol/kg SO_4, and, at 350 °C, is in equilibrium with 3 mmol of anhydrite and 0.51 mmol haematite per kg fluid. Thus 83% of the Ca is present in anhydrite and 17% is in solution (or $F = 0.17$). Sr in the solution, calculated using equation (3), is 7.5 µmol/kg, assuming a D_{Sr} of 0.6, and Sr/Ca in the anhydrite is 7.4 mmol/mol, considerably higher than the Sr/Ca calculated for anhydrite precipitating from a simple mixture of 20% black smoker fluid and 80% sea water of 6.3 mmol/mol. As in the replenished system, the rate of fluid evolution depends on the D_{Sr} value and the composition of the fluid introduced to the batch mixing process, however closed-system evolution is much more effective in generating the high and varied Sr/Ca values observed in the upper TAG mound.

This scenario of fluid evolution is shown schematically in figure 7. Sea water is entrained, mixes with black smoker fluid in a ratio of 40:60, and precipitates anhydrite. The resultant 'evolved fluid', undersaturated in anhydrite principally because of a lack of SO_4, rises within the mound until it encounters more entrained sea water. The additional SO_4 (and Ca and Sr) results in deposition of more anhydrite with a higher Sr/Ca.

Clearly the scenarios discussed, both open and closed system, are over simplifications, but they serve as examples of the possible complexities involved in sea water entrainment, fluid mixing and evolution of fluid compositions within the mound. They also provide an explanation for the observed increase in Sr/Ca ratios in anhydrite at shallower levels in the mound. A more detailed study, utilizing Sr isotope

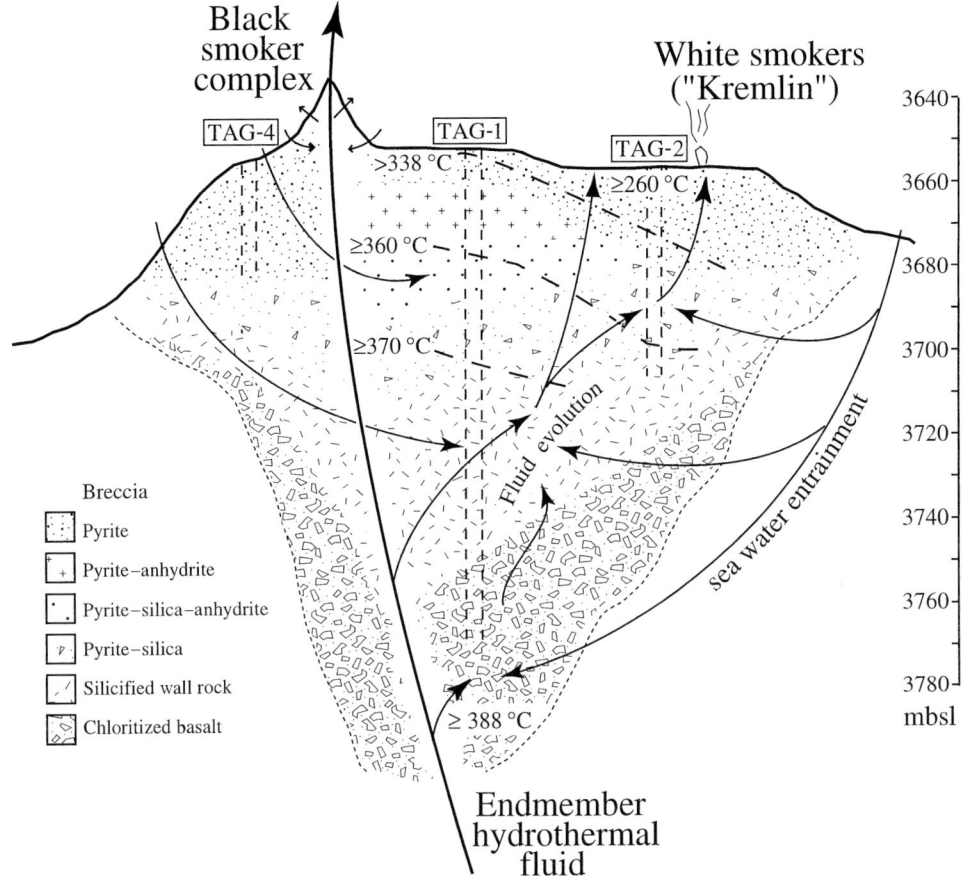

Figure 7. Schematic depiction of sea water entrainment and fluid evolution within the TAG mound as inferred from this study. Arrows indicate sea water penetration into the mound and fluid pathways. Note that 'evolved fluids' that result from mixing of sea water and hydrothermal fluid at depth in the mound will continue to rise, encountering more entrained sea water at shallower levels. Dashed lines show isotherms inferred from fluid inclusion trapping temperatures.

data, Sr/Ca ratios and REE concentrations, is planned to further examine the complexities of fluid evolution within the TAG mound.

(b) Conductive heating

The discussion above considers mixtures of variable amounts of black smoker fluid and sea water at 350 °C. Evidence for consistently high temperatures (> 337 °C and up to 388 °C in the TAG-1 area) is provided by fluid inclusion trapping temperatures (figures 2 and 5). Sea water and sea water/black smoker fluid mixtures can attain these high temperatures by a combination of mixing and conductive heating. Table 2 shows the temperatures that would be attained from mixtures of sea water and hydrothermal fluid in proportions indicated by Sr isotope systematics considering the heat capacity of sea water at differing temperatures (Bischoff & Rosenbauer, 1985). Extensive conductive heating is required to raise temperatures from those inferred from mixing to the maximum values observed from fluid inclusion trapping temperatures (figure 8). A simple exercise can be carried out to examine whether these temperatures can be reasonably explained by conductive heating, and to esti-

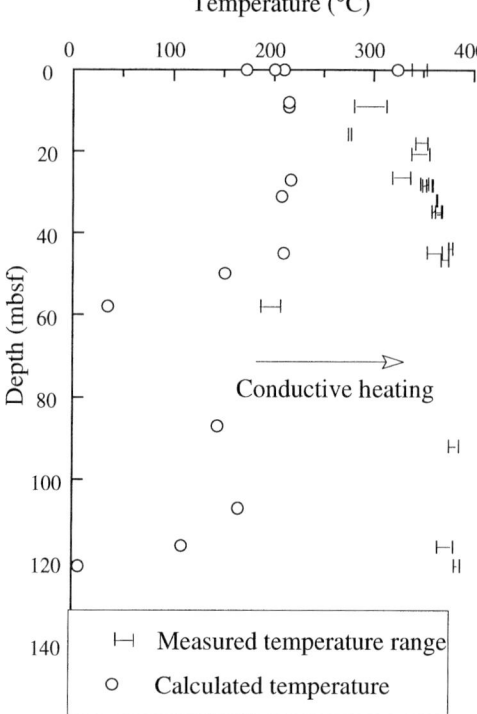

Figure 8. Comparison of trapping temperature ranges observed in TAG anhydrite samples and temperatures calculated from Sr isotope mass balance considerations. The arrow represents the extensive conductive heating of fluids.

mate possible rates of flow of entrained sea water. Calculations are carried out by considering heat gain by fluid flowing through a hollow cylinder, assuming that the outer layer is maintained at a constant temperature, T_0. The temperature of the fluid exiting the end of the cylinder of length L is given by:

$$T_{\text{Fluid}} = \left[(T_i - T_0)e^{-\left(\frac{2\pi kL}{fC_p \ln(r_2/r_1)}\right)}\right] + T_0 \qquad (4)$$

where T_i is the initial temperature of the sea water, or hydrothermal fluid/sea water mix entering the cylinder, k is thermal conductivity of the wall of the cylinder, f is fluid flow rate, C_p is heat capacity of the fluid, and r_2 and r_1 are the radii of the outer and inner walls of the cylinder, respectively (Seewald & Seyfried, 1990; Tivey et al. 1995). The value used for thermal conductivity of the inner walls of the cylinder is an average value based on measured thermal conductivities of breccia samples from the mound that range from 10.2 to 15.0 W/mK (Rona et al. 1998), and extrapolate to values of 6.1 to 8.7 W/mK at 350 °C (Tivey, 1998). Using a value for k of 8 W/mK and for C_p of 5200 J/kg/K (an average value for sea water from 2 to 390 °C; Bischoff & Rosenbauer, 1985), flow rates required to heat the entrained fluid to the observed temperature of 388 °C can be calculated from equation (4) over a variety of length scales. With values of T_i, T_0, L, r_2 and r_1 of 2 °C, 390 °C, 100 m, 1 m and 0.01 m respectively, the maximum flow rate allowed is 0.04 kg/s to attain 388 °C. Fluid velocity is related to flow rate by $F = V\pi(r_1)^2 \rho$, where ρ is the density of the fluid (830 kg/m^3 at 275 °C and 400 bars (Bischoff & Rosenbauer, 1985), and time (t) is related to velocity by $t = L/V$. Thus a flow rate of 0.04 kg/s corresponds to a velocity of 15 cm/s, and the time required for the fluid to attain the high temperatures observed would be 11 minutes. If a value of r_2 of 0.1 m is used instead, flow rates and velocities can be as high as 0.08 kg/s and 31 cm/s and still attain the observed heat gain in 5.4 mins (Tivey et al. 1998). Mixtures of sea water and hydrothermal fluid can likewise be heated over similar distances. using values of T_1, T_0, C_p, L and R_1 of 200 °C, 370 °C, 6300 J/kg/K (value for sea water at −360 °C and 40 MPa (Bischoff & Rosenbauer, 1985), 100 m and 0.01 m results in a final temperature (T_f of 369 °C being attained in 21 mins for an r_2 of 1 m (with $F < 0.03$ kg/s and $V < 8$ cm/s) or in 6.2 mins for $r_2 = 0.1$ m (with $F < 0.07$ kg/s and $v < 26$ cm/s).

Given these flow rates, Tivey et al. (1998) calculated the amount of anhydrite that would precipitate in the mound per year as sea water/hydrothermal fluid mixtures flow through such channels. For one channel carrying a mix of 20% hydrothermal fluid and 80% sea water (that will precipitate ~ 0.0143 moles of anhydrite per kg fluid) at a flow rate of 0.05 kg/s, a total volume of ~ 1 m^3 would be deposited in a year (molar volume of anhydrite is 4.6×10^{-5} m^3; Robie et al. 1978). We can also make a rough estimate of how much sea water can be entrained and heated by considering estimates of heat flux from the mound, and that sea water or hydrothermal fluid/sea water mixtures must be heated to > 338°C. Rona et al. (1993b) calculated a convective heat flux from the black smoker complex of 225 MW, and Rudnicki & Elderfield (1992) estimated a flux from the entire mound of ~ 500 to 1000 MW. If we use 5% of the black smoker heat flux (or 1 to 2% of the entire mound heat flux) as an estimate of the energy being used to heat sea water from 2 to ~ 360°C, and an average heat capacity of sea water over this temperature range of 5250 J/kg/K (Bischoff & Rosenbauer 1985), then ~ 6 kg/s can be heated. This corresponds to 120 separate channels or veins each carrying ~ 0.05 kg/s. The amount of anhydrite that would precipitate annually in the mound from this rate of sea water entrainment, assuming precipitation of ~ 0.0143 mol anhydrite per kg fluid is ~ 120 m^3. Similarly, if either 1% or 10% of 225 MW is used to heat sea water, the amounts of fluid heated are 1.2 and 12 kg/s respectively and the amounts of anhydrite precipitated per year are 24 m^3 and 240 m^3 respectively. Considering these deposition rates it would take 80–800 years to accumulate the $\sim 2 \times 10^4$ m^3 of anhydrite currently present within the mound (estimated by assuming that the anhydrite-rich zone is equivalent to a 50 m diameter cylinder, 30 m thick, containing roughly 1/3 anhydrite). These estimates suggest that anhydrite deposition is rapid and much of the anhydrite inventory at TAG is recently precipitated.

The combination of analyses carried out on anhydrite recovered from both the surface and within the mound, to depths of 121 mbsf, provides information on depths of entrainment and on the complexities of entrainment, mixing and precipitation. Data indicate that the internal fluid regime is complex and involves deep-seated recharge and pooling and evolution of fluids at shallow depths within the mound.

5. Conclusions

Data from fluid inclusions in anhydrite provide evidence for high temperatures (> 337 °C) throughout the central and southeastern quadrants of the TAG mound, and suggest that temperatures at depths of greater than 100 mbsf within the mound are currently in excess of 380 to 390 °C. The salinity range of fluids within the TAG mound provides evidence for phase separation at depth below the seafloor. Sea water is entrained into the mound to a depth of at least 120 m and Sr isotopic data indicate that anhydrite recovered from the stockwork zone is formed from conductively heated sea water with little to moderate hydrothermal fluid influence.

Both Sr and Mg substitute into the anhydrite lattice, though Mg partitioning is minimal due to mismatch in cationic radius. High Mg/Ca ratios in near-surface samples have been attributed to the coexistence of a Mg-bearing crystalline phase observed in thin section. Sr/Ca ratios vary downcore but this does not arise from the temperature control on Sr partitioning into the crystalline phase. Instead there is significant fluid evolution as anhydrite precipitates with $D_{Sr} < 1$ and the resultant fluid mixes within the mound. The fluid evolution can be modelled as either a replenished system with continued replenishment of fluid within the mound or a closed system with batch mixing and precipitation followed by further mixing. The TAG system probably exhibits both types of behaviour as fluid flow pathways clog during mineralization and reopen later with continued fluid flow. The result is high and varied Sr/Ca ratios particularly in near-surface anhydrite.

A general trend of decreasing amounts of hydrothermal fluid in mixtures with depth is consistent with a corresponding decrease in permeability. The upper, anhydrite-rich portion of the mound is dominated by evolved fluids forming from mixtures of hydrothermal fluid and sea water in a zone of high permeability. The subsurface permeability must control sea water ingress and the extent of mixing and heat exchange within the mound. The discrepency between temperatures calculated from fluid composition inferred from Sr isotopic systematics and the temperatures measured as fluid inclusion trapping temperatures is due to conductive heating of the resultant fluid. Consideration of estimated fluid flow rates and the heat flux for the TAG mound suggest that the entire anhydrite inventory inferred from ODP drilling could precipitate in 80–800 years.

Predrilling and postdrilling submersible surveys of the mound, in particular the Kremlin area where black smoker fluids are now venting, suggest that venting styles exhibit long-term variation caused by changes in fluid circulation pathways (BRAVEX 1995; Schultz et al. 1996). Such observations suggest that the subsurface fluid pathways are constantly modified as mineralization and dissolution occur and as larger scale structural controls on the subsurface flow modify the circulation. Whether or not such entrainment, which not only results in deposition of anhydrite, but also in the generation of acidic fluids and subsequent metal remobilization, is peculiar to the TAG active hydrothermal mound, or occurs to various degrees at all vent sites, needs to be investigated.

Thanks are extended to Harry Elderfield and Bob Nesbitt for supplying laboratory facilities, Darryl Green for carrying out analyses and to Rex Taylor for TIMS support. P. A. Rona is thanked for providing Mir anhydrite samples. MKT thanks Kathy Gillis for help and guidance in performing fluid inclusion analyses, and for use of her laboratory facilities. M. Sulanowska and K. Davis helped with drafting of figures. Helpful discussions with Greg Ravissa and Wolfgang Bach improved this manuscript. RAM was supported by NERC grant GST/02/0976; BRIDGE and United Kingdom ODP contributed to her participation in the Leg 158 post-cruise meeting. Support for MKT was provided by JOI-USSSP Account No. 158-20889.

References

Albarede, F., Michard, A., Minster, J. F. & Michard G. 1981 $^{87}Sr/^{86}Sr$ ratios in hydrothermal waters and deposits from the East Pacific Rise at 21 °N. *Earth Planet. Sci. Lett.* **55**, 229–236.

Becker, K. & Von Herzen, R. P. 1996 Pre-drilling observations of conductive heat flow at the TAG active mound using DSV Alvin. In *Proceedings of the Ocean Drilling Program, Initial Report* **158** (ed. Humphris, S. E., Herzig, P. M., Miller, D. J. & Zierenberg, R.), 15–29. College Station, Tx (Ocean Drilling Program).

Becker, K., Von Herzen, R., Kirklin, J., Evans, R., Kadko, D., Kinoshita, M., Matsubayashi, O., Mills, R., Schultz, A. & Rona, R. 1996 Conductive heat flow at the TAG active hydrothermal mound: Results from 1993–1995 submersible surveys. *Geophys. Res. Lett.* **23**, 3463–3466.

Bischoff, J. L. & Rosenbauer, R. J. 1985 An empirical equation of state for hydrothermal sea water (3.2 percent NaCl). *Am. J. Sci.* **285**, 725–763.

Bodnar, R. J. 1993 Revised equation and table for determining the freezing point depression of H_2O-NaCl solutions. *Geochim. Cosmochim. Acta* **57**, 683–684.

BRAVEX 1995. The effects of ODP drilling on the TAG hydrothermal mound: the TAG post-drilling survey. *BRIDGE Newsletter* **8**, 7–11.

Brett, R., Evans, H. T., Jr., Gibson, E. K., Jr., Hedenquist, J. W., Wandless, M. V. & Sommer, M. A. 1987 Mineralogical studies of sulphide samples and volatile concentrations of basalt glasses from the southern Juan de Fuca Ridge. *J. Geophys. Res.* **92**, 11373–11379.

Butterfield, D. A. & Massoth, G. J. 1994 Geochemistry of north Cleft segment vent fluids: temporal changes in chlorinity and their possible relation to recent volcanism. *J. Geophys. Res.* **99**, 4951–4968.

Campbell, A. C., Palmer, M. R., Klinkhammer, G. P., Bowers, T. S., Edmond, J. M., Lawrence, J. R., Casey, J. F., Thompson, G., Humphris, S., Rona, P. & Karson, J. A. 1988 Chemistry of hot springs on the Mid-Atlantic Ridge. *Nature* **335**, 514–519.

Charlou, J. L., Donval J. P., Jean-Baptiste, P., Dapoigny, A. & Rona P. 1996 Gases and helium isotopes in high-temperature solutions sampled before and after ODP Leg 158 drilling at TAG hydrothermal field (26 °N, MAR). *Geophys. Res. Lett.* **23**, 3491–3494.

Chiba, H., Uchiyama, N. & Teagle, D. A. H. 1998 Stable isotope study of anhydrite and sulphide minerals at the TAG hydrothermal mound, Mid-Atlantic Ridge 26 °N, ODP Leg 158. In *Proceedings of the Ocean Drilling Program, Scientific Results* **158** (ed. Humphris, S. E., Herzig, P. M., Miller, D. J. & Zierenberg R.) (In the press). College Station, TX (Ocean Drilling Program).

Constantinou, G. & Govett, G. J. S. 1973 Geology and geochemistry and genesis of Cyprus sulphide deposits. *Econ. Geol.* **68**, 843–858.

Ding, K. & Seyfried, W. E., Jr. 1992a Experimental determination of Cu–Cl speciation at the T–P conditions relevant to ridge crest hydrothermal activity: Implications to $\log(fO_2)$ in the hot spring fluids (abstract). *Eos, Trans. Am. Geophys. U.* **73**, 254.

Ding, K. & Seyfried, W. E., Jr. 1992b Determination of Fe–Cl complexing in the low pressure supercritical region (NaCl fluid): Iron solubility constraints on pH of subseafloor hydrothermal fluids. *Geochim. Cosmochim. Acta* **56**, 3681–3692.

Edmond, J. M., Campbell, A. C., Palmer, M. R., Klinkhammer, G. P., German, C. R., Edmonds, H. N., Elderfield, H., Thompson, G. & Rona, P. 1995 Time series studies of vent fluids from the TAG and MARK sites (1986, 1990) Mid-Atlantic Ridge: a new solution chemistry model and a mechanism for Cu/Zn zonation in massive sulphide orebodies. In *Hydrothermal Vents and Processes* (ed. L. M. Parson, C. Walker & D. R. Dixon), **87**, 77–86. London: Geol. Soc. Spec. Publ.

Edmonds, H. N. & Edmond, J. M. 1995 A three-component mixing model for ridge-crest hydrothermal fluids. *Earth Planet. Sci. Lett.* **134**, 53–67.

Edmonds, H. N., German, C. R., Green, D. R. H., Huh, Y., Gamo, T. & Edmond, J. M. 1996 Continuation of the hydrothermal fluid chemistry time series at TAG, and the effects of ODP drilling. *Geophys. Res. Lett.* **23**, 3487–3490.

Elderfield, H., Mills, R. A. & Rudnicki, M. D. 1993 Geochemical and thermal fluxes, high-temperature venting and diffuse flow from mid-ocean ridge systems: the TAG hydrothermal field, Mid-Atlantic Ridge 26 °N. *Geol. Soc. Spec. Publ.* **76**, 295–307.

Gamo, T., Chiba, H., Masuda, H., Edmonds, H. N., Fujioka, K., Kodama, Y., Nanba, H. & Sano, Y. 1996 Chemical characteristics of hydrothermal fluids from the TAG mound of the Mid-Atlantic Ridge in August 1994: Implications for spatial and temporal variability of hydrothermal activity. *Geophys. Res. Lett.* **23**, 3483–3486.

Humphris, S. E., Herzig, P. M., Miller, D. J., Alt, J. C., Becker, K., Brown, D., Brugmann, G., Chiba, H., Fouquet, Y., Gemmell, J. B., Guerin, G., Hannington, M. D., Holm, N. G., Honorez, J. J., Iturrino, G. J., Knott, R., Ludwig, R., Nakamura, K., Petersen, S., Reysenbach, A. L., Rona, P. A., Smith, S., Sturz, A. A., Tivey, M. K. & Zhao, X. 1995 The internal structure of an active sea-floor massive sulphide deposit. *Nature* **377**, 713–716.

James, R. H. & Elderfield, H. 1996 The chemistry of ore-forming fluids and mineral precipitation rates in an active hydrothermal sulphide deposit on the Mid-Atlantic Ridge. *Geology* **24**, 1147–1150.

James, R. H. & Elderfield, H. 1998 Addendum to Chemistry of ore-forming fluids and mineral formation in an active hydrothermal sulfide deposit on the Mid-Atlantic Ridge. *Geology* (In the press), **25**, 480.

Janecky, D. R. & Shanks, W. C. III 1988 Computational modeling of chemical and sulfur isotopic reaction processes in seafloor hydrothermal systems: chimneys, massive sulphides, and subjacent alteration zones. *Can. Mineral.* **26**, 805–825.

Johnson, J. W., Oelkers E. H. & Helgeson, H. C. 1992 SUPCRT92: A software package for calculating the standard molal thermodynamic properties of minerals, gases, aqueous species, and reactions from 1–5000 bars and 0–1000 °C. *Comput. Geosci.* **18**, 899–947.

Kelley, D. S. & Delaney, J. R. 1987 Two-phase separation and fracturing in mid-ocean ridge gabbros at temperatures greater than 700 °C. *Earth Planet. Sci. Lett.* **83**, 53–66.

Kelley, D. S. & Robinson, P. T. 1990 Development of a brine-dominated hydrothermal system at temperatures of 400–500 °C in the upper level plutonic sequence, Troodos ophiolite, Cyprus. *Geochim. Cosmochim. Acta* **54**, 653–661.

Kelley, D. S., Gillis, K. M. & Thompson, G. 1993 Fluid evolution in submarine magma-hydrothermal systems at the Mid-Atlantic Ridge. *J. Geophys. Res.* **98**, 19 579–19 596.

Kelley, D. S. & Malpas, J. 1996 Melt-fluid evolution in gabbroic rocks from Hess Deep. In *Proceedings of the Ocean Drilling Program, Scientific Results* **147** (ed. Mevel, C., Gillis, K. M., Allan, J. F. & Meyer, P. S.) pp. 213–226. College Station, TX (Ocean Drilling Program).

Kerridge, J. F., Haymon R. M. & Kastner, M. 1983 Sulfur isotope systematics at the 21° N site, East Pacific Rise. *Earth Planet. Sci. Lett.* **66**, 91–100.

Kleinrock, M. & Humphris S. 1996 Structural control on sea-floor hydrothermal activity at the TAG active mound. *Nature* **382**, 149–153.

Kusakabe, M., Chiba, H. & Ohmoto, H. 1982 Stable isotopes and fluid inclusion study of anhydrite from the East Pacific Rise at 21° N. *Geochem. J.* **16**, 89–95.

Kushnir, J. 1982 The partitioning of sea water cations during the transformation of gypsum to anhydrite. *Geochim. Cosmochim. Acta* **46**, 433–446.

LeBel, L. & Oudin, E. 1982 Fluid inclusion studies of deep-sea hydrothermal sulphide deposits on the East Pacific Rise near 21° N. *Chem. Geol.* **37**, 129–136.

Mills, R. A. & Elderfield, H. 1995 Rare earth element geochemistry of hydrothermal deposits from the active TAG mound, 26° N Mid-Atlantic Ridge. *Geochim. Cosmochim. Acta* **59**, 3511–3524.

Mills, R. A., Alt, J. C. & Clayton, T. 1996 Low-temperature fluid flow through sulfidic sediments from TAG: modification of fluid chemistry and alteration of mineral deposits. *Geophys. Res. Lett.* **23**, 3495–3498.

Mills, R. A., Teagle, D. A. H. & Tivey, M. K. 1998 Fluid mixing and anhydrite precipitation within the TAG mound. In *Proceedings of the Ocean Drilling Program, Scientific Results* **158** (ed. Humphris, S. E., Herzig, P. M., Miller, D. J. & Zierenberg, R.), (In the press). College Station, TX (Ocean Drilling Program).

Nehlig, P. 1991 Salinity of oceanic hydrothermal fluids: a fluid inclusion study. *Earth Planet. Sci. Lett.* **102**, 310–325.

Peter, J. M., Goodfellow, W. D. & Leybourne, M. I. 1994 Fluid inclusion petrography and microthermometry of the Middle Valley hydrothermal system, northern Juan de Fuca Ridge. In *Proceedings of the Ocean Drilling Program, Scientific Results* **115** (ed. Mottl, M. J., Davis, E. E., Fisher, A. T. & Slack, J. F.), pp. 411-428. College Station, TX (Ocean Drilling Program).

Petersen, S., Herzig, P. M. & Hannington, M. D. 1998 Fluid inclusion studies as a guide to the temperature regime within the TAG hydrothermal mound, 26° N, Mid-Atlantic Ridge. In *Proceedings of the Ocean Drilling Program, Scientific Results* **158** (ed. Humphris, S. E., Herzig, P. M., Miller, D. J. & Zierenberg, R.), (In the press). College Station, TX (Ocean Drilling Program).

Robie, R. A., Hemingway, B. S. & Fisher, J. R. 1978 Thermodynamic properties of minerals and related substances at 298.15 K and 1 bar (10^5) pascals) pressure and at higher temperatures. *US Geol. Surv.* **1452**, 1–456.

Roedder, E., 1984 Fluid inclusions. *Rev. Mineral.* **12**.

Rona, P. A., Hannington, M. D., Raman, C. V., Thompson, G., Tivey, M. K., Humphris, S. E., Lalou, C. & Petersen, S. 1993a. Active and relict sea-floor hydrothermal mineralization at the TAG hydrothermal field, Mid-Atlantic Ridge. *Econ. Geol.* **88**, 1989–2017.

Rona, P. A., Bogdanov, Y. A., Gurvich E. G., Rimski-Korsakov, N. A., Sagalevitch, A. M. & Hannington, M. D. 1993b Relict hydrothermal zones in the TAG hydrothermal field, Mid-Atlantic Ridge, 26° N, 45° W. *J. Geophys. Res.* **98**, 9715–9730.

Rona, P. A., Davis, E. E. & Ludwig, R. 1998 Thermal properties of TAG hydrothermal precipitates, Mid-Atlantic Ridge, and comparison with Middle Valley, Juan de Fuca Ridge. In *Proceedings of the Ocean Drilling Program, Scientific Results* **158**, (ed. Humphris, S. E., Herzig, P. M., Miller, D. J. & Zierenberg, R.) (In the press.) College Station, TX (Ocean Drilling Program).

Rudnicki, M. D. & Elderfield, H. 1992 Theory applied to the Mid-Atlantic Ridge hydrothermal plumes: the finite difference approach. *J. Volcanol. Geotherm. Res.* **50**, 163–174.

Saccocia, P. J. & Gillis, K. M. 1995 Hydrothermal upflow zones in the oceanic crust. *Earth Planet. Sci. Lett.* **136**, 1–16.

Schultz, A., Delaney, J. R. & McDuff R. E. 1992 On the partitioning of heat flux between diffuse and point source seafloor venting. *J. Geophys. Res.* **97**, 12299–12314.

Schultz, A., Dickson, P. & Elderfield, H. 1995 Temporal variations in diffuse hydrothermal flow at TAG. *Geophys. Res. Lett.* **23**, 3471–3474.

Seewald, J. S. & Seyfried, W. E., Jr. 1990 The effect of temperature on metal mobility in subseafloor hydrothermal systems: constraints from basalt alteration experiments. *Earth Planet. Sci. Lett.* **101**, 388–403.

Shannon, R. D. 1976 Revised effective ionic radii and systematic studies of interatomic distances in halides and chalcogenides. *Acta Cryst.* **A32**, 751–767.

Shikazono, N. & Holland, H. D. 1983 The partitioning of strontium between anhydrite and aqueous solutions from 150 °C to 250 °C. *Econ. Geol. Mono.* **5**, 320–328.

Shipboard Scientific Party 1996, Site 957. In *Proceedings of the Ocean Drilling Program, Scientific Results* **158** (ed. Humphris, S. E., Herzig, P. M., Miller, D. J. & Zierenberg, R.), pp. 65–226. College Station, TX (Ocean Drilling Program).

Styrt, M. M., Brackman, A. J., Holland, H. D., Clark, B. C., Pisutha-Arnold, V., Eldridge, C. S. & Ohmoto, H. 1981 The mineralogy and the isotopic composition of sulfur in hydrothermal sulfide/sulfate deposits on the East Pacific Rise, 21° N latitude. *Earth Planet. Sci. Lett.* **53**, 382–390.

Sverjensky, D. A., Hemley, J. J. & d'Angelo, W. M. 1991 Thermodynamic assessment of hydrothermal alkali feldspar–mica–aluminosilicate equilibria. *Geochim. Cosmochim. Acta* **55**, 989–1004.

Teagle, D. A. H., Alt, J. C., Chiba, H. & Halliday, A. N. 1998 Dissecting an active hydrothermal deposit: the strontium and oxygen isotopic anatomy of the TAG hydrothermal mound, Part II – Anhydrite. In *Proceedings of the Ocean Drilling Program, Scientific Results* **158** (ed. Humphris, S. E., Herzig, P. M., Miller, D. J. & Zierenberg, R.), (In the press.) College Station, TX (Ocean Drilling Program).

Thompson, G., Humphris, S. E., Schroeder, B., Sulanowska, M. & Rona, P. A. 1988 Active vents and massive sulphides at 26° N (TAG) and 23° N (Snakepit) on the Mid-Atlantic Ridge. *Can. Mineral.* **26**, 697–711.

Tivey, M. K. 1998 In *Proceedings of the Ocean Drilling Program, Scientific Results* **158** (ed.

Humphris, S. E., Herzig, P. M., Miller, D. J. & Zierenberg, R.), (In the press.) College Station, TX (Ocean Drilling Program).

Tivey, M. K., Humphris, S. E., Thompson, G., Hannington, M. D. & Rona, P. A. 1995 Deducing patterns of fluid flow and mixing within the TAG active hydrothermal mound using mineralogical and geochemical data, *J. Geophys. Res.* **100**, 12 527–12 555.

Tivey, M. K., Mills, R. A. & Teagle, D. A. H. 1998 Temperature and salinity of fluid inclusions in anhydrite as indicators of sea water entrainment and heating in the TAG active mound. In *Proceedings of the Ocean Drilling Program, Scientific Results* **158** (ed. Humphris, S. E., Herzig, P. M., Miller, D. J. & Zierenberg, R.), (In the press.) College Station, TX (Ocean Drilling Program).

Tivey, M. K. & McDuff, R. E. 1990 Mineral precipitation in the walls of black smoker chimneys: A quantitative model of transport and chemical reaction. *J. Geophys. Res.* **95**, 12 617–12 637.

Vanko, D. A., Griffith, J. D. & Erickson, C. L. 1992 Calcium-rich brines and other hydrothermal fluids in fluid inclusions from plutonic rocks, Oceanographer Transform, Mid-Atlantic Ridge. *Geochim. Cosmochim. Acta* **56**, 35–47.

Von Damm, K. L. 1988. Systematics and postulated controls on submarine hydrothermal solution chemistry. *J. Geophys. Res.* **93**, 4551–4561.

Von Damm, K. L. 1990 Seafloor hydrothermal activity: black smoker chemistry and chimneys. *Ann. Rev. Earth Planet. Sci.* **18**, 173–204.

Von Damm, K. L., Oosting, S. E., Kozlowski, R., Buttermore, L. G., Colodner, D. C., Edmonds, H. N., Edmond, J. M. & Grebmeier, J. M. 1995 Evolution of East Pacific Rise hydrothermal vent fluids following a volcanic eruption. *Nature* **375**, 47–50.

Westall, J. C., Zachary, J. L. & Morel, F. M. M. 1976 MINEQL, a computer program for the calculation of chemical equilibrium composition of aqueous systems. Tech. Note 18, Dept. Civil Eng., MIT, Cambridge, MA.

Zhang, Y. & Frantz, J. D. 1987 Determination of the homogenization temperatures and densities of supercritical fluids in the system $NaCl-KCl-CaCl_2-H_2O$ using synthetic fluid inclusions. *Chem. Geol.* **64**, 335–350.

Thermocline penetration by buoyant plumes

By Kevin G. Speer

Laboratoire de Physique des Océans IFREMER/CNRS,
B.P. 70, 29280 Plouzané, France

Plumes of buoyant fluid rise in a stratified environment until their buoyancy with respect to the environment reverses, they become heavier than their surroundings and gravitational forces bring them to a halt. Obstacles to turbulent plume rise, occur in the form of external stratification and two-component mixing, which changes the buoyancy of the plume. Volcanic eruptions introduce large amounts of heat to the water column, and the question arises as to whether or not such eruptions can drive plumes up to the sea surface, and create a significant sea surface temperature anomaly.

A turbulent plume model is used to estimate the magnitude of an eruption which might be capable of driving a plume across the ocean's thermocline, which poses a substantial barrier to vertical motion—more so, for instance, than the tropopause with respect to atmospheric plumes. The confining effect of Earth's rotation helps to maintain stronger anomalies in the horizontal spreading phase of the motion at the sea surface. Plumes which cannot attain the surface may also have substantial temperature and salinity anomalies if these quantities vary in the the source or water column through which the plume rises and entrains water.

1. Introduction

The effect of marine geothermal sources is mostly hidden from view, occuring for the most part on deep mid-ocean ridges where new crust is formed, or in basins where old crust is sinking into the Earth. Often the sources are thousands of metres deep, and require large institutional efforts merely to access them. While their effect can be dramatic on the deep ocean environment, as they are capable of sustaining life and generating ocean circulation, a whole new class of environmental effects is implied if these geothermal sources should somehow modify significantly the sea surface temperature (SST). Air–sea interaction in the form of heat and moisture fluxes from the ocean to the atmosphere strongly depends on the sea surface temperature, as phenonema ranging from the El Niño-southern oscillation to hurricanes are at least partly controlled by sea surface temperature anomalies of a few degrees Celsius.

Volcanoes can modify the sea surface temperature directly when they happen near the surface, on seamounts, for instance, or when lava from terrestrial eruptions flows into the sea. Large eruptions presumably generate significant SST anomalies, but the direct heat input from magma has to compete with secondary effects such as anomalous winds, ash clouds and aerosols affecting the Sun's energy input and particulates in the surface layer which modify absorption. The physics of this complex surface layer activity clearly involves atmospheric circulation as well as oceanic circulation, and has been studied from the atmospheric perspective especially in terms of the large-scale influence of aerosol injections on global temperature. The *indirect* heating of the sea surface by seafloor geothermal sources projected to the surface in hydrothermal plumes is considered here. The focus is on the rising and spreading plume generated by the geothermal sources; further interaction of hot plumes with the atmosphere has been suggested, among other things, to be partly responsible for triggering El Niño (Walker 1995) and, in extreme cases, large hurricane-like storms (Emanuel et al. 1995).

How strong does the heat source on the seafloor have to be in order to generate a buoyant plume capable of reaching the surface? To answer this question the full water column density (buoyancy) structure, or stratification must be taken into account. Over most of the ocean, away from high latitudes and outside of marginal seas, the stratification can be characterized by a deep, subthermocline density gradient of -10^{-4} kg m^{-3} per metre, and a shallow thermocline gradient of -10^{-2} kg m^{-3} per metre. Corresponding buoyancy frequencies (N, table 1) are $N = 10^{-3}$ s^{-1} in deep water, and $N = 10^{-2}$ s^{-1} in the thermocline. A turbulent buoyant jet is characterized (see, for example, Turner 1986) by its initial volume flux Q (units: $L^3 T^{-1}$), momentum source M ($L^4 T^{-2}$), and buoyancy source F ($L^4 T^{-3}$). From these quantities, two length scales may be constructed representing the distance to which source geometry and initial momentum influence plume evolution. A 1 MW hydrothermal vent discharging at 1 ms^{-1} through a 10 cm diameter orifice has, roughly, $Q = 10^{-2}$ m^3 s^{-1}, $M = 10^{-2}$ m^4 s^{-2}, and $F = 10^{-3}$ m^4 s^{-3}; at distances greater than $l_M = M^{3/4} F^{-1/2} = 1$ m and $l_Q = Q^{3/5} F^{-1/5} = 0.3$ m the orifice geometry and momentum source are unimportant and the resulting plume may be characterized by F alone, emanating from a virtual point source about 1 m below the orifice. These distances and orifice size are all small compared to rise height, even for the strong sources to be considered here. (Little et al. (1987) provide a detailed discussion of source characteristics for a particular vent site).

From N and F, rising plume scales may be constructed for height H_N, velocity W_N, and buoyancy g'_N:

$$h_N = (F/N^3)^{1/4}, \qquad (1.1)$$

$$W_N = (FN)^{1/4}, \qquad (1.2)$$

$$g'_N = (FN^5)^{1/4}. \qquad (1.3)$$

From the definition of N, the buoyancy jump across the thermocline is $N^2 H$, where H is the vertical thickness scale of the thermocline, roughly 500 m. Setting the plume buoyancy equal to that of the thermocline, $(N^5 F)^{1/4} = N^2 H$, we find that $F = N^3 H^4$. With thermocline values for N and H, $F \sim 10^5$ m^4 s^{-3} represents the source strength needed to penetrate the thermocline. Up to the base of the thermocline the stratification is weak by comparison and F is approximately conserved in the plume. This source has to operate over a time period of at least one buoyancy period N^{-1} if the plume is to reach the surface; stronger sources for proportionally shorter periods generate thermals with the same penetration scale H.

The associated heat source is approximately $(\rho C_p/g\alpha) F$, or about 10^7 MW, for $\rho = 500$ kg m^{-3}, $C_p = 6000$ J kg^{-1} °C^{-1} and $\alpha = 3 \times 10^{-3}$ (values relevant to high-temperature seawater). The transfer of this much buoyancy to cooler ambient temperature seawater by conduction at the surface of freezing magma requires a somewhat higher heat source, of order 10^8 MW, mainly because α is smaller at lower temperatures. This source strength is far greater than that from a vent or cluster of vents,

Table 1. *Definition of basic parameters, where T is temperature, S is salinity, ρ is density, ρ_{amb} is the ambient density, ρ_o is a constant reference density and g is gravity*

parameter	symbol	units
Coriolis parameter	f	s^{-1}
buoyancy frequency	$N = \left(-g\dfrac{\partial \rho_{\text{amb}}}{\partial z}\right)^{1/2}$	s^{-1}
buoyancy	$b = -g\dfrac{(\rho - \rho_{\text{amb}})}{\rho_o}$	m s^{-2}
reduced gravity	$g' = -b$	m s^{-2}
thermal expansion	$\alpha = -\dfrac{1}{\rho}\dfrac{\partial \rho}{\partial T}$	$^\circ\text{C}^{-1}$
haline contraction	$\beta = +\dfrac{1}{\rho}\dfrac{\partial \rho}{\partial S}$	10^{-3}
specific heat	C_p	$\text{J kg}^{-1}\,^\circ\text{C}^{-1}$
vertical velocity	W	m s^{-1}
specific mass transport	Q	$\text{m}^3\,\text{s}^{-1}$
momentum transport	M	$\text{m}^4\,\text{s}^{-2}$
heat flux	H	W m^{-2}
excess salt flux	$W(S - S_{\text{amb}})$	$\text{m }\permil\,\text{s}^{-1}$
heat transport	$\mathcal{H}$	W
salt transport	$\mathcal{S}$	kg s^{-1}
buoyancy flux	$B = \dfrac{g\alpha \mathcal{H}}{\rho C_p} + g\beta W(S - S_{\text{amb}})$	$\text{m}^2\,\text{s}^{-3}$
buoyancy transport	$F = \displaystyle\int_{\text{area}} B\,\mathrm{d}A$	$\text{m}^4\,\text{s}^{-3}$

Table 2. *Plume regimes based on order of magnitude values of source strengths $\mathcal{H}$ and F, and associated velocity and transport scales in the flow above the source*

	$\mathcal{H}$ (MW)	F (m^4 s^{-3})	W (m s^{-1})	Q (m^3 s^{-1})
vent	10	10^{-2}	10^{-1}	10^{-2}
mound	10^2	10^{-1}	1	1
megaplume	10^5	10	1	10^2
hyperplume	10^8	10^5	10	10^6

and roughly one thousand times greater than even that thought to be responsible for the largest observed plumes, or megaplumes (Baker *et al.* 1987). For convenience, plumes formed by sources of this magnitude may be termed hyperplumes (table 2).

An immediate question arises concerning the plausibility of supplying this amount of heat by magma outflow and subsequent cooling. A magma supply of roughly 10^4–10^5 m^3 s^{-1} appears to be needed to be cooled to ambient temperature, transferring heat to the seawater. A further question about the *mechanism* of this transfer is pertinent, because lava flows tend to cap themselves with an insulating ceramic lid, and simple scaling with plausible conductivity coefficients suggests that huge areas would be needed if thicknesses more than a few metres or so are built up by the outflow. Thus a turbulent mixing between the magma and seawater, as might occur in explosive eruptions, is necessary to obtain such heat transfer rates. Whether or not these conditions are satisfied by an immense eruption (high pressure at the seafloor would reduce gas expansion) or bollide impact, for instance, is hard to estimate; but heat sources of comparable strength have been documented in the largest terrestrial eruptions (Wilson et al. 1978)—though of short duration.

It is assumed here that very strong sources of heat arise and last for a period of hours to days. Numerous phenomena that could permit warm water to reach the surface more readily are neglected, such as bubbles or buoyant particles, or vortex rings (which trap fluid; Turner 1960), in favour of concentrating on the situation that seems to require the least special circumstances. Rise height in other geometries, such as a line source of length L or a surface flux over an area L^2, depend more strongly on the buoyancy source strength $((F/LN^3)^{1/3}$ and $(F/L^2N^3)^{1/2})$, and might be expected to attain the surface more easily. However, to obtain these geometrical limits practically requires $L \gg H$ in which case the buoyancy source per unit length or per unit area is much reduced. In the limit $F \to \infty$ the ratio of line source penetration to point source penetration $(h_N/L)^{1/3}$ grows large since L is fixed, but for values such that $h_N \sim H$ and $L \gg H$ the line source penetration is smaller. When the ocean depth and source size are similar ($L \sim H$), details of the source geometry are likely to be important to the vertical structure of the plume. The point source model is used here only as a guide to the relation between source strength and penetration.

Weak oceanic stratification at higher latitudes obviously allows even modest hydrothermal sources to reach the surface; on the other hand the sea surface anomalies would also be modest, and moreover the air-sea interactions described above are tropical, and do not operate at higher latitudes.

To understand how a plume behaves when it encounters a thermocline it is helpful to refer to the laboratory experiment of Kumagai (1984). He studied the evolution and eventual penetration of a buoyant plume across a finite density or buoyancy jump. There is a downward buoyancy flux as the plume impinges on the interface and erodes the stratification, over a time scale which can be longer than the time scale N^{-1} implied by the above plume scaling. Relatively weak plumes can thus penetrate the thermocline, though the resulting anomaly will be weak since the plume is always near its equilibrium density. With Earth's rotation, this process might happen more readily since energy is confined; on the other hand for some parameter ranges baroclinic instability could break up the spreading plume and re-establish the thermocline stratification, prolonging or suppressing the process.

A simple plume model is used next to determine the penetration height as a function of the source strength F in the presence of a realistic thermocline. At some point the plume is buoyant enough to reach the surface. Further increases in source strength produce stronger sea surface buoyancy, or equivalently, temperature anomalies in the one component system. The spreading phase is considered in an idealized model designed to examine the tendency for Earth's rotation to confine the plume fluid, and maintain a significant anomaly in the presence of mixing. Finally, a case relevant to the release of the Red Sea hot brine reservoir is described, to highlight salinity effects.

2. An exponential thermocline

A standard plume model (Turner 1986; Little et al. 1987; Speer & Rona 1989) was used to find the vertical structure of the plume for a variety of source strengths. The model equations are:

$$Q_z = 2\pi RW, \tag{2.1}$$
$$M_z = \lambda^2 \pi R^2 b, \tag{2.2}$$
$$F_z = -\pi R^2 W N^2(z), \tag{2.3}$$

where W is vertical velocity, R is the radial scale for the assumed Gaussian profile. An ambient stratification N must be specified, and based on exponential fits to data from the central subtropical Pacific Ocean (similar to the Atlantic Ocean), a simple form:

$$T(z) = T_{00} + T_0 e^{\delta z}, \quad S(z) = S_0, \tag{2.4}$$

appeared to be adequate to represent $N(z)$. For the runs described here, $T_{00} = 1.0$, $T_0 = 0.01$, $S_0 = 35$ for $0 \leqslant z \leqslant 2000$ m. The thickness scale $1/\delta = 250$ m. An equation relating temperature and salinity to density is required. For temperatures below $40\,°$C, and with S within the range of normal seawater, the full UNESCO equation of state was used. At higher temperatures Chen's (1981) result for pressure near 200 bars and constant salinity was substituted.

Initial values for Q, M and F were based essentially on the following megaplume scenario and segment scenario. From chemical and geophysical observations the megaplume was thought to form from high-temperature water venting at roughly $1\,\mathrm{m\,s^{-1}}$ along a fissure several hundred metres long, and less than 1 m wide (Baker et al. 1987). If the temperature were about $300\,°$C, typical values would be $Q = 125\,\mathrm{m^3\,s^{-1}}$, $M = 62.5\,\mathrm{m^4\,s^{-2}}$, $F = 50\,\mathrm{m^4\,s^{-3}}$. The source F is similar to the total source strength $39\,\mathrm{m^4\,s^{-3}}$ used by Lavelle (1995) in model studies of the megaplume (integrating over a 1200 m long line source). In the segment scenario the fissure occupies the entire segment, multiplying the total area of venting fluid by a thousand or so, and magnifying the source by a similar amount.

The standard formula for rise height $Z^* = 3.8(F/N^3)^{1/4}$ applies to a constant N or uniformly stratified environment. With arbitrary N one usually must solve numerically for the rise height (figure 1a); however, substituting the exponential form $N = N_0 e^{\delta Z^*/2}$ and solving for Z^* turns out to give a reasonable agreement with model results (figure 1b). Thus the effective stratification felt by the rising plume is essentially that near the final equilibrium level. A simple approximate solution to the equation that results from the above substitution is

$$Z^* = \frac{2}{3\delta} \ln\left(\frac{F}{F_0}\right), \tag{2.5}$$

valid when the rise height is close to the full water depth, showing the logarithmic dependence This form gives heights approaching those of the plume model near the surface (figure 1c; $F_0 = 1\,\mathrm{m^4\,s^{-3}}$).

When the strength is of order $10^4\,\mathrm{m^4\,s^{-3}}$ the plume reaches the surface. Inertia carries the plume past its equilibrium point, though, so that it arrives at the surface with a negative buoyancy or temperature anomaly (figure 2). This would give it an appearance similar to other eddy-like structures such as the megaplume or cold-core Gulf Stream rings, with a positive upper density anomaly and negative lower density anomaly, centered on its equilibrium level.

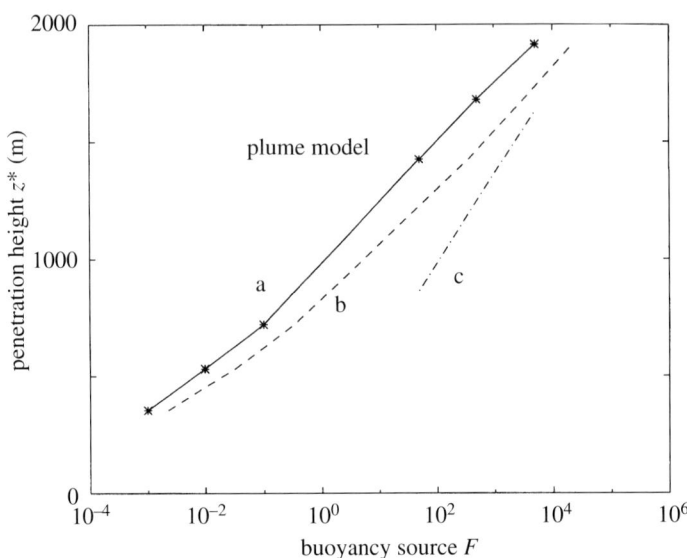

Figure 1. Plot of plume penetration height Z^* versus source strength F, results from plume model (a). Also shown are approximations based on the equation for rise height: valid for all heights (b), and simplified for heights close to the full ocean depth (c).

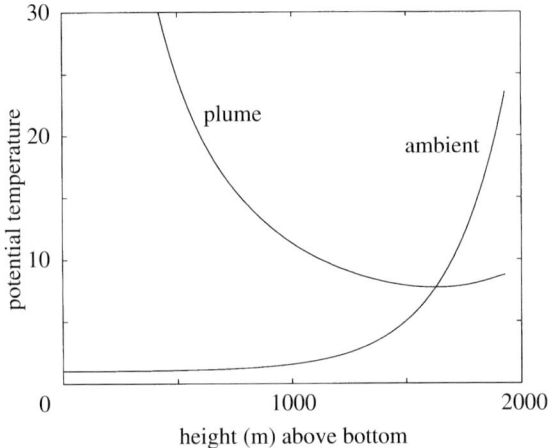

Figure 2. Plume temperature as a function of height for a source just adequate to reach the surface layer ($F = 5 \times 10^3$ m^4 s^{-3}). The initial temperature is 365 °C. The plume rises at about 5 m s^{-1} over most of the water column. Also shown is the exponential ambient temperature representing the thermocline. The top of the plume is cool relative to ambient surface water.

Of particular interest is the possibility that the plume could raise the sea surface temperature. Increasing the source strength raises the centreline temperature to surface values at about $F = 3 \times 10^5$ m^4 s^{-3} (figure 3). Beyond this point, SST rises slowly, as expected from plume scaling for buoyancy, giving rise to anomalies of several degrees. At values of 6×10^6 m^4 s^{-3} the SST is potentially high enough to have important local air–sea interaction effects (e.g. Emanuel et al. 1995). However, such effects are only plausible if the anomaly covers a significant area of the ocean, larger than that of the plume, and if it lasts for a period of time sufficient for the

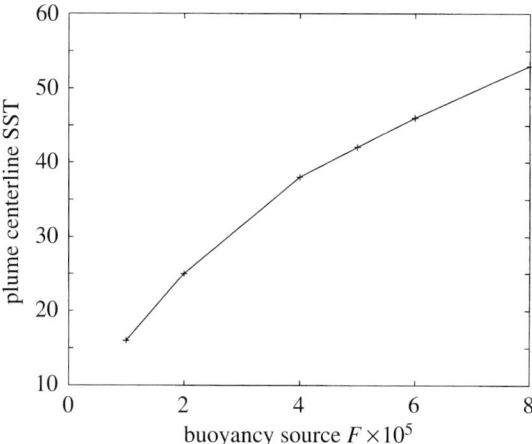

Figure 3. Plume temperature at the surface for various source strengths ($m^4 s^{-3}$). The ambient SST is about $30\,°C$. Only sources stronger than about 3×10^5 $m^4 s^{-3}$ actually raise SST for the present choice of thermocline structure.

atmosphere to react, of the order of one day. A simple model of the spreading phase is presented next to address these issues.

3. A circular slab model of spreading

The time evolution of lenses of fluid, either stratified or homogeneous, is the subject of much research; in different contexts they are commonly observed in the ocean, and their dynamics includes many interesting processes (McWilliams 1985). Here, a drastically simplified set of equations is obtained by focusing on bulk balances in a circular slab, and the evolution of its perimeter.

The plume fluid is assumed to spread out on the surface axisymmetrically to radius R, in a layer with mean thickness h and buoyancy g' (figure 4). The surrounding water has a constant density and zero buoyancy. Conservation of mass or volume $V_{OL} = \pi R^2 h$ gives, in the presence of mixing represented by a vertical entrainment velocity W_e (see later):

$$\frac{dV_{OL}}{dt} = Q + W_e \pi R^2, \quad (3.1)$$

with Q the amount of plume fluid feeding the slab.

The equation for the bulk balance of buoyancy in the slab is

$$\frac{d(g'V_{OL})}{dt} = Qg'_0, \quad (3.2)$$

where g'_0 is the buoyancy of the entering plume fluid.

A reduced-gravity model with motion only in the buoyant layer is commonly used in the study of fronts and vortices (e.g. Cushman-Roisin et al. 1985). Often, strong assumptions about the spatial structure of the vortex are made to simplify the problem. Here, we ignore the spatial structure completely and focus on the evolution of the rim, or radius R. The specification of the pressure gradient at the rim, which normally involves the spatial structure, is contrived to produce a simplified set of equations that mimic the behaviour of the more complicated model. The equations of motion of a parcel at R, together with bulk mass and buoyancy conservation, are

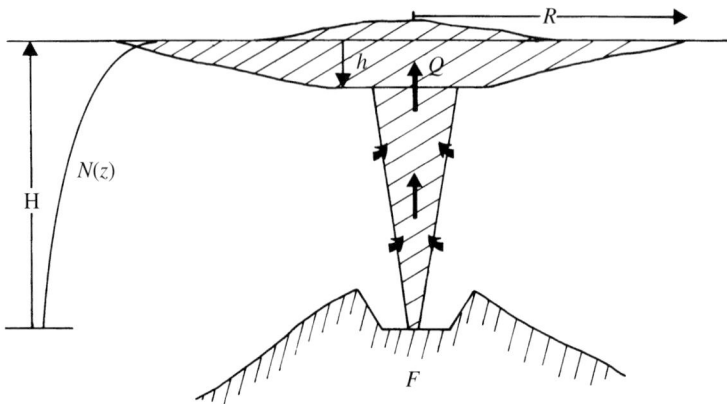

Figure 4. Schematic of a plume resulting from a major eruption on a ridge. The plume rises through a thermocline $N(z)$ and spreads out laterally on the surface, with a continuous supply Q of fluid from below.

written as:

$$\dot{R} = u,$$

$$\dot{u} = g'h/R + fv - \frac{W_e}{h}u + \frac{v^2}{R},$$

$$\dot{v} = -fu - \frac{W_e}{h}v - \frac{uv}{R},$$

$$\dot{h} = \frac{Q - 2\pi Rhu}{\pi R^2} + W_e,$$

$$\dot{g'} = \frac{Q}{V_{OL}}(g'_0 - g') - \frac{W_e}{h}g'.$$

All radial structure is ignored, and the radial pressure gradient has been replaced by the form $g'h/R$, simulating the hydraulic pressure drop across the rim, of reduced amplitude at large radius or small thickness. These simplifications eliminate many processes, including wave radiation, but the primary interest here is the effect of mixing. Exact particular solutions to *inviscid* equations for the time-dependent motion of a lens of fluid at the surface were obtained by Cushman–Roisin et al. (1985), showing the basic modes of motion of circular and elliptic vortices. In the above system, a mixing term in the form of a vertical entrainment velocity is included:

$$W_e = E_0\, Ri^{-1}\, s, \quad s = \sqrt{u^2 + v^2}, \tag{3.3}$$

parametrized here as a function of a Richardson number $Ri = g'h/s^2$, which is small (and mixing large) when the scale for inertial forces is stronger than that for buoyancy forces (Turner 1986). Values for E_0 are zero (no mixing) or 10^{-3}. This parametrization is meant only to capture roughly the changes in properties due to mixing with the surrounding water as the plume spreads.

Initial conditions are set at a radius $R_0 = 5000$ m, by which point the plume's radial expansion has decreased to values more appropriate for the above system.

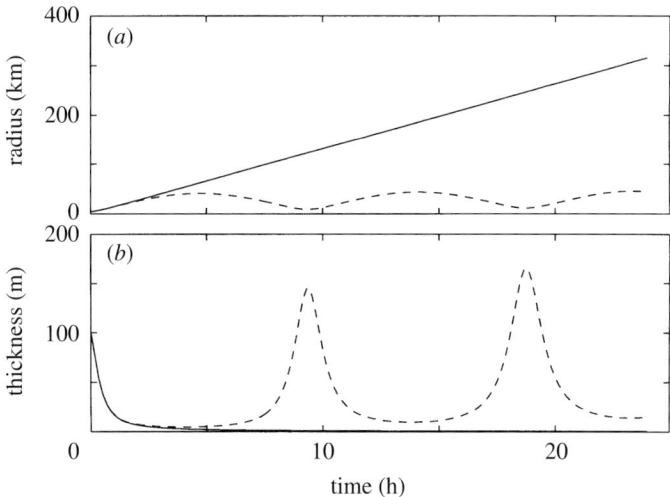

Figure 5. Slab model solutions without mixing. Radius versus time (upper panel) without rotation (solid), and with rotation (dashed). Thickness versus time (lower panel) without (solid) and with (dashed) rotation.

They are: $h = 100$ m, $g' = 0.1$ m s^{-2}, $u = 1$ m s$^{-1} \sim \sqrt{g'h}$, and $v = 0$. The other parameters are $f = 10^{-4}$ s^{-1} and $Q = 10^6$ m^3 s^{-1}. Starting at smaller radii and higher velocity exaggerates the behaviour described here, but does not produce new behaviour. Initial thickness was chosen arbitrarily; initial buoyancy corresponds to a temperature anomaly of tens of degrees. The radial velocity for small time is $g'ht/R_0$, and it increases to values ca. $\sqrt{g'h}$ if the slab starts from rest.

Overall, the magnitude of the (specific) angular momentum for a circular disk $L = V_{OL}Rv/2$ increases owing to lateral expansion and Earth's rotation, spinning up anticyclonic motion in the slab, similar to the case for normal nonbuoyant hydrothermal lenses (Lavelle & Baker 1994; Speer & Marshall 1995). The centrifugal force term v^2/R plays a small role and u and v are basically out of phase. Note that the slab can regain buoyancy lost to mixing, since the plume resupplies it at the initial value g'_0.

Solutions with no mixing show the basic behaviour with and without rotation (figure 5). Without rotation the slab expands and thins monotonically; with rotation, the expansion is arrested after one quarter of an inertial period $(2\pi/f)$, and the plume begins to pulsate. A similar pulsation of a circular lens was described by Cushman–Roisin et al. (1985). Thickness grows rapidly near the minimum radius owing both to contraction and filling (Q).

Ignoring the nonlinear velocity terms and mixing, the equations for u and v can be combined into one equation:

$$\ddot{u} + f^2(1 + (3R_{\rm d}^2/R^2))u = 0, \quad (3.4)$$

$$R_{\rm d}^2 = g'h/f^2, \quad (3.5)$$

where $R_{\rm d}$ is the deformation radius. Substituting $u = u_0 e^{i\omega t}$ gives $\omega^2 = f^2(1 + 3R_{\rm d}^2/R^2)^{1/2}$ showing free oscillations analogous to Poincaré waves, for which $\omega^2 = f^2 + \kappa_{\rm H}^2 g'H$, where $\kappa_{\rm H}$ is the horizontal wavenumber. For g' and f as above, and with $h = 10$ m at $R = 25$ km, the frequency is about $1.2f$. In the linear, horizontally uniform limit $(R \to \infty)$ the effect of mixing is only to dampen the oscillations, and not to modify their frequency; solutions with and without mixing tended to show a spectral peak near $1.5f$, qualitatively consistent with the linear limit.

The damping effect is illustrated in solutions with mixing (figure 6). The initial pulsation period has increased by about 25%, measured by the time to the first thickness maximum, but further oscillations are only about 10% slower than the case without mixing. Velocities (figure 6b) show the adjustment to geostrophic-like flow with negative azimuthal flow v, and radial velocity oscillating about zero. Froude number $Fr = \sqrt{Ri}^{-1}$ is a gauge of mixing since it is large when mixing is large. Rotation tends to diminish mixing by preserving thickness. In these solutions, the Burger number R_d/R rapidly approaches a value of roughly 0.5, passing through a minimum near $tf \sim 3$, but remaining close to 0.5 for longer times. Finally, a key result is the stronger buoyancy or temperature anomaly with rotation. Confinement and reduced mixing allows the central plume to recharge the surface layer and maintain buoyancy. There is little dissipation even after 100 h (figure 7). By this time ($tf \sim 10$–100) baroclinic instability is likely to be developing, causing a horizontal displacement of the entire slab away from the source (Helfrich & Speer 1995).

4. A Red Sea brine plume

At the bottom of the Red Sea, pools of hot brine have been formed by the exchange of heat and chemicals between seawater and the seafloor, with temperatures above 60 °C and salinities of roughly 250‰ (Ross 1983), giving a density of about 1200 kg m^{-3}. The overlying deep water, at 22 °C and 40.5‰, has a density close to 1030 kg m^{-3}. To make the dense brine buoyant, its temperature must be raised several hundred degrees by volcanic activity.

The standard plume equations were modified to deal with heat and salinity transports separately:

$$H_z = -\lambda_f^{-1}\rho_0 C_p Q \frac{\partial T_{\mathrm{amb}}}{\partial z}, \qquad (4.1)$$

$$S_z = -\lambda_f^{-1}\rho_0 Q \frac{\partial S_{\mathrm{amb}}}{\partial z}10^{-3}, \qquad (4.2)$$

where $H = \lambda_f^{-1}\rho_0 C_p Q(T - T_{\mathrm{amb}})$, $S = \lambda_f^{-1}\rho_0 Q(S - S_{\mathrm{amb}})$ and $\lambda_f^{-1} = (1 + \lambda^2)/\lambda^2$. Now the buoyancy transport $F = \lambda_f^{-1} Q b$, with b calculated from an equation of state. Salinity was accomodated in the simplest possible way at high temperature by setting the haline contraction coefficient $\beta = 0.9 \times 10^{-3}$ for $T > 40$ °C. The resulting potential density (figure 8) has curvature due to the temperature dependence alone.

This diagram helps to illustrate how dense brine might become buoyant. All water to the right of about 1028.5 kg m^{-3}, the density of the bottom water, is nonbuoyant. Heating, to increase temperature, or mixing with overlying water, to reduce salinity, is required to generate a buoyant plume. Thus dense brine in the lower right-hand part of the diagram migrates up (with heating) and to the left (with mixing) to become finally buoyant. Then it rises off the seafloor and entrains the fluid surrounding it. To represent this entrainment temperature and salinity profiles were chosen to simulate the Red Sea (figure 9). Only a finite volume on the order of several cubic kilometres of brine is available at the source; the vertical transport in the plume drains this volume in a period of hours to days (table 3).

Two solutions are plotted (figure 10) in order to make two points. The first is that if the buoyancy source is strong enough, then the plume can reach the surface

Thermocline penetration by buoyant plumes 259

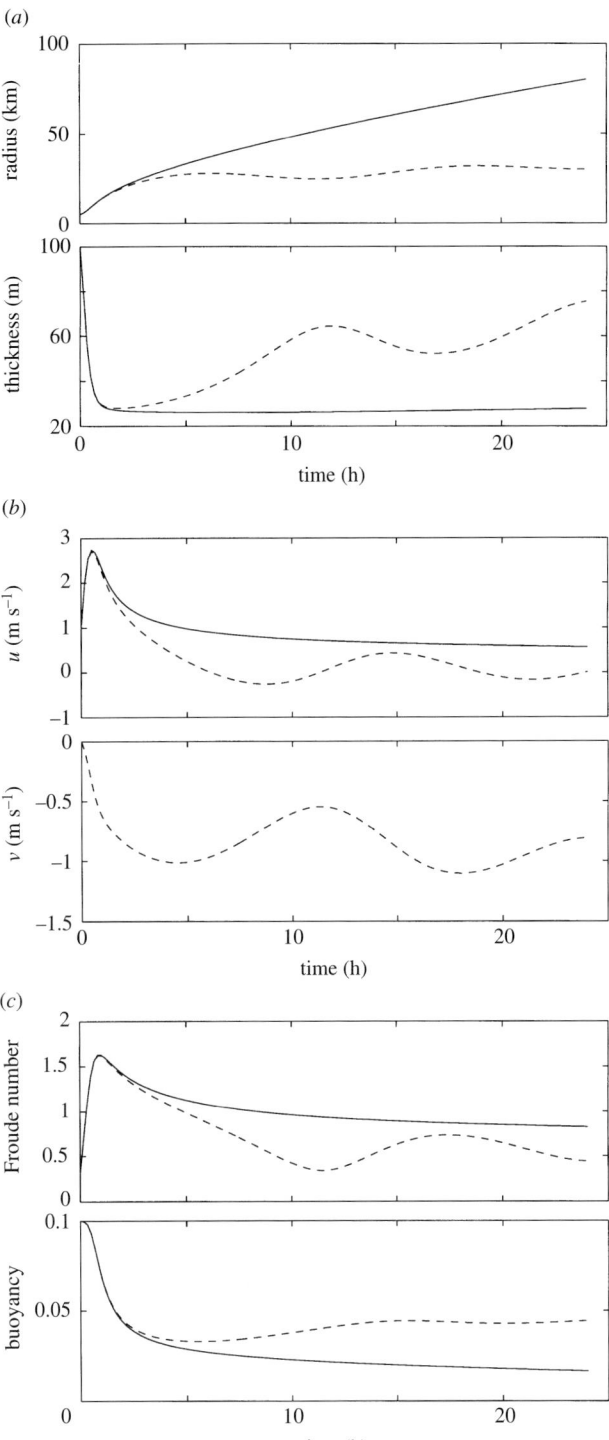

Figure 6. Slab model solutions with mixing. (a) Radius (upper) and thickness (lower) without rotation (solid), and with rotation (dashed). (b) Radial (upper) and azimuthal (lower) velocity. Azimuthal velocity is zero without rotation. (c) Froude number (upper) and buoyancy (lower).

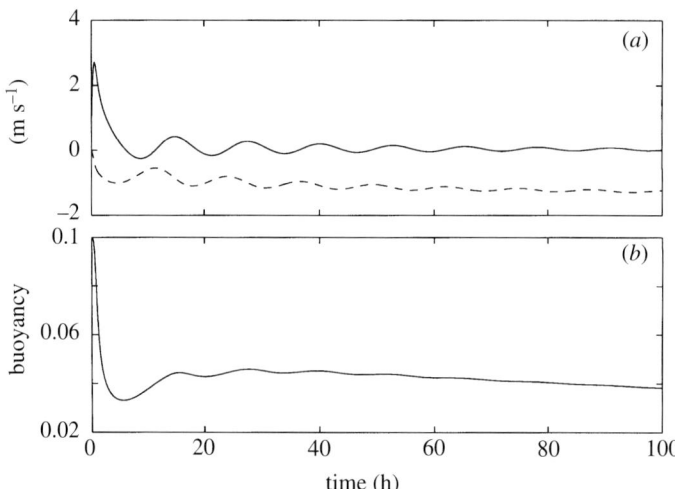

Figure 7. Longer slab model run with rotation. Upper panel shows radial (solid) and azimuthal (dashed) velocity; lower panel shows nearly constant buoyancy after one day.

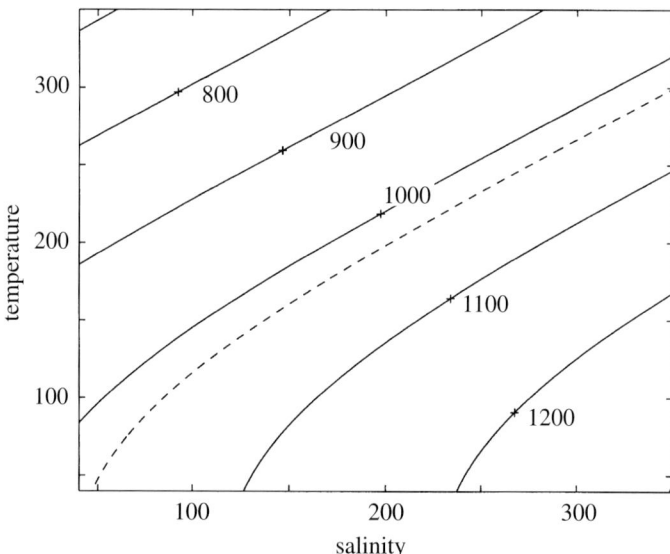

Figure 8. Density (kg m^{-3}) from a simplified high temperature and salinity equation of state. Bottom density (dashed) separates buoyant from nonbuoyant source conditions. Note that on this scale, the full density range (1024–1028.5 kg m^{-3}) of the idealized Red Sea lies nearly along the curve for bottom density.

as before. Here, in order to be that strong, it seems to be necessary to reduce the salinity of the brine quite substantially. Presumably the motion generated by any major eruption could do so by stirring up water near the seafloor. Second, mixing between the rising plume and the interior or ambient seawater drives T and S toward ambient values, but because of the curved isopycnals, the *density* of the plume can become equal to that of the surrounding water at some intermediate level. This cabelling effect (Sato 1972) halts the plume at mid-depth, despite the absence of any stratification ($N_T = N_S = N = 0$) at these levels. Subsequent turbulent mixing is reduced as a lens is formed, isolating a core of the plume fluid from its surroundings.

Thermocline penetration by buoyant plumes

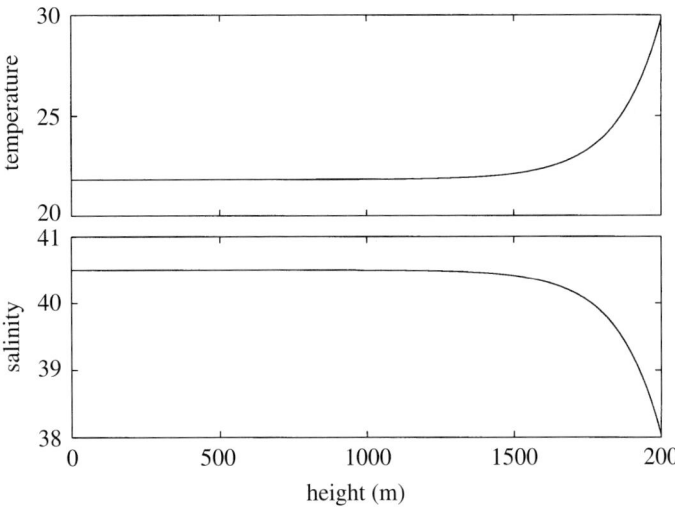

Figure 9. Idealized ambient temperature and salinity versus height for the Red Sea.

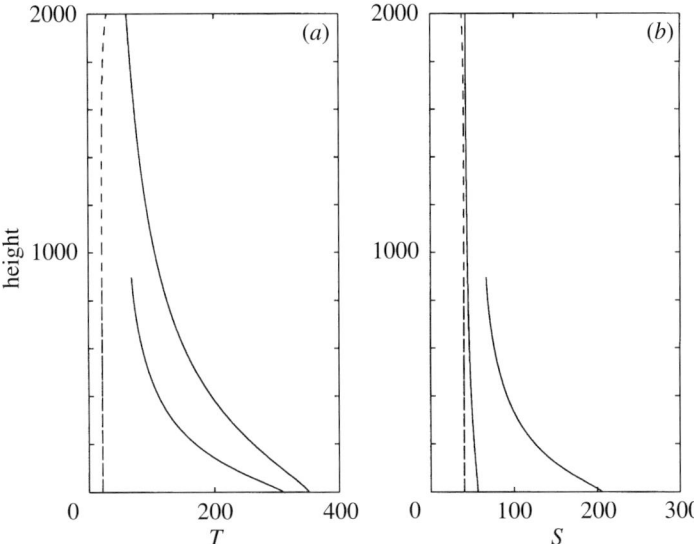

Figure 10. Plume solutions for the Red Sea case with ambient structure (dashed). The plume starting at higher temperature and lower salinity reaches the surface; another plume reaches its equilibrium level at mid-depth.

Table 3. *Source conditions for Red Sea runs*

penetration	$\mathcal{H}$ (MW)	$\mathcal{S}$ (kg s^{-1})	F (m^4 s^{-3})	Q (m^3 s^{-1})	T (°C), S (‰)
surface	8×10^8	1×10^7	1.9×10^6	1×10^6	350, 57
mid-depth	7×10^7	1×10^7	8×10^4	1×10^5	309, 205

5. Discussion

Buoyant plumes generated by geothermal sources typically rise in a stratified environment, ultimately due to solar heating or chemical components in oceans, lakes and atmospheres. Only a single case among the tremendous variety of possibilities has been investigated here, to focus on what appears to be a ubiquitous problem: plume penetration through a thermocline, or more generally, a pycnocline. Many other phenomena such as seamount volcanism or air–sea interaction are capable of modifying the surface layer properties of the ocean as well, without the burden of crossing a thermocline.

Some aspects of the solutions presented here apply more generally to hydrothermal plumes, such as the expected oscillation of the lens, and a possible two-component mixing behaviour. An interesting area for future study is the initial supercritical expansion of the plume fluid, whether on the surface or internally, as it spreads laterally. Also, two-component effects, such as double diffusion, with a broad definition of salinity to include diverse chemical compositions, might turn out to be quite relevant to normal hydrothermal plumes, which tend to create T–S anomalies by their nature (see also Campbell et al. 1984). The complicated chemical mixture that makes up some geothermal sources could add new effects to plume dynamics; thus far, however, promising avenues for such effects seem to be the generation of vapour and brine phases, thought now to be a rather generic aspect of ridge systems, as well as the release of bubbles in shallow seafloor systems.

The very large transfer of heat required to create thermocline-penetrating plumes must make them rare events. The circumstances of the eruption process leading, moreover, to such transfers remains unclear. Studies of heat transfer in shallow underwater eruptions might be feasible and helpful in this regard. At a fundamental level, it is the rate of heat exchange that drives hydrothermal circulation, whether this takes the form of rising plumes or slow seepage through porous crust. Better constraints on heat flow or the circulation driven by it at any length and time scale are valuable.

Support is provided by the CNRS. The international RIDGE community supports a forum for interesting interdisciplinary exchanges, from which I have benefited. K. Emanuel restimulated interest in getting a plume to the surface, and the Royal Society generously provided the occasion; thoughtful comment on the subject from K. Helfrich is appreciated.

References

Baker, E. T., Massoth, G. J. & Feely, R. A. 1987. Cataclysmic hydrothermal venting on the Juan de Fuca Ridge. *Nature* **329**, 149–171.

Campbell, I. H., McDougall, T. J. & Turner, J. S. 1984 A note on the fluid dynamic processes which can influence the deposition of massive sulfides. *Econ. Geol.* **79**, 1905–1913.

Chen, C.-T. A. 1981 Geothermal systems at 21° N. *Science* **211**, 298.

Cushman-Roisin, B., Heil, W. H. & Nof, D. 1985 Oscillations and rotations of elliptical warm-core rings. *J. Geophys. Res.* **90**, 11 756–11 764.

Emanuel, K. A., Speer, K., Rotunno, R., Srivastava, R. & Molina, M. 1995 Hypercanes: a possible link in global extinction scenarios. *J. Geophys. Res.* **100**, 13 755–13 765.

Helfrich, K. R. & Speer, K. G. 1995 Oceanic hydrothermal circulation: mesoscale and basin-scale flow. In *Seafloor hydrothermal systems, physical, chemical, biological and geological interactions* (ed. S. Humphries, R. Zierenberg, L. Mullineaux & R. Thomson), pp. 347–356. Geophysical Monograph 91, American Geophysical Union.

Kumagai, M. 1984 Turbulent buoyant convection from a source in a confined two-layered region. *J. Fluid Mech.* **147**, 105–131.

Lavelle, J. W. 1995 The initial rise of a hydrothermal plume from a line segment source: results from a three-dimensional numerical model. *Geophys. Res. Lett.* **22**, 159–162.

Lavelle, J. W. & Baker, E. T. 1994 A numerical study of local convection in the benthic ocean induced by episodic hydrothermal discharges. *J. Geophys. Res.* **99**, 16 065–16 080.

Little, S. A., Stolzenbach, K. D. & Von Herzen, R. P. 1987. Measurements of plume flow from a hydrothermal vent field. *J. Geophys. Res.* **92**, 2587–2596.

McWilliams, J. C. 1985 Submesoscale, coherent vortices in the ocean. *Rev. Geophys.* **23**, 165–182.

Ross, D. A. 1983 The Red Sea. In *Estuaries and enclosed seas* (ed. B. H. Ketchum), pp. 293–307. Amsterdam: Elsevier.

Sato, T. 1972 Behaviours of ore-forming solutions in seawater. *Mining Geol.* **22**, 31–42.

Speer, K. G. & Marshall, J. 1995 The growth of convective plumes at seafloor hot springs. *J. Mar. Res* **53**, 1025–1057.

Speer, K. G. & Rona, P. A. 1989 A model of an Atlantic and Pacific hydrothermal plume. *J. Geophys. Res.* **94**, 6213–6220.

Turner, J. S. 1960 A comparison between buoyant vortex rings and vortex pairs. *J. Fluid Mech.* **33**, 639–656.

Turner, J. S. 1986 Turbulent entrainment: the development of the entrainment assumption and its application to geophysical flows. *J. Fluid Mech.* **173**, 431–471.

Walker, D. 1995 More evidence indicates links between El Niños and seismicity. *EOS Transactions*, vol. 76. Washington, DC: American Geophysical Union.

Wilson, L., Sparks, R. S. J., Huang, T. C. & Watkins, N. D. 1978 The control of eruption column heights by eruption energetics and dynamics. *J. Geophys. Res.* **83**, 1829–1836.

Crustal accretion and the hot vent ecosystem

By S. Kim Juniper[1] and Verena Tunnicliffe[2]

[1] Centre de recherche en géochimie isotopique et en géochronologie (GEOTOP) and Département des Sciences Biologiques, Université du Québec à Montréal, C.P. 8888, Succ. Centre-Ville, Montréal (Québec), Canada H3C 3P8
[2] School of Earth and Ocean Sciences and Department of Biology, University of Victoria, P.O. Box 1700, Victoria (British Columbia), Canada V8W 2Y2

We examine evidence for links between seafloor spreading rate and properties of vent habitat most likely to influence species diversity and other ecosystem properties. Abundance of vent habitat along spreading centres appears positively related to spreading rate while habitat stability shows an opposite relationship. Habitat heterogeneity is lowest at faster spreading ridges. Limited data indicate an increasing species diversity with spreading rate, complicated by historical factors. Ecosystem productivity and efficiency of resource utilisation may also reflect diversity differences.

1. Introduction

In explaining organism distribution or in quantifying flows of energy and materials, ecologists are confronted with changing communities whose component populations shift as do interactions with the environment. Fitting patterns to observed changes and formulating predictive models are important elements of contemporary ecology. How we examine a novel system, like that of deep-sea hydrothermal vents, depends partly on our *a priori* beliefs on what the important controls will be. Explanations of community composition and structure include three fundamental view-points: (1) control through biological interactions such as competition and predation; (2) control through environmental limitations such as energy supply or growing conditions; and (3) influence of historical events (see Real & Brown 1991). Their relative importance seems to vary depending upon the system under study, as well as the spatial and temporal scales being examined.

The hydrothermal vent ecosystem requires the chemical energy present in discharging fluids. Distribution of vent communities is thus controlled by the mantle and crustal processes that determine the nature and distribution of hydrothermalism. Even vent community composition is subject to geological and geophysical controls, at all levels: from locally through hydrology (Grehan & Juniper 1996), to globally, through the historic relations of tectonic plates (Tunnicliffe & Fowler 1996). While every ecosystem is shaped by the environment in which it is found, physical (environemental) controls are often complicated, if not obscured, by interaction with biotic processes. Vent ecosystems appear to provide a prime example of accommodation to the physical environment, where geological and geophysical controls are clearly evident across a broad spectrum of space and time scales. Hot vent faunas are influenced

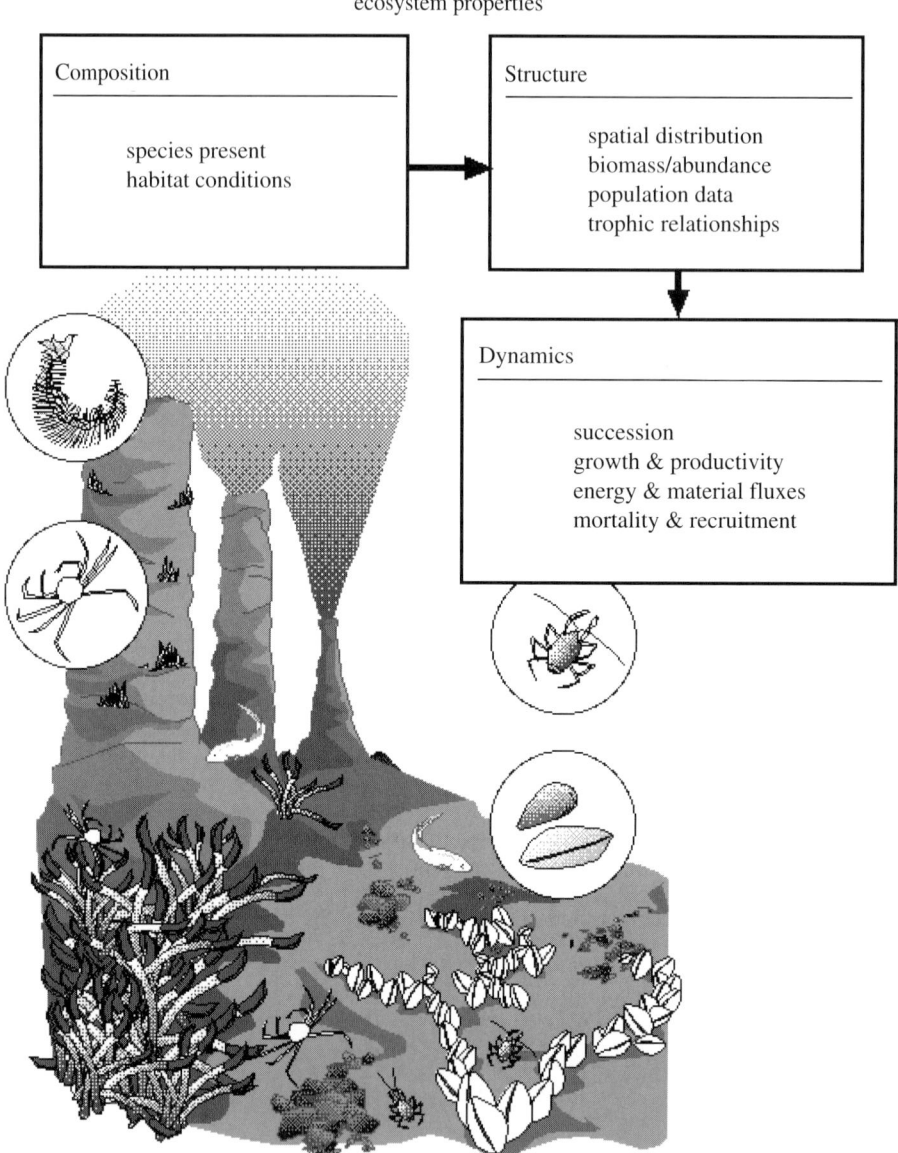

Figure 1. Major ecosystem properties in relation to information required to understand them.

by the tectonic and volcanic processes of crustal accretion because hydrothermalism dictates the habitat. Numerous papers discuss the link between tectonism and volcanism and venting; a most complete recent review is that of Fornari & Embley (1995). It is a small step to make a further link to the biological system. The present challenge is to identify those processes with the greatest control on ecosystem development and functioning and to evaluate whether or not these relationships are amenable to predictive models.

Ecosystem properties can be seen to exist at three conceptual levels that require different levels of knowledge of organisms and environment (figure 1). A first level of understanding is essentially an inventory of the species present and a description of their habitat derived from collections and simple field observations. At the next level, to understand ecosystem structure and identify functional links, quantitative information on species abundance and biomass and knowledge of trophic relationships is required. Finally, measurement of biological rate processes (metabolism, growth, reproduction, etc.), time series observations (colonization, succession) and experimentation enable ecologists to develop models of ecosystem dynamics. Dynamic models require equivalent environmental information for consideration of the physical forces that can drive ecosystem processes.

While the hierarchical representation in figure 1 implies a progressive path of information sophistication and depth of understanding, present knowledge of vent ecosystems tends to reflect the logistic realities of working in the deep sea more than it does any model of how ecological research is conducted. Until recently, vent ecologists were primarily limited to data describing ecosystem elements. Quantitative sampling at vents with submersibles has proved technically difficult and the few estimates that exist do not allow any consideration of environmental influences on biomass or abundance. Knowledge of feeding relationships and trophic structure is incomplete, and the contribution of chemosynthetically produced organic matter remains unquantified (Karl 1995). We know that chemosynthetic microbes provide at least some food for the macrofauna either in symbiosis or through ingestion of free living cells (see conceptual model in Tunnicliffe (1991)). Vent fluids provide the compounds that microorganisms oxidize to generate energy for organic carbon synthesis. Dissolved O_2 for oxidation processes at vents originates from photosynthetic sources. Hydrogen sulphide abundance and the energetics of the oxidation reaction result in a predominance of sulphide-oxidizing microbes at vents (Jannasch 1985, and this volume). Beyond this, we have only a rudimentary grasp of vent ecosystem structure and its accommodation to physical controls.

Vent ecosystem properties are thus understood in an unequal and incomplete manner, imposing definite constraints on the questions that can be addressed in this paper. One of the few comparative pieces of information available on global vent communities is the type and number of species present. Naming and classifying species has represented a major effort partly because of the unusual nature of many of these animals and the fact that more than 90% are species new to science (Tunnicliffe 1991). Numbers of species vary dramatically across the ecosystems of our planet and among similar ecosystems. The major explanations for this variation include habitat stability, level of productivity, habitat heterogeneity (Brown 1988) and geological age (Sanders 1968). We will consider different spatial and temporal properties of the Mid-Ocean Ridge (MOR) vent environment and examine evidence for influences across the three levels of ecosystem properties outlined in figure 1. Reliance on species distribution information limits the scope of our evaluation, since occurrence data are not powerful indicators of community structure or dynamics. Precise locations of all sites mentioned can be found in Fornari & Embley (1995) and Hannington *et al.* (1995).

2. Distribution of vent habitat

Vent habitat has a highly irregular distribution in an approximately linear trend. Communication between vents becomes erratic or intermittent beyond the turbu-

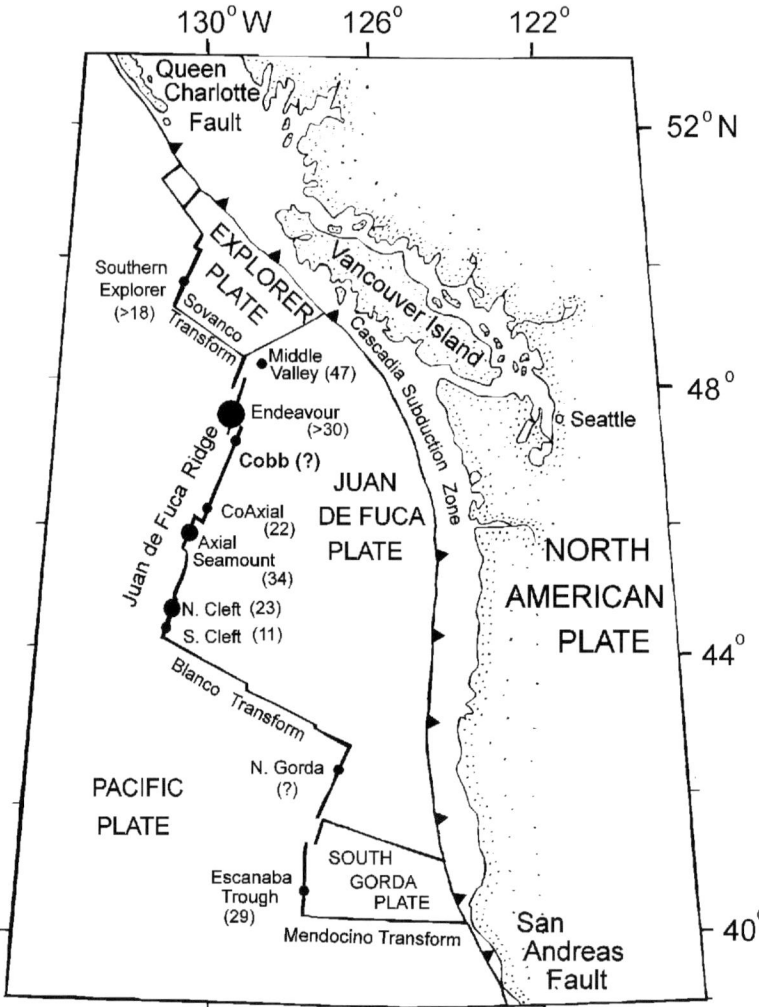

Figure 2. Sites of known vent fields in the northeast Pacific. Size of dot indicates the relative size of the vent field. Numbers in brackets indicate numbers of species sorted from site collections.

lence and fluid dispersion of a single field. The next vent field may be located on the next ridge segment or further. The concept of island stepping stones and the application of island biogeography theory have been suggested several times for use in vent distributions and diversity analysis. However, the theory was developed using a 'source' or continental mass from which species would migrate to offshore islands (MacArthur & Wilson 1967). For vents, the total 'island' fauna is the source and the more the islands the higher the diversity. Abundant small islands will hold more diversity among them than a few larger ones (Rey & Strong 1983). We sampled 16 vents on a 2.2 km length of north Cleft Segment, Juan de Fuca. These young vents held 23 species among them but a maximum of only 75% of them were found at any one vent (Milligan 1993).

With the linear distribution of vents, one can expect to see either a clinal distribution in community and population characters or a complete interchange over the entire region. Such a comparative analysis has not been completed for total faunal

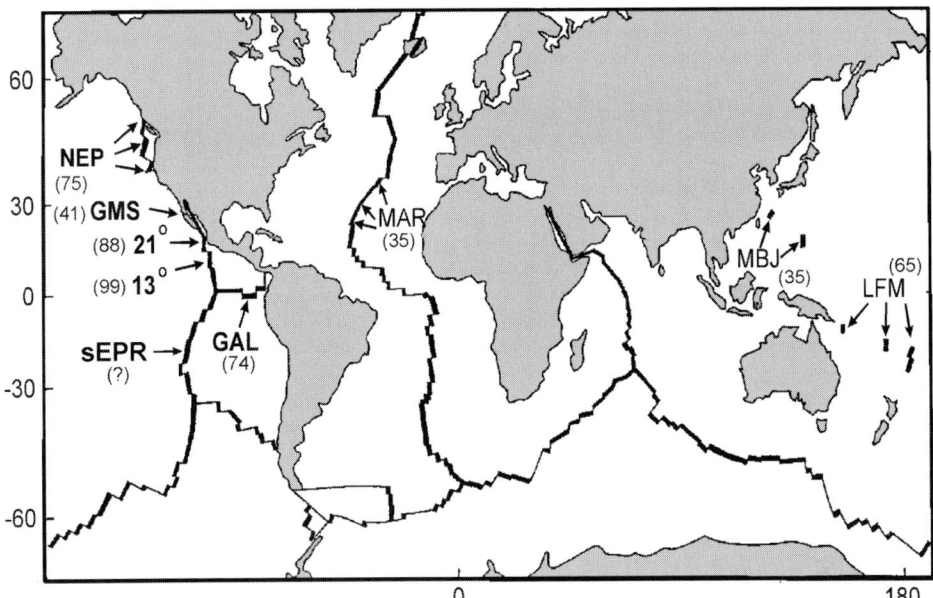

Figure 3. Locations of global sites from which faunal collections have been analysed. Numbers indicate the species presently known from these sites. After Tunnicliffe et al. (1997). See figure 4 for relative spreading rates of ridges.

diversity along ridge segments. However, the genetic structure of populations of several species has been examined. Two species, a polychaete on EPR (Jollivet et al. 1994) and a tubeworm on Juan de Fuca Ridge (Southward et al. 1996) both show extensive communication and mixing: there is no clinal variation over 600–1000 km. The EPR tubeworm (Riftia) does show some genetic differences along 5000 km (Black et al. 1994) but major differentiation across transform faults was not present to the extent reported by France (1992) for amphipods. There is a tendency for extensive mixing of larvae in a common pool over mid-ocean ridges. Models addressing larval entrainment in hydrothermal plumes (Kim et al. 1994; Mullineaux & France 1995) describe the mixing and transport of larvae in distinct water masses above the ridge crest.

Figure 2 illustrates the occurrence of vent sites along the Northeast Pacific Ridges. Within each site indicated, the number of vent openings is highly variable (numbering from hundreds at Endeavour to just a few on Cobb Segment and CoAxial) so that the within-site pool of species should vary just as the within-regional pool will vary. For the present, the larger vent fields remain undersampled and the influence of vent field size on species abundance cannot be critically evaluated. Van Dover (1995) compares spacing of vents on the Mid-Atlantic Ridge (100s of kilometres) with those on northern East Pacific Rise (as low as 5 km) (figure 3). She notes that local endemism should be much higher in the first case where exchange among sites will be limited.

Generally, if there is more habitat, more species can fit even if the system is in a state of flux; pioneer species may live at one site while succeeding species dominate another. The distribution of hydrothermal plumes over active ridges approximates vent distribution (Baker et al. 1995). Plume occurrence increases with spreading rate (figure 4) to the extreme that 60% of surveyed segments of the southern EPR have

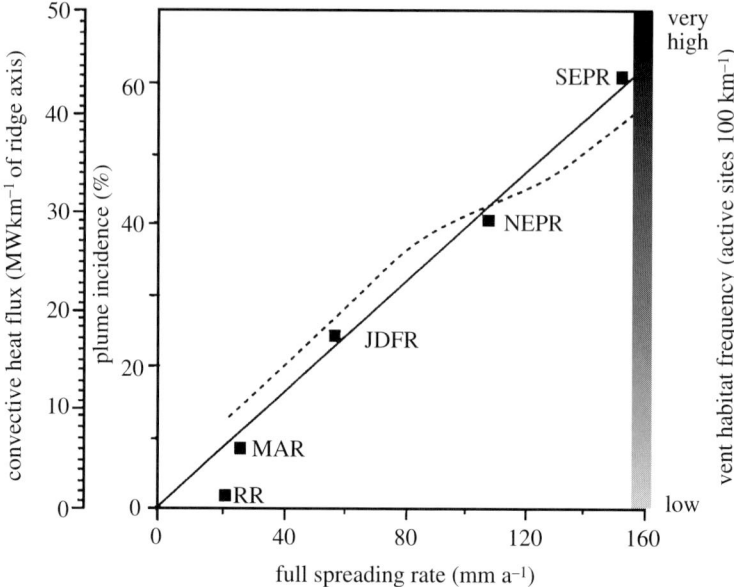

Figure 4. Proposed relationship of vent habitat abundance in relation to the plume incidence, convective heat flow and seafloor spreading rate. Horizontal and left vertical axes adapted from Baker et al. (1995, fig. 22b).

hydrothermal activity (Urabe et al. 1995). Size of regional species pools should relate to abundance of the habitat along a ridge – and thus the spreading rate (figure 3). Diversity data from the Mid-Atlantic Ridge, the northern EPR and the Juan de Fuca regions appear to support the predicted trend (figure 3), although sampling effort is uneven. The relationship of segment scale plume incidence to area and spacing of vent habitat needs to be better defined.

3. Habitat stability

In this section, we propose that annual to decadal scale stability of vent habitat relates to ridge spreading rate through magmatic activity. Thus the stability and predictability of the vent habitat varies among ridges and we could see this variability reflected in community diversity – and possibly structure and dynamics. For example, habitat stability will interact with reproductive strategy, which is highly variable among vent taxa (Mullineaux & France 1995), in determining species presence within a vent field or region.

Ecologists have long asserted that the type, frequency and/or intensity of disruption of a community influences diversity (see Connell 1978). Disturbance that occurs at frequencies on the order of organism generation times can interfere with biological patterns and the gradual adjustment of community components to each other. The relative importance of habitat stability compared to other factors is debatable (see Huston 1979; Ricklefs 1987), as is the relevance of discussions of equilibrium states in natural communities. Many ecosystems may be out of equilibrium much of the time (Wiens 1984; Sousa 1979). In the non-equilibrium state, species tend to respond to environmental variations independently of one another and exploit resources oppor-

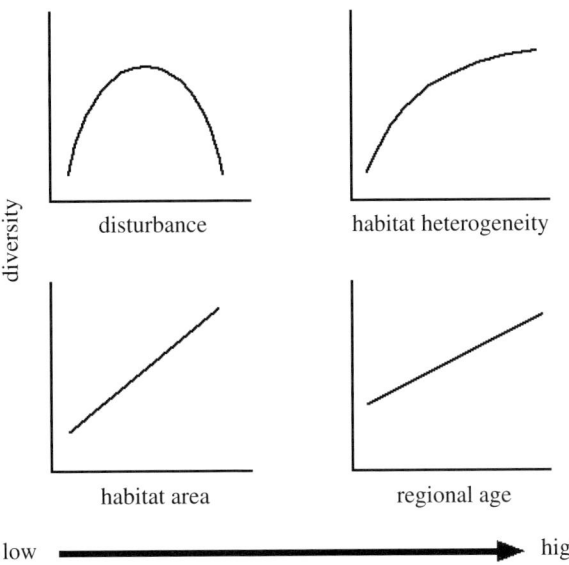

Figure 5. Summary representation of influence of major physical controls on diversity of natural communities. See text for explanation.

tunistically, and habitats can be undersaturated with species and individuals (Wiens 1984).

Vent community stability relates most strongly to constancy of fluid supply. Evidence for the longevity of vent sites indicates a range from as short as two years (CoAxial; Tunnicliffe et al. 1997) to at least 80 years (TAG; Lalou et al. 1993). Controls on fluid supply are complex but can be related to surface deposit characters, subsurface conduits and heat source behaviour (Alt 1995).

To consider surface controls first: Venting through sedimentary deposits is likely to be stable in the medium-term. However, sites like Middle Valley and Escanaba Trough are overlain by turbidite sequences deposited over hundreds of years (Davis et al. 1987), thus catastrophic deposition and community annihilation must occur at intervals. Habitat stability on active sulphide chimneys and mounds varies with the space and time scales under consideration. While sulphide accretion can result in rapid habitat alteration at decimeter to metre scales, chimneys represent an expanding substratum in a space-limited system. The long-lived hydrothermal mounds such as those on the Mid-Atlantic Ridge or Explorer Ridge contain a large area of habitat within which small scale changes are frequent (Tunnicliffe et al. 1986; Karson & Brown 1988; Segonzac 1992). Tunnicliffe & Juniper (1990) document the effects of chimney collapse including mortality of proximal animals. Fustec (1987) has mapped faunal changes on a growing chimney over two years. Modern sulphides frequently contain engulfed vent animals (Koski et al. 1984; Hannington & Scott 1988; Cook & Stakes 1995). The rapid adjustments of flow, sulphide deposition and fauna represent a fertile area of investigation.

Secondly, stability of habitat is influenced by subsurface crustal features (figure 5). Deposition of sulphide within the stockwork (Fouquet et al. 1993; Hannington et al. 1995) may be a common mechanism of influencing quality and constancy of the overlying habitat. Intrusive magma emplacement or diking is another mechanism. The former will be more common in slower spreading situations that foster sulphide accumulation and the latter in ridges with high magmatic activity (Embley & Fornari 1995). Individual tectonic faulting events can disturb vent habitat while longer term faulting processes can be critical to maintaining hydrothermal circulation. Fujioka (1995) reports a seismic event at TAG that caused slumps and animal deaths. Yet, at sites with a less robust magma supply, sustained hydrothermal activity may depend on deep faulting to provide access to a diminishing heat source, as has been proposed for TAG and other Mid-Atlantic Ridge sites (Fouquet et al. 1993; Rona et al. 1993) and Endeavour Segment, Juan de Fuca (Delaney et al. 1992). Even on fast and superfast spreading segments of the East Pacific Rise, long-lived and high temperature vents are typically located along major faults (Fustec et al. 1987; Haymon et al. 1993; Auzende et al. 1996).

Thirdly, heat source can have a wide influence on venting in terms of both vent distribution and stability. Eruptive activity appears to correlate with spreading rate (Fornari & Embley 1995). Creation of vent habitat occurs after an eruption; high frequency of magmatic activity also must alter the pattern of subsurface hydrothermal circulation to cause frequent vent closure. Recent observations of eruptions at ridges are expanding our view of crustal dynamics and faunal responses.

We have followed post-eruption development of communities on Cleft and CoAxial Segments (Juan de Fuca Ridge) and work proceeds on the eruptive site at 9° N East Pacific Rise (Lutz et al. 1994). In terms of stability, the most prominent features are: the immediacy of the food supply, the swift colonization response by vent animals and the high rate of change in these new communities. Magma movement stimulates and/or flushes subsurface microbial production (Holden 1996; Juniper et al. 1995) providing an instant food source and probably 'seeds' initial seafloor production. Dense white mats of (presumably) sulphide-oxidizing bacteria appeared within weeks/months of eruption (Haymon et al. 1993, Delaney et al. 1994; Tunnicliffe et al. 1997). On CoAxial, we know the date of the eruption (Embley et al. 1995) and that no vent fauna existed previously in the area. After one year, five vent species were observed; by two years, 22 species were present (Tunnicliffe et al. 1997). After two years, these vent communities were still changing, with some having disappeared with short-lived venting. After three years, known vents were inactive and their communities were essentially dead.

Biological observations commenced on Cleft Segment in 1988, two years after a large megaplume was observed over the site (Baker et al. 1987). While the exact nature of previous venting remains obscure, it was evident the biological communities were in a state of change. The biomass was high but diversity was low (Milligan & Tunnicliffe 1994). Over the subsequent years, we observed the contraction of venting and diminution of biota. By 1993, 15 km of venting had contracted to under 5 km; large numbers of deep-sea predators were observed. These observations fit the evolution of H_2S/heat ratios that peaked shortly after eruption (Butterfield et al., this volume). Both events show rapid cooling. Between the short time-frame and changing fluid conditions, neither habitat nor community will come to equilibrium.

4. Spatial heterogeneity

Within any given habitat, competition for resources, predation and disease can limit the number of species that coexist and thus restrict diversity and food web complexity. A more heterogeneous physical environment can enhance diversity by offering more opportunities for specialized primary producers and dependent consumers, and a broader resource base for mobile predators and omnivores. Habitat heterogeneity can thus interact with input of immigrants from a regional species pool in determining local diversity (Ricklefs 1987). In this section we will consider how crustal accretion might influence the composition of vent communities by controlling spatial heterogeneity of habitat at local and regional scales.

Within a vent field, habitat heterogeneity is augmented by variations in flow rate, fluid chemistry and substratum type. Although habitat requirements of most vent species remain poorly defined, several descriptive models relate the fine scale distribution of larger organisms to temperature and sulphide concentrations (reviewed by Juniper & Sarrazin 1995). While the complete range of conditions from smokers to weak diffuse flows can be observed at most sites, high temperature chimneys and associated organisms can be locally and even regionally absent, such as on the present day Galapagos Ridge.

Venting through sediments and sulphide deposits creates heterogeneity of flow and fluid chemical characteristics. The large number of species found at Middle Valley (Juan de Fuca Ridge) vents has been attributed to the greater diversity of habitat created by the presence of a sediment cover (Juniper et al. 1992). Variability in faunal assemblages colonizing actively-forming sulphide deposits has been described by Fustec et al. (1987), Tunnicliffe & Juniper (1990) and Segonzac et al. (1993). Venting through large sulphide deposits at sites such as Endeavour Segment (Juan de Fuca Ridge) and TAG (MAR) can produce broad patterns of habitat zonation as hydrothermal activity waxes and wanes in different areas of the deposit. At a finer scale, spatial heterogeneity can result from processes such as infilling and local reductions in porosity that alter fluid supply to different areas of chimney surfaces. New opportunities for colonization are created by chimney structural failure and abrupt fluid flow variations (Tunnicliffe 1991; Gaill & Hunt 1991; Hannington et al. 1995). Large sulphide deposits can represent long lived habitats that feature a great deal of heterogeneity in both the spatial and time domains. Opportunities for increased species diversity are extensive.

The importance of regional heterogeneity of habitat is understood in a theoretical sense, in that it should permit a province to hold a larger pool of species. On the Juan de Fuca/Explorer Ridge system, venting through sulphides and sediments in the north gives way in the south to a more robust magma supply and venting through small sulphide edifices and basaltic substrata. This contrasts with the northern EPR where more homogeneous substratum and fluid conditions are encountered. Yet the northern EPR pool of vent species is larger than that of the northeast Pacific (figure 3). Other influences, such as habitat abundance and spreading history probably interact with large scale habitat diversity. As more of the global ridge system is systematically surveyed it may be possible to separately evaluate the influence of regional habitat diversity on species pool size.

5. Historical factors

The diversity and distribution of marsupial mammals in Asia, the Americas and Australia were difficult to understand before reference to historical processes and plate tectonics (Marshall et al. 1982). In the same manner, given their position on spreading ridges, vent communities are likely to carry a strong imprint of tectonic history. Comparison of the faunas of different regions of the world (figure 3) can be deceptive using only modern proximity. The ridges themselves have been the major pathways for distribution of the global fauna and past positions and juxtapositions of the ridges are important to understanding relative diversity and composition (Tunnicliffe & Fowler 1996). Factors that influence the accumulation of species in a region include: age of the region, history of connectedness to another region and proximity to a diverse region. Figure 3 demonstrates considerable differences in diversity (= species richness). Caution must be used in interpreting differences as sampling methods and intensity varies dramatically. In particular, the western Pacific and Atlantic have seen much less biological sampling. Nonetheless, it is likely the pattern presented will remain. The following examples illustrate the different ages of global vent provinces.

The East Pacific Rise is a traceable ridge from the early Mesozoic. Northern EPR presently holds the greatest species diversity. Since it is of similar age, the poorly sampled southern EPR could provide equal diversity. The Galapagos Rift began propagating about 25 Ma BP and was probably populated from the EPR although many endemic species are now found at Galapagos (see Tunnicliffe et al. (1996) for details). The Gorda/Juan de Fuca/Explorer region at the northern end of the original ridgeway was isolated when North America overran the ridge some 30 Ma BP, thus separating part of the ancestral fauna. That separation may have taken only part of the original fauna (Tunnicliffe 1988) leaving a lower diversity community that embarked on a diverging evolutionary path from EPR. In contrast, the Lau/Fiji/Manus area and the Atlantic are spreading ridges of younger age that formed conjoining the East Pacific Rise. They did not form independent faunas but show great similarity to the eastern Pacific in higher taxonomic structure. Van Dover (1995) presents the spreading history of the Atlantic suggesting the earlier opening northern Atlantic (175 Ma BP) may have been colonized via a shallow link through the Caribbean region. At present, faunal similarities suggest the Indian Ocean as the most likely conduit (Tunnicliffe & Fowler 1996). Perhaps successive extinctions and recolonizations occurred in the Atlantic. The greatly different character of the Atlantic vent fauna is not well explained by biogeography. Better information from the Indian and south Atlantic ridges is required.

6. Discussion

We have identified several physical factors that influence vent faunas. It is much more difficult to identify biological controls on these communities. Competition for nutrients may explain the compositional shift seen at Rose Garden, Galapagos (Hessler et al. 1988) or the size shift in worms one to two years after the CoAxial eruption (Tunnicliffe et al. 1997); Jollivet (1993) suggests that crab predation can be important especially as crabs often select dominant organisms. Much more work is needed. The energy supply at vents has a variable and often unpredictable behaviour, a phenomenon fundamentally different from solar-powered systems. Relative stability of the vent site or region is a potentially important factor affecting species accumulation, succession and the trophic structure and energy flow that ensue. Habitat abundance may counter instability at regional scales and we suspect that the history of spreading in a region is important to faunal accumulation through both immigration and speciation. Faunal origins will also dictate the initiating fauna in a region and the pattern of subsequent species interactions.

Figure 5 illustrates the expected response of species diversity to the various phenomena discussed in this paper. The peak of diversity at moderate levels of disturbance follows the intermediate disturbance hypothesis of Connell (1978). Increasing diversity with habitat heterogeneity on both local and regional scales has been verified in many ecosystems (Ricklefs 1987). Diversity increase with habitat area is adapted from island biogeography theory (MacArthur & Wilson 1967) while the influence of regional age is presented by Valentine (1971) among others. The net result of all these factors will vary depending on their relative importance. In the following section we consider these influences in relation to species occurrence data known from the major spreading centres.

(a) Slow spreading: the Mid-Atlantic Ridge

The MAR appears to be a region of low frequency of major disturbance, moderate/high local habitat heterogeneity with low regional heterogeneity, small habitat area, and high age. The relatively low species diversity seems consistent but remains a puzzle to explain in relation to the Pacific. Endosymbioses are generally rare, many taxa are absent, yet consumer biomass appears high. Stochastic extinction processes may be very important here. Dating by Lalou et al. (1993) indicates episodicity of venting at TAG and Snake Pit. At widely spaced vents, if one should shut down, local extinction has a high probability (Van Dover 1995). If several adjacent areas cease for several thousand years then the probability of regional species extinction becomes much higher. The modern MAR community may a product of widespread extinctions in quiescent periods. A possible test may be to examine local to regional diversity ratios: the value should be higher in areas where venting persistence has supported species diversification. However, further work on phylogenetic relations of separate groups is necessary to track the probably origins of the individual components of the MAR fauna. The utilization of alternative food supplies may be crucial to the persistence of species between episodes of venting. Phytoplankton derived lipids are abundant in postlarval MAR vent shrimp, and present even in adult animals (Dixon & Dixon 1996; Dixon et al. 1995). As well, MAR shrimp have been observed on inactive, weathering sulphides where films of mineral degrading bacteria (Wirsen et al. 1993) may provide sustenance for deposit feeders. Abundant methane emissions resulting from reaction of seawater with outcropping ultramafic rock (Bougault et al. 1993) could provide a stable refuge for MAR mussels, the only endosymbiosis known from vents in the region. These mussels harbour both sulphide oxidizing and methanotrophic bacteria in their gill tissues (Cavanaugh et al. 1992; A. Fiala-Médioni, personal communication).

(b) Intermediate spreading: Juan de Fuca area

In this region major disturbances should be moderately frequent and perhaps more varied than on the MAR. Local habitat heterogeneity varies depending on substratum and there is a high regional heterogeneity. Total habitat area is moderate (low on Explorer and Gorda), regional age high but there may be a mixed spreading history (slow spreading on Gorda). Observed lower diversity (versus EPR) may have some element of stochastic events in that extinctions over the last 20 million years could not be easily reseeded from EPR. We see less development in trophic structure in this region compared to the EPR (large endemic predators missing, planktonic grazers missing, peripheral suspension-feeders poorly developed).

(c) Lau/Fiji back-arc basins

There has been little collection here but the diversity appears high. Descriptions by Desbruyères et al. (1994) and Galkin (1992) suggest that species diversity may be favoured by a diverse habitat, augmented by spatial separation (i.e. many separated small sites) and a long history of spreading activity in the southwestern Pacific. Like the Juan de Fuca, there are discrete areas of venting through sediments, through sulphide deposits and through bare basalts. Site speciation may be occurring, judging from the numbers of new taxa of known genera that are reported.

(d) Fast spreading, the northern EPR

The high species diversity observed on the northern EPR corresponds to a moderate/high disturbance rate, moderate habitat heterogeneity, moderate/high habitat area, high regional age and diversity (9° vs 13° vs stable 21° vs sedimented Guaymas). Niche diversity and many aspects of trophic structure appear to be the most developed here: three tube worm and two bivalve symbioses, endemic predators both small and large, abundant peripheral suspension feeders and detrivores in addition to the large biomass in symbiont hosts. The latter observation raises the issue of relative productivity of different hydrothermal systems. Is the development of a peripheral fauna resource limited in other areas? Or have the combined physical and historic features of northern EPR provided a unique setting for the development of a vent dependent suspension feeding fauna? Why is the giant tube worm *Riftia pachyptila* so massive compared to its northeast Pacific counterpart *Ridgeia piscesae*?

(e) Ultra-fast spreading, the southern EPR

Very high disturbance, low heterogeneity, high habitat area and high regional age combine at this little explored end of the spreading rate spectrum. Our discussion above suggests that the net effect is likely to be a moderate diversity, perhaps lower than northern EPR. The proliferation of venting provides tremendous opportunities for fast growing, rapidly reproducing animals that directly exploit chemosynthesis. At the same time, the high frequency of eruption must continually destabilize incipient communities. Dives in 1984 indicated a mostly mobile vent fauna (Renard et al. 1985; Juniper et al. 1990) and low diversity, as would be predicted for such ephemeral conditions. However, more extensive surveys in 1993 between 17° S and 19° S documented a wide range of venting conditions (Fouquet et al. 1994; Auzende et al. 1996), with most major EPR macrofaunal species being present (Geistdoerfer et al. 1995), although communities appeared to be evolving rapidly. Extensive biological work is required in this area to better understand the opposing influences of habitat abundance and instability.

7. Conclusion

Vent habitat is inherently unstable at many scales. Magmatic activity – reflecting spreading rate – is probably the ultimate determinant of both vent distribution and stability. The clearest biological expression of spreading rate is likely to be found in the composition of regional species pools. Since diversity can be linked to ecosystem productivity and efficiency (Darwin 1859; May 1973; Tilman et al. 1996) these ecosystem properties merit comparative study at regional scales. Fornari & Embley (1995) caution against simplistic, global correlations of spreading rate and vent behaviour. Biologists must await the time when the relations among spreading rate, magmatic activity, tectonism, sulphide accumulation and venting are clearer to the geologists. The inaccessibility of vents leaves us with limited data to formulate ecological theories. It remains important in these early stages to adopt a 'fluid' approach and continue to test different theories. However, no matter how sophisticated the ecological model undertaken, we still require a fundamental understanding of what the creatures are that inhabit the system. 'The value of model-making is that it provides us with a series of possibilities... however, a wide and deep understanding of organisms, past and present, is a basic requirement...' (Hutchinson 1975, p. 515).

Financial support has been provided by NSERC Canada research grants to S.K.J. and V.T. Laurel Franklin and Enriette Gagnon prepared figures. The Royal Society and the discussion meeting organizers are thanked for the invitation that led to the writing of this paper. We thank Paul Tyler, Alan and Eve Southward, Chuck Fisher and Robert Ricklefs for their thoughtful reviews of an earlier version of this manuscript.

References

Auzende, J.-M., Ballu, V., Batiza, R., Bideau, D., Charlou, J.-L., Cormier, M. H., Fouquet, Y., Geistdoerfer, P., Lagabrielle, Y., Sinton, J. & Spadea, P. 1996 Recent tectonic, magmatic, and hydrothermal activity on the East Pacific Rise between 17° S and 19° S: submersible observations. *J. Geophys. Res.* B **101**, 17995–18010.

Baker, E. T, German, C. R. & Elderfield, H. 1995 Hydrothermal plumes over spreading-centre axes: global distributions and geological inferences. In *Seafloor hydrothermal systems: physical, chemical, biological and geological interactions* (ed. S. E. Humphris, R. A. Zierenberg, L. S. Mullineaux & R.E. Thomson), pp. 47–70. (Geophysical Monograph 91.) Washington, DC: AGU.

Baker, E. T., Massoth, G. J. & Feely, R. A. 1987 Cataclysmic hydrothermal venting on the Juan de Fuca Ridge. *Nature* **329**, 149–151.

Black, M. B., Lutz, R. A. & Vrijenhoek, R. C. 1994 Gene flow among vestimentiferan tube worm (Riftia pachyptila) populations from hydrothermal vents of the eastern pacific. *Mar. Biol.* **120**, 33–39.

Bougault, H. et al. 1993 Fast and slow spreading ridges: structure and hydrothermal activity, ultramafic topographic highs, and CH_4 output. *J. Geophys. Res.* B **98**, 9643–9651.

Brown, J. H. 1988 Species diversity. In *Analytical biogeography* (ed. A. A. Myers & P. S. Giller), pp. 57–89. London: Chapman & Hall.

Cavanaugh, C., Wirsen, C. O. & Jannasch, H. W. 1992 Evidence for methylotrophic symbionts in a hydrothermal vent mussel (Bivalvia: Mytilidae) from the Mid-Atlantic Ridge. *Appl. Environ. Microb.* **58**, 3799–3803.

Childress, J. J. & Mickel, T. J. 1985 Metabolic rates of animals from the hydrothermal vents and other deep-sea habitats. *Biol. Soc. Wash. Bull.* **6**, 249–260.

Connell, J. H. 1978 Diversity in tropical rain forests and coral reefs. *Science* **199**, 1302–1310.

Cook, T. L. & Stakes, D. S. 1995 Biogeological mineralization in deep-sea hydrothermal deposits. *Science* **267**, 1975–1979.

Darwin, C. 1859 *On the origin of species by means of natural selection*. London: John Murray.

Davis, E. E., Goodfellow, W. D., Bornhold, B. D., Adshead, J., Blaise, B., Villinger, H. & Le Cheminant, G. M. 1987 Massive sulfides in a sedimented rift valley, northern Juan de Fuca Ridge. *Earth Planet. Sci. Lett.* **82**, 49–61.

Delaney, J. R., Baross, J. A., Lilley, M. D., Kelley, D. S. & Embley, R. W. 1994 Is the quantum event of crustal accretion a window into a deep hot biosphere? *Eos* **75**, 617.

Delaney, J. R., Robigou, V., McDuff, R. E. & Tivey, M. K. 1992 Geology of a vigorous hydrothermal system on the Endeavour Segment, Juan de Fuca Ridge. *J. Geophys. Res.* **97**, 19663–19682.

Desbruyères, D. *et al.* 1994 Deep-sea hydrothermal communities in the Southwestern Pacific back-arc basins (the North Fiji and Lau Basins): composition, microdistribution and food web. In *The North Fiji Basin (SW Pacific)* (ed. J. M. Auzende & T. Urabe). *Mar. Geol. STARMER Special Issue* **116**, 227–242.

Desbruyères, D. 1995 Biological studies at the ridge crest: temporal variations of deep-sea hydrothermal communities at 13° N/EPR. *InterRidge News* **4**, 6–10. (University of Durham, UK.)

Dixon, D. R. & Dixon, L. R. J. 1996 Results of DNA analyses conducted on vent shrimp postlarvae collected above the Broken Spur vent field during the CD95 cruise, August 1995. *BRIDGE Newsletter* **11**, 9–15. BRIDGE Office, Department of Earth Sciences, University of Leeds, Leeds, UK.

Dixon, D. R., Jollivet, D. A. S. B., Dixon, L. R. J., Nott, J. A. & Holland, P. W. H. 1995 The molecular identification of early life-history stages of hydrothermal vent organisms. In *Hydrothermal vents and processes* (ed. L. M. Parsons, C. L. Walker & D. R. Dixon) *Geol. Soc. Lond. Spec. Publ.* **87**, 343–350.

Embley, R. W., Jonasson, I. R., Perfit, M. R., Franklin, J. M., Tivey, M. A., Malahoff, A., Smith, M. F. & Francis, T. J. G. 1988 Submersible investigation of an extinct hydrothermal system on the Galapagos Ridge: sulfide mounds, stockwork zone, and differentiated lavas. *Canad. Mineral.* **26**, 517–539.

Embley, R. W., Chadwick Jr, W. W., Jonasson, I. R., Butterfield, D. A. & Baker, E. T. 1995 Initial results of the rapid response to the 1993 CoAxial event: relationships between hydrothermal and volcanic processes. *Geophys. Res. Lett.* **22**, 143–146.

Fornari, D. J. & Embley, R. W. 1995 Tectonic and volcanic controls on hydrothermal processes at the mid-ocean ridge: an overview based on near-bottom and submersible studies. In *Seafloor hydrothermal systems: physical, chemical, biological and geological interactions* (ed. S. E. Humphris, R. A. Zierenberg, L. S. Mullineaux & R. E. Thomson), pp. 1–46. (Geophysical Monograph 91.) Washington, DC: AGU.

Fouquet, Y., Wafik, A., Cambon, P., Mevel, C., Meyer, G. & Gente, P. 1993 Tectonic setting and mineralogical and geochemical zonation in the Snake Pit sulfide deposit (Mid-Atlantic Ridge at 28° N). *Econ. Geol.* **88**, 2018–2036.

France, S. C., Hessler, R. R. & Vrijenhoek, R. I. 1992 Genetic differentiation between spatially-disjunct populations of the deep-sea, hydrothermal vent-endemic amphipod Ventiella sulfuris. *Mar. Biol.* **114**, 551–559.

Fujioka, K. 1995 TAG hydrothermal mound of the Mid-Atlantic ridge, its evolution and long-term change. *Eos* **76**, 574.

Fustec, A., Desbruyères, D. & Juniper, S. K. 1987 Deep-sea hydrothermal vent communities at 13° N on the East Pacific Rise: microdistribution and temporal variations. *Biol. Oceanogr.* **4**, 121–164.

Geistdoerfer, P., Auzende, J. M., Batiza, R., Bideau, D., Cormier, M. H., Fouquet, Y., Lagabrielle, Y., Sinton, J. & Spadea, P. 1995 Hydrothermalisme et communautés animales associées sur la dorsale du Pacifique Oriental entre 17° S et 19° S (Campagne Naudur, Décembre 1993). *C.R. Acad. Sci. Paris* (IIa) **320**, 47–54.

Grehan, A. J. & Juniper, S. K. 1996 Clam distribution and subsurface hydrothermal processes at Chowder Hill (Middle Valley), Juan de Fuca Ridge. *Mar. Ecol. Prog. Ser.* **130**, 105–115.

Hannington, M. D. & Scott, S. D. 1988 Mineralogy and geochemistry of a hydrothermal silica-sulfide-sulfate spire in the caldera of Axial Seamount, Juan de Fuca Ridge. *Can. Mineral.* **26**, 603–625.

Hannington, M. D., Jonasson, I. R., Herzig, P. M. & Peterson, S. 1995 Physical and chemical processes of seafloor mineralization at mid-ocean ridges. In *Seafloor hydrothermal systems: physical, chemical, biological and geological interactions* (ed. S. E. Humphris, R. A. Zierenberg, L. S. Mullineaux & R. E. Thomson), pp. 115–157. (Geophysical Monograph 91.) Washington, DC: AGU.

Haymon, R. M., Fornari, D. J., Von Damm, K. L., Lilley, M. D., Perfit, M. R., Edmond, J. M., Shanks III, W. C., Lutz, R. A., Grebmeier, J. M., Carbotte, S., Wright, D., McLaughlin, E., Smith, M., Beedle, N. & Olson, E. 1993 Volcanic eruption of the mid-ocean ridge along the East Pacific Rise crest at 9°45.52′ N: direct submersible observations of sea floor phenomena associated with an eruption event in April, 1991. *Earth Planet. Sci. Lett.* **119**, 85–101.

Hessler, R. R., Smithey, W. M., Boudrias, M. A., Keller, C. H., Lutz, R. A. & Childress, J. J. 1988 Temporal change in megafauna at the Rose Garden hydrothermal vent (Galapagos Rift; eastern tropical Pacific). *Deep-Sea Res.* **35**, 1681–1709.

Hessler, R. R., Smithey, W. M. & Keller, C. H. 1985 Spatial and temporal variation of giant clams, tube worms and mussels at deep-sea hydrothermal vents. *Biol. Soc. Wash. Bull.* **6**, 411–428.

Holden, J. F. 1996 Ecology, diversity, and temperature-pressure adaptation of the deep-sea hyperthermophilic archaea thermococcales. Ph.D. thesis, University of Washington, USA.

Huston, M. 1979 A general hypothesis of species diversity. *Am. Nat.* **113**, 81–101.

Hutchinson, G. E. 1975 Variations on a theme by Robert MacArthur. In *Ecology and evolution of communities* (ed. M. L. Cody & J. M. Diamond), pp. 492–521. Cambridge, MA: Belknap Press.

Jannasch, H. W. 1985 The chemosynthetic support of life and the microbial diversity at deep-sea hydrothermal vents. *Proc. R. Soc. Lond.* B **225**, 277–297.

Jollivet, D. 1993 Distribution et evolution de la faune associée aux sources hydrothermales profondes a 13° N sur la dorsale du Pacifique Oriental: le cas particulier des polychetes alvinellidae. Ph.D. thesis, Université de Bretagne Occidentale, France.

Jollivet, D., Desbruyères, D., Bonhomme, F. & Moraga, D. 1995 Genetic differentiation of deep-sea hydrothermal vent alvinellid populations (Annelida: Polychaeta) along the East Pacific Rise. *Heredity* **74**, 376–391.

Juniper, S. K., Martineu, P., Sarrazin, J. & Gélinas, Y. 1995 Microbial-mineral floc associated with nascent hydrothermal activity on CoAxial Segment, Juan de Fuca Ridge. *Geophys. Res. Lett.* **22**, 179–182.

Karson, J. A. & Brown, J. R. 1988 Geologic setting of the Snake Pit hydrothermal site: an active vent field on the Mid-Atlantic Ridge. *Mar. Geophys. Res.* **10**, 91–108.

Kim, S. L., Mullineaux, L. S. & Helfrich, K. R. 1994 Larval dispersal via entrainment into hydrothermal vent plumes. *J. Geophys. Res.* C **99**, 12 655–12 665.

Koski, R. A., Clague, D. A. & Oudin, E. 1984 Mineralogy and chemistry of massive sulfide deposits from the Juan de Fuca Ridge. *Geol. Soc. Am. Bull.* **95**, 930–945.

Lalou, C., Reyss, J.-L., Brichet, E., Arnold, M., Thompson, G., Fouquet, Y. & Rona, P. A. 1993 New Age dating for Mid-Atlantic Ridge hydrothermal sites: TAG and Snake Pit chronology revisited. *J. Geophys. Res.* **98**, 9705–9713.

Lalou, C., Reyss, J.-L., Brichet, E., Rona, P. A. & Thompson, G. 1995 Hydrothermal activity on a 105-year scale at a slow-spreading ridge, TAG hydrothermal field, Mid-Atlantic Ridge 26° N. *J. Geophys. Res.* **100**, 17 855–17 862.

Large, R. R. 1992 Australian volcanic-hosted massive sulfide deposits: features, styles and genetic models. *Econ. Geol.* **87**, 471–510.

Lutz, R. A. 1991 The biology of deep-sea vents and seeps. *Oceanus* **34**, 75–83.

Lutz, R. A., Fritz, L. W. & Cerrato, R. M. 1988 A comparison of bivalve (Calyptogena magnifica) growth at two deep-sea hydrothermal vents in the eastern Pacific. *Deep-Sea Res.* **35**, 1793–1810.

Lutz, R. A., Shank, T. M., Fornari, D. J., Haymon, R. M., Lilley, M. D., Von Damm, K. L. & Desbruyères, D. 1994 Rapid growth at deep-sea vents. *Nature* **371**, 663–664.

MacArthur, R. H. & Wilson, E. O. 1967 *The theory of island biogeography*. Princeton University Press.

Marshall, L. G., Webb, S. D., Sepkoski, J. J. & Raup, D. M. 1982 Mammalian evolution and the great American interchange. *Science* **215**, 1351–1357.

May, R. M. 1973 *Stability and complexity in model ecosystems*. Princeton University Press.

Milligan, B. 1993 Geological effects on faunal distributions on the Cleft Segment, Juan de Fuca Ridge. Masters Thesis, University of Victoria, Canada.

Milligan, B. N. & Tunnicliffe, V. 1994 Vent and nonvent faunas of Cleft Segment, Juan de Fuca Ridge, and their relations to lava age. *J. Geophys. Res.* **99**, 4777–4786.

Mullineaux, L. S. & France, S. C. 1995 Dispersal mechanisms of deep-sea hydrothermal vent fauna. In *Seafloor hydrothermal systems: physical, chemical, biological and geological interactions* (ed. S. E. Humphris, R.A. Zierenberg, L.S. Mullineaux & R. E. Thomson), pp. 408–424. (Geophysical Monograph 91.) Washington, DC: AGU.

Odum, E. P. 1971 *Fundamentals of ecology*, 3rd edn. Philadelphia, London and Toronto: W. B. Saunders Co.

Real, L. A. & Brown, J. H. (eds) 1991 *Foundations of ecology*. Chicago University Press.

Renard, V., Hekinian, R., Francheteau, J., Ballard, R. D. & Backer, H. 1985 Submersible observations at the axis of the ultra-fast-spreading East Pacific Rise (17°30′ to 21°30′ S). *Earth Planet. Sci. Lett.* **75**, 339–353.

Ricklefs, R. E. 1987 Community diversity: relative roles of local and regional processes. *Science* **235**, 167–171.

Robigou, V., Delaney, J. R. & Stake, D. S. 1993 Large massive sulfide deposits in a newly discovered active hydrothermal system, the High-Rise Field, Endeavour Segment, Juan de Fuca Ridge. *Geophys. Res. Lett.* **20**, 1887–1890.

Rona, P. A., Hannington, M. D., Raman, C. V., Thompson, G., Tivey, M. K., Humphris, S. E., Lalou, C. & Petersen, S. 1993 Active and relict seafloor hydrothermal mineralization at the TAG hydrothermal field, Mid-Atlantic ridge. *Econ. Geol.* **88**, 1989–2017.

Sanders, H. L. 1968 Marine benthic diversity: a comparative study. *Am. Nat.* **102**, 243–282.

Segonzac, M. 1992 Les peuplements associés a l'hydrothermalisme océanique du Snake Pit (dorsale medio-Atlantique; 23° N, 3480 m): composition et microdistribution de la mégafaune. *C.R. Acad. Sci. Paris* **314**, 593–600.

Southward, E. C., Tunnicliffe, V. & Black, M. 1995 Revision of the species of Ridgeia from northeast Pacific hydrothermal vents, with a redescription of *Ridgeia piscesae* Jones (Pogonophora: Obturata = Vestimentifera). *Can. J. Zool.* **73**, 282–295.

Southward, E. C., Tunnicliffe, V., Black, M. B., Dixon, D. & Dixon, L. 1997 Ocean ridge segmentation and vent tubeworms (Vestimentifera) in the NE Pacific. In *Tectonic, magmatic, hydrothermal and biological segmentation of mid-ocean ridges* (ed. C. J. MacLeod, P. Tyler & C. L. Walker). London: Geol. Soc. Lond. Spec. Publ. **118**, 211–224.

Tilman, D., Wedin, D. & Knops, J. 1996 Productivity and sustainability influenced by biodiversity in grassland ecosystems. *Nature* **379**, 718–720.

Tunnicliffe, V. 1988 Biogeography and evolution of hydrothermal vent fauna in the eastern Pacific Ocean. *Proc. R. Soc. Lond.* B **233**, 347–366.

Tunnicliffe, V. 1991 The biology of hydrothermal vents: ecology and evolution. *Oceanogr. Mar. Biol. A. Rev.* **29**, 319–407.

Tunnicliffe, V. & Fowler, C. M. R. 1996 Influence of sea-floor spreading on the global hydrothermal vent fauna. *Nature* **379**, 531–533.

Tunnicliffe, V. & Juniper, S. K. 1990 Dynamic character of the hydrothermal vent habitat and the nature of sulphide chimney fauna. *Prog. Oceanog.* **24**, 1–14.

Tunnicliffe, V., Botros, M., De Burgh, M. E., Dinet, A., Johnson, H. P., Juniper, S. K. & McDuff, R. E. 1986 Hydrothermal vents of Explorer Ridge, northeast Pacific. *Deep-Sea Res.* **33**, 401–412.

Tunnicliffe, V., Fowler, C. M. R. & McArthur, A. 1997 Plate tectonic history and hot vent biogeography. In *Tectonic, magmatic, hydrothermal and biological segmentation of mid-ocean ridges* (ed. C. J. MacLeod, P. Tyler & C. L. Walker). London: Geol. Soc. Lond. Spec. Publ. **118**, 225–238.

Urabe, T. *et al.* 1995 The effect of magmatic activity on hydrothermal venting along the superfast-spreading East Pacific Rise. *Science* **269**, 1092–1095.

Valentine, J. W. 1971 Plate tectonics and shallow marine diversity and endemism, an actualistic model. *Syst. Zool.* **20**, 253–264.

Van Dover, C. L. 1995 Ecology of Mid-Atlantic Ridge hydrothermal vents. In *Hydrothermal vents and processes* (ed. L. M. Parson, C. L. Walker & D. R. Dixon), pp. 257–294. London: Geol. Soc. Lond. Spec. Publ. 87.

Biocatalytic transformations of hydrothermal fluids

By Holger W. Jannasch

Woods Hole Oceanographic Institution, Woods Hole, MA 02543, USA

The occurrence of copious animal populations at deep-sea vents indicates an effective microbial chemosynthetic biocatalysis of hydrothermal fluids on their emission into oxygenated ambient seawater. The large metabolic and physiological diversity of microbes found at these sites, including anaerobic and aerobic hyperthermophiles, reflects an even higher variety of biocatalytic or enzymatic reactions that greatly influence deep-sea hydrothermal geochemistry.

1. Introduction

The extent of biocatalysis in geochemical transformations is still an enigmatic issue. Although it is well understood that processes such as microbial photosynthesis, sulphate reduction, denitrification, etc., do not occur in the absence of the responsible organisms, geochemists generally are often unclear as to the effectiveness of biocatalytic transformations, where they occur, where they cannot occur, and how can they be predicted. This dilemma is due, in part, to an incompatibility between quantitative approaches in geochemical and microbiological studies. The unexpected discovery of high chemosynthetic bacterial production of biomass near deep-sea vents has resulted in studies that led to the following clarifying remarks for geochemists on biocatalytic transformation of hydrothermal emissions.

The observations contain that certain reduced inorganic compounds may serve as biocatalytically utilizable 'sources of energy' wherever they occur under conditions that allow the production of enzymes through microbial growth. During the Earth's history, biochemical–physiological evolution has resulted in large numbers of biogeochemical processes that are mediated by a large microbial and, thereby, biocatalytic diversity. Many microbes contributed to determining the composition of the Earth's atmosphere, hydrosphere and upper crust during the period of about two billion years before higher forms of life appeared, including the onset of oxygen formation by cyanobacteria (Cloud 1976; Schidlowski 1984; Cohen *et al.* 1984). While today 'green plants' are limited to photosynthesis and 'animals' to heterotrophy (see later), all of their primordial forms have died out. In contrast, many of the early microbes are probably identical, at least phenotypically, with those that we find today wherever conditions are similar to those of the early Earth, especially in hydrothermal marine and terrestrial environments. The fact that microbes live in microenvironments and can survive long periods of time in a dormant stage—possibly millions of years (Cano & Borucki 1995)—without genetic change, may turn a few cm^3 of ancient reheated smoker wall material into a microbial 'jurassic park'.

2. Definitions

Organic catalysts or enzymes are complex proteins with characteristic metal-containing subunits or co-enzymes. Imprinted as 'genes' on DNA (the reproducing genetic storage for each species), individual enzymes can be identified in known organisms or even on strands of DNA extracted from water or sediment samples (Pace et al. 1986; Muyzer et al. 1993; Burggraf et al. 1994). Biocatalytic processes are principally autocatalytic as enzymes are continuously produced during growth and multiplication of the organism.

'Lithotrophic' microbes are able to use electrons from reduced inorganic compounds in oxidations of the electron acceptor with a gain of energy, e.g. the methanogenic bacteria:

$$\tfrac{1}{4}CO_2(aq) + H_2(aq) \longrightarrow \tfrac{1}{4}CH_4(aq) + \tfrac{1}{2}H_2O \quad (\Delta G° = -48). \tag{2.1}$$

Since the free energies ($\Delta G°$ in kJ mol^{-1} of the electron donor) are calculated for standard conditions with respect to temperature and pressure at pH 7, they are useful for comparative purposes only. They cannot be applied directly to the natural environment, especially to the unknown microenvironment of the metabolically active cell.

Most lithotrophic microbes are also 'autotrophic', i.e. they are able to couple the energy gained from lithotrophic oxidations to the reduction of CO_2 to organic carbon:

$$\tfrac{1}{3}CO_2(aq) + H_2(aq) \longrightarrow \tfrac{1}{6}[CH_2O] + \tfrac{1}{6}CH_4(aq) + \tfrac{1}{2}H_2O \quad (\Delta G° = -39). \tag{2.2}$$

In contrast, the anaerobic lithotrophic oxidation of H_2S is endergonic (requiring free energy) and requires light as an external source of energy as in pigmented bacteria that conduct an anoxic photosynthesis ('photosystem I', Pagan & Cohen 1982):

$$\tfrac{1}{2}CO_2(aq) + H_2S(aq) \longrightarrow \tfrac{1}{2}[CH_2O] + S^0 + \tfrac{1}{2}H_2O \quad (\Delta G° = +26). \tag{2.3}$$

When 'photosystem II' evolved within the photosynthetic cyanobacteria, the electron donor H_2S was replaced by H_2O:

$$CO_2(aq) + H_2O \longrightarrow [CH_2O] + O_2(aq) \quad (\Delta G° = +487). \tag{2.4}$$

This enabled these early photosynthetic organisms to spread much further in the environment but also resulted in the production of free oxygen (instead of sulphur or sulphate) as a byproduct. Because of the formation of the toxic peroxide, oxygen acts as a 'poison' to all anaerobic life forms. Subsequently, after the evolution of H_2O_2-detoxifying biocatalysts, free oxygen became a highly efficient metabolic oxidant:

$$H_2S(aq) + 2O_2(aq) \longrightarrow HSO_4^- + H^+ \quad (\Delta G° = -801), \tag{2.5}$$

empowering aerobic lithoautotrophic bacteria to metabolize independently of light:

$$CO_2(aq) + H_2S(aq) + H_2O + O_2(aq) \longrightarrow [CH_2O] + HSO_4^- + H^+ \quad (\Delta G° = -314), \tag{2.6}$$

and widely available in the early atmosphere and hydrosphere. The resulting oxidized forms of sulphur became, in turn, readily available electron acceptors for other microbes such as the autotrophic and heterotrophic sulphate and sulphur reducing bacteria.

Aerobic biocatalysis can also use H_2:

$$H_2(aq) + \tfrac{1}{2}O_2(aq) \longrightarrow H_2O \quad (\Delta G° = -263), \tag{2.7}$$

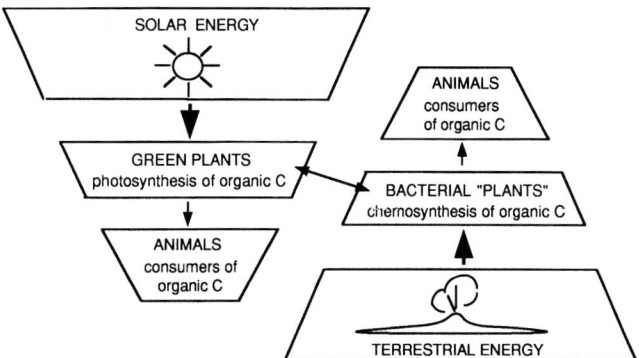

Figure 1. Analogy between the photo- and chemosynthetic food chains (plants = Calvin–Bensen cycle enzymes reducing inorganic to organic carbon).

or NH_3, NO_2^-, CH_4, Fe^{2+}, Mn^{2+} and possible other reduced metals as electron donors. In other words, specific microbes potentially chemosynthesize and grow wherever relevant reduced inorganic compounds reach free oxygen, particularly in the marine sediment–water interface.

Organic, i.e. reduced, carbon is also an electron donor readily available for biocatalytic oxidations. This 'heterotrophic' decomposition of organic matter depends on even more complex biocatalytic systems in a variety that is as large as the number of possible organic compounds produced by plants, animals or microbes themselves. Without showing the coupling to biosynthesis, one example is the aerobic oxidation of hexose (glucolysis):

$$C_6H_{12}O_6 + 6O_2(aq) \longrightarrow 6CO_2(aq) + 6H_2O \quad (\Delta G^\circ = -2920), \tag{2.8}$$

or the much less energy-efficient anaerobic oxidation (fermentation) of acetate and ethanol to butyrate:

$$CH_3COO^- + CH_3CH_2OH \longrightarrow CH_3CH_2CH_2COO^- + H_2O \quad (\Delta G^\circ = -38.7). \tag{2.9}$$

Many of the enzymatic reactions are reversible, i.e. they occur in 'anabolic' biosynthesis as well as in 'catabolic' respiration.

3. Aerobic chemosynthesis

Pfeffer (1897), impressed by Winogradsky's discovery of microbial 'chemo-autotrophy' in *Beggiatoa* (1887), coined the term 'chemosynthesis' in analogy to photosynthesis. It is now often used in place of chemolithoautotrophy. The process is driven by energy derived from biocatalytically mediated chemical oxidations and transferred to the Calvin–Bensen cycle enzymes that synthesize organic carbon from CO_2. Because this system is also found in green plants, the aerobic chemosynthetic bacteria can be described as 'plants' that fix CO_2 in the dark (figure 1). Until the discovery of deep-sea 'oases', it had never been imagined that chemosynthesis could serve as the food chain for whole animal communities.

While many chemosynthetic bacteria are living as symbionts in the most prominent vent animals, the population of free-living microorganisms, invisible to the naked eye, is probably even more extensive. This is indicated by scanning electronmicrographs of bacterial mats covering basalt lava surfaces (Jannasch & Wirsen 1981; Jannasch 1985). The emissions of dense bacterial, often flocculent, suspensions from 'warm' vents (less than approximately 40 °C) indicate productive growth chambers and mat formation in the layers of porous lava below the sea floor where entrained oxygenated seawater mixes with hydrothermal fluid. Here the aerobically chemolithoautotrophic genus *Thiomicrospira* appears to be prevalent (Ruby et al. 1981; Muyzer et al. 1995).

Large masses of white material were also observed to occur just after new volcanic activity at Juan de Fuca and 9° N East Pacific Rise vent sites (Haymon et al. 1993). From examining samples that we obtained we believe that the bulk of this material constitutes mineral encrusted sheaths of *Leptothrix*-like aerobic metal-oxidizing bacteria. They might have grown in shallow subseafloor lava pockets and been dislodged during the new volcanic activity. The sheaths did not contain cells any longer, which is not unusual for this type of organism that autolyses readily while the sheaths stay behind (D. Emerson, personal communication). Both the enzymatic activity and concentration of organic carbon were low relative to the amount of sample material. Juniper et al. (1995) report similar observations. This supports the assumption that populations of these normally slow-growing organisms led to a subsurface accumulation of metabolic products, such as the iron and silica-encrusted sheaths, over some period of time.

The chemosynthetic carbon fixation in bacterial suspensions or in microbial mats can be quantitatively determined by the uptake of $^{14}CO_2$ (Wirsen et al. 1993; Jannasch 1995) and showed that the organisms are mesophilic (optimal activity between 25–35 °C) and hardly affected by the *in situ* pressure of 260 atm (Ruby & Jannasch 1982). In contrast to the occurrence of anaerobic hyperthermophilic bacteria at hydrothermal vents, no comparable aerobic chemosynthetic organisms have yet been found (Wirsen et al. 1993). At sediment-covered vent sites, chemosynthetic activity takes place in yellow (cytochrome-containing) *Beggiatoa* mats that may be several cm thick (Nelson et al. 1989).

The visibly highest productivity of biomass is carried out by chemosynthetic bacteria—so far uncultured—that live in symbiosis with novel invertebrates, especially the large white clams (*Calyptogena*) and vestimentiferan tube worms (*Riftia*). It is remarkable that the chemosynthetically based food chain (figure 1) gives rise to animal populations that are more dense and productive, though locally restricted, than any comparable populations maintained by photosynthesis. At cold seeps and other benthic environments where methane is vented, methane-oxidizing symbionts occur in mussels and smaller vestimentiferans.

The argument has been raised, for example, in readers' comments to a paper by Jannasch & Mottl (1985), that the aerobic chemosynthesis is not entirely independent of photosynthesis since it requires free oxygen. In other words, the production is not 'primary' as in photosynthesis, but secondary.

4. Anaerobic chemosynthesis and high-temperature biocatalysis

This is not true for the strictly anaerobic chemosynthesis which was probably one of the earliest biocatalyses in the Earth's history. The exergonic (free energy producing) anaerobic reduction of CO_2 in hydrothermal fluid (equations (2.1) and (2.2)) requires hydrogen. This biocatalysis was first observed in a *Methanococcus* isolated from a deep-sea vent site by Jones et al. (1983) and growing at 90 °C. Its formation of methane, as dependent on temperature and pressure, was later studied by Miller et al. (1988) and the purification and properties of some of the biocatalysts involved by Shah & Clark (1990). The methanogenic new genus, *Methanopyrus* (Huber et al.

Table 1. *Energy-yielding reactions catalysed by some lithoautotrophic marine hyperthermophilic vent isolates*

electron donor	electron acceptor	reaction	organisms
H_2	CO_2	$4H_2 + CO_2 \longrightarrow CH_4 + 2H_2O$	*Methanopyrus, Methanococcus*
H_2	S^0	$H_2 + S^0 \longrightarrow H_2S$	*Pyrodictium*[a]
H_2	$SO_4^{2-}, S_2O_3^{2-}$	$H_2 + H_2SO_4 \longrightarrow H_2S + 4H_2O$	*Archaeoglobus*[a]
$S^{2-}; S^0$	O_2	$H_2 + \frac{1}{2}O_2 \longrightarrow H_2O$, $2S^0 + 3O_2 + 2H_2O \longrightarrow 2H_2SO_4$	*Aquifex Sulfolobus*[b]

[a] Facultatively heterotrophic
[b] Not yet isolated from marine hydrothermal sources

1989; Kurr *et al.* 1991) grew at temperatures up to 110 °C. This hyperthermophilic anaerobic chemosynthesis represents a truly 'primary' production which is, however, not catalysed by the Calvin–Bensen enzyme system but by a variety of new metabolic pathways summarized by Schönheit & Schäfer (1995).

Over the years hydrothermal vents yielded a large number of other anaerobic autotrophic and heterotrophic types of microbes (table 1) containing biocatalysts that are active at temperatures between 80 and 110 °C (Stetter *et al.* 1990; Ply *et al.* 1991; Jannasch *et al.* 1992). So far the hydrostatic pressures have little affect on their activity and most of the hyperthermophilic microbes described from deep-sea vents also occur in shallow marine hot springs. Nelson *et al.* (1992) observed a stabilizing influence of pressure on some biocatalytic activities. How protein and nucleic acids are protected from denaturation at these high temperatures is not yet fully understood. Most hyperthermophilic microbes belong to the kingdom of the archaea. What is an archaeum?

When Woese *et al.* (1990) based modern phylogeny, the evolutionary relationships between organisms, on the particular nucleotide sequences of the 16S ribosomal RNA, three domains emerged (figure 2): the *Bacteria*, the—also microbial—*Archaea*, and the *Eucaria* (all higher forms of life with nuclei-containing cells). Members of the archaea include many of those organisms that live under 'extreme' conditions such as temperatures of up to 110 °C, pH as low as 2 or as high as 11, and concentrated brines. These 'extremophiles' were thought to have lived under conditions of the early Earth.

Originally it appeared that all hyperthermophilic microbes had to be placed into the first 'kingdom' of the archaeal domain, the Crenarchaeota, with only two genera, *Thermococcus* and *Methanococcus*, in the Euryarchaeota (figure 2). Then two hyperthermophilic genera were found that clearly belonged to the bacterial domain, *Thermotoga* and *Aquifex*. The fact that hyperthermophiles occupy the lower branches in both the bacterial and the archaeal domains (figure 2), has been interpreted to the effect that life may have originated within a high temperature regime (Woese *et al.* 1990; Woese 1991). In fact, the discovery of mid ocean hydrothermal ridges has rejuvenated interest in theoretical and experimental studies on 'the origin of life' (Holm 1992). The first hyperthermophilic organism isolated from a deep-sea vent,

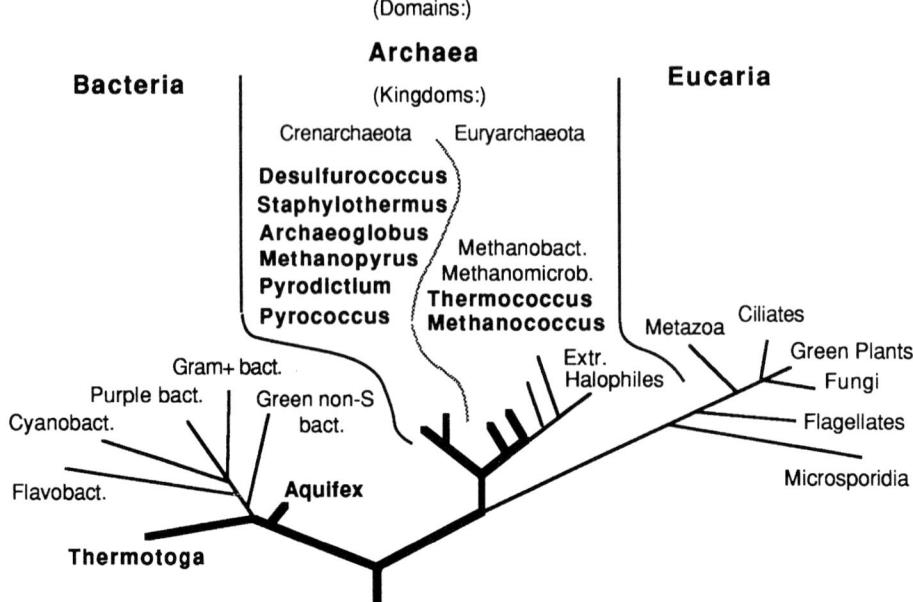

Figure 2. The major marine genera of hyperthermophilic archaea and bacteria, isolated from hydrothermal vents, superimposed (in bold) on the 16S rRNA-based phylogenetic tree as proposed by Woese *et al.* 1990 (modified from Jannasch 1995).

the previously mentioned *Methanococcus* (Jones *et al.* 1983), recently also became the first archaeum of which the entire genome has been sequenced. This involved the decoding of 1.7×10^6 base pairs and substantiating the uniqueness of the archaeal domain (news item 1996 *Science* **271**, 1061).

Aerobic hyperthermophiles were recently also discovered in both microbial domains. One of them is the bacterial genus *Aquifex* (the 'water maker', equation (2.7) and table 1, Huber *et al.* 1992). The facultatively aerobic hyperthermophilic archaeal *Pyrobaculum* (Völkl *et al.* 1993) is able to catalyse the reduction of nitrate to dinitrogen (denitrification) if oxygen is absent. More and more cases emerge indicating that physiological characteristics of phenotypes are not as easily predictable from their positioning on the phylogenetic tree as anticipated. A number of biocatalysts active at high temperatures have been identified and purified (figure 3, table 2). Some of them are of high commercial value, especially the DNA cleaving polymerases that are used in molecular biology.

During one of our earlier dive series at the sediment-covered Guaymas Basin spreading centre in the Gulf of California, Jørgensen *et al.* (1990) observed in sediment cores sulphate reducing activity with maxima around 90 °C. From collected material a new sulphate reducing archaeum was isolated (*Archaeoglobus*) that grew at temperatures of up to 90 °C (Burggraf *et al.* 1990). As electron donors, this organism uses hydrogen for autotrophic and acetate (or other organic substrates) for heterotrophic metabolism (table 1). Besides sulphate, electron acceptors can be sulphite and thiosulphate. Speculations on the existence of substantial and metabolically active populations of microbes in deeper parts of the globe's crust (Deming & Baross 1993) must also be based on an anaerobic chemosynthetic catalysis at extremely high temperatures, largely involving methane production.

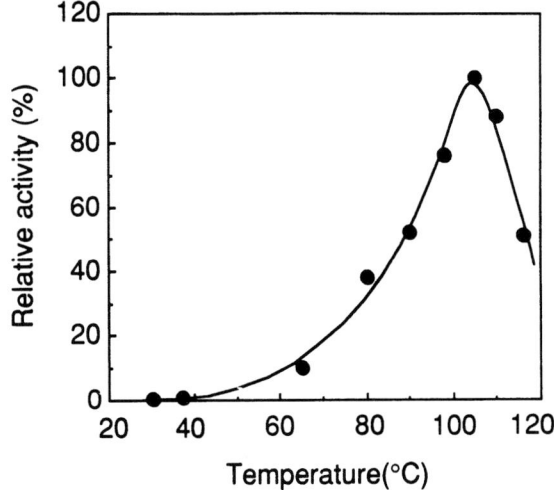

Figure 3. Temperature–activity plot of a protease purified from *Desulfurococcus* strain SY:200ng enzyme tested in 0.1 M HEPES buffer and 1% Azocasein for 10 min above and 30 min below 60 °C (from Hanzawa et al. 1996).

Table 2. *Examples of purified hyperthermophilic biocatalysts (enzymes) (from Adams (1993))*

enzyme	catalytic activity	temp. (°C)	organism
protease	peptide hydrolysis	100	*Pyrococcus*
amylase	starch hydrolysis	100	*Pyrococcus*
sulphide hydrogenase	sulphur reduction	> 95	*Pyrococcus*
hydrogenase	hydrogen generation	> 95	*Pyrococcus*
α-galactidase	galactomannan hydrolysis	> 90	*Thermotoga*
xylanase	xylan hydrolysis	105	*Thermotoga*
glucose isomerase	glucose isomeration	105	*Thermotoga*
amylopollulanase	starch degradation	118	*Thermococcus*
DNA polymerase[a]	DNA synthesis	> 95	*Thermococcus*

[a]NE Biolabs Catalog 1995

5. Discussion

From a geochemical point of view it remains enigmatic that many of the various microbial processes have been identified by bacterial isolations only and, therefore, have to be considered as 'potential'. How important such transformations may actually be in a natural setting depends on conditions that are conducive for growth of the responsible organisms, i.e. production of the biocatalyst. In other cases activities were measured but the organisms not yet isolated. For instance, in geothermal Guaymas Basin sediments, rates of sulphate reduction were 19–61 µM SO_4^{2-} day^{-1} (Jørgensen et al. 1992) which is unusually high for normal deep-sea conditions and resemble rates that are found in coastal waters (Jørgensen 1983). The maximum temperature of this biocatalysis is 110 °C (optimum at 103–106 °C). This approaches the lower temperature limit for thermolytic sulphate reduction (Krouse et al. 1988) and

Table 3. *Pyrite formation as hypothetical source of energy for biocatalytical formation of organic carbon as proposed by Wächtershäuser et al. (1988, 1990)*

reaction	free energy ($\Delta G°$ in kJ mol^{-1})
(I) FeS + H$_2$S (aqueous) → FeS$_2$ (crystalline) + H$_2$	$\Delta G^0 = -41.9$
(II) CO$_2$ (aqueous) + H$_2$ → HCOOH	$\Delta G^0 = +30.2$
(III) FeS + H$_2$S + CO$_2$ → FeS$_2$ + H$_2$O + HCOOH	$\Delta G^0 = -11.7$

might become 'important for an interpretation of the formation of sulphide deposits and their sulphur isotope distribution' (Jørgensen et al. 1992). The isolation of the sulphate reducer growing optimally at 106 °C has not yet been successful.

A quantitative estimate of the overall chemosynthesis of organic carbon at deep-sea vents can be made by using the figure for total annual seawater percolation through the oceanic crust at tectonic spreading centres, expressed as sulphate entrainment, of 1.2×10^8 tonnes (Edmond et al. 1982). Three quarters of this amount are estimated to be deposited as polymetallic sulphides, and one quarter emitted as dissolved sulphide by diffusive and hot venting. Assuming further that half of this deep-sea sulphide emission is used for chemosynthesis (the stoichiometric C:S proportion being 1:1 and the molecular weight proportion 1:3), the annual production of organic carbon of 5×10^6 tonnes would just amount to 0.03% of the 18×10^9 tons produced by oceanic photosynthesis (Woodwell et al. 1978). Since, however, only approximately 1% of this photosynthetic organic matter reaches the deep-sea (Honjo & Manganini 1993), it can be estimated that about 3% of all organic carbon found in the deep-sea derives from chemosynthesis at hydrothermal vents. The assumption that half of the emitting sulphide is used for chemosynthesis may be too generous, but is based on the predominant occurrence of diffuse flow at hydrothermal vent sites (Shultz et al. 1992; Rona & Trivett 1992; Ginster et al. 1994) where the conditions for biocatalysis are most favourable. The visible animal biomass is, most likely, much smaller than that of the non-visible bacteria in suspension and in mats covering all solid surfaces within vent regions (Jannasch & Wirsen 1981).

As well as the oxidation of sulphur by microbial metabolism, the reduction of oxidized sulphur compounds to sulphide can also be biocatalyzed by a number of microbes as indicated in figure 4. Elemental sulphur is actively reduced by almost all anaerobic hyperthermophilic archaea, partly as a detoxification reaction. The accumulation of molecular or ionic hydrogen during fermentative microbial metabolism quickly reaches inhibitory concentrations. The reduction of elemental sulphur (added in the laboratory to the growth media for this purpose) eliminates hydrogen by the formation of hydrogen sulphide. Sulphate is reduced by *Archaeoglobus* and thiosulphate by *Thermotoga* (Ravot et al. 1995) as well as by *Archaeoglobus* (Burggraf et al. 1990).

Considering the tremendous amount of polymetallic sulphides available at mid-ocean ridges, Wächtershäuser's suggestion (1988, 1990) that an energetically feasible biogeochemical mechanism for the earliest appearance of life on Earth might result from oxidation of primary precipitates of pyrrhotite (FeS) to form pyrite (FeS$_2$) seems reasonable. In table 3 the overall reaction III would be favoured by the insolubility

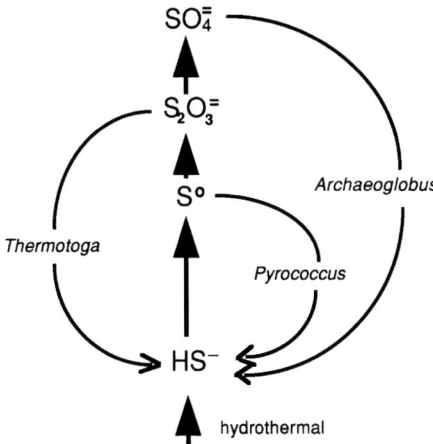

Figure 4. Hydrothermal and biocatalytic origin of sulphide in Guaymas Basin geothermal sediment (modified and updated from Jannasch 1995).

of pyrite at a low pH. Only reaction I, which is enhanced by elevated temperatures, has so far been verified experimentally (Drobner et al. 1990).

Modern approaches to the *in situ* identification of organisms as carriers and producers of specific biocatalysts have been developed through the use of 'molecular probes' (Pace et al. 1986; Muyzer et al. 1993; Burggraf et al. 1994): particular nucleotide sequences from pure culture isolates are compared to matching sequences within DNA extractions from samples such as sediment, mat material, smoker wall scrapings, etc. For biotechnological purposes, genes of certain commercially useful biocatalysts or enzymes (table 2), such as highly thermostable lipases, proteases (figure 3) or amylases, may be extracted and PCR-amplified. Aiming at geochemically more relevant enzymes, such approaches could be combined with microbiological studies in order to make quantitative estimates of biocatalytic contributions to geochemical transformations. Hydrothermal vent environments will provide a rich source of such materials.

I thank Professor J. Cann for the opportunity to attend the excellent meeting held at the Royal Society in London. I am also indebted to B. Sulzberger, U. Jans and H. Kramer (EAWAG, Dübendorf, Switzerland) for the calculation of free energies (standard reaction enthalpies) listed in this paper. The work was funded by the US National Science Foundation (OCE 92-00458) and carries the WHOI Contribution number 9205.

References

Adams, M. W. W. 1993 Enzymes and proteins from organisms that grow near and above 100 °C. *A. Rev. Microbiol.* **47**, 627–658.

Burggraf, S., Jannasch, H. W., Nicolaus, B. & Stetter, K. O. 1990 *Archaeoglobus profundus* sp. nov., represents a new species within the sulfate reducing archaebacteria. *Syst. Appl. Microbiol.* **13**, 24–28.

Burggraf, S., Mayer, T., Amann, R., Schadhauser, S., Woese, C. R. & Stetter, K. O. 1994 Identifying members of the domain Archaea with rRNA-targeted oligonucleotide probes. *Appl. Environ. Microbiol.* **60**, 3112–3119.

Cano, R. J. & Borucki, M. K. 1995 Revival and identification of bacterial spores in 25- to 40-million-year-old dominicam amber. *Science* **268**, 1060–1064.

Cloud, P. E. 1976 Beginnings of biospheric evolution and their biogeochemical consequences. *Paleobiology* **2**, 351–387.

Cohen, Y., Castenholz, R. W. & Halvorson, H. (eds) 1984 *Microbial mats: stromatolites*. New York: Alan R. Liss Inc.

Deming, J. W. & Baross, J. A. 1993 Deep-sea smokers: windows to a subsurface biosphere? *Geochim. Cosmochim. Acta.* **57**, 3219–3230.

Drobner, E., Huber, H., Wächtershäuser, G., Rose, D. & Stetter, K. O. 1990 Pyrite formation linked with hydrogen evolution under anaerobic conditions. *Nature* **346**, 742–744.

Edmond, J. M., Von Damm, K. L., McDuff, R. E. & Measures, C. I. 1982 Chemistry of hot springs on the East Pacific Rise and their effluent dispersal. *Nature* **297**, 187–191.

Ginster, U., Mottl, M. J. & Von Herzen, R. P. 1994 Heat flux from black smokers on the Endeavour and Cleft segments, Juan de Fuca Ridge. *J. Geophys. Res.* **99**, 4937–4950.

Hanzawa, S., Hoaki, T., Jannasch, H. W. & Maruyama, T. 1996 An extremely thermostable serine protease from a hyperthermophilic archaeum, Desulfurococcus strain SY isolated from a deep-sea hydrothermal vent. *J. Mar. Biotech.* **4**, 121–126.

Haymon, R. M., Fornari, D. J., Von Damm, K. L., M. D., Lilley, M. D., Perfit, M. R., Edmond, J. M., Shanks III, W. C., Lutz, R. A., Grebmeier, J. M., Carbotte, S., Wright, D., McLaughlin, E., Smith, M., Beedl, N. & Olsen, E. 1993 Volcanic eruption of the mid-ocean ridge along the East Pacific Rise crest at 9° 45–52' N: direct submersible observations of seafloor phenomena associated with an eruption event in April 1991. *Earth Planet. Sci. Lett.* **119**, 85–101.

Holm, N. G. (ed.) 1992 Marine hydrothermal systems and the origin of life. In *Origins of life and the evolution of the biosphere*, vol. 22, pp. 1–241. Dordrecht: Kluwer.

Honjo, S. & Manganini, S. J. 1993 Annual biogenic particle fluxes to the interior of the North Atlantic Ocean; studied at 34° N 21° W and 48° N 21° W. *Deep-Sea Res.* **40**, 587–607.

Huber, R., Kurr, M., Jannasch, H. W. & Stetter, K. O. 1989. A novel group of abyssal methanogenic archaebacteria (*Methanopyrus*) growing at 110 °C. *Nature* **342**, 833–834.

Huber, R., Wilharm, T., Huber, D., Trincone, A., Burggraf, S., König, H., Rachel, R., Rockinger, I., Fricke, H. & Stetter, K. O. 1992 *Aquifex pyrophilis* gen. nov. sp. nov., represents a novel group of marine hyperthermophilic hydrogen-oxidizing bacteria. *Sys. Appl. Microbiol.* **15**, 340–351.

Jannasch, H. W. 1985 The chemosynthetic support of life and the microbial diversity at deep sea hydrothermal vents. *Proc. R. Soc. Lond.* B **225**, 277–297.

Jannasch, H. W. 1995 Microbial interactions with hydrothermal fluids. In *Seafloor hydrothermal systems* (ed. S. E. Humphris, R. A. Zierenberg, L. S. Mullineaux & R. E. Thomson). Geophys. Monogr. no. 91, pp. 273–296. Washington, DC: AGU.

Jannasch, H. W. & Mottl, M. J. 1985. Geo-microbiology of deep sea hydrothermal vents. *Science* **229**, 717–725 (Readers' comments in 1985 *Science* **230**, 496).

Jannasch, H. W. & Wirsen C. O. 1981 Morphological survey of microbial mats near deep sea thermal vents. *Appl. Environ. Microbiol.* **41**, 528–538.

Jannasch, H. W., Wirsen, C. O., Molyneaux, S. J. & Langworthy, T. A. 1992 Comparative physiological studies on hyperthermophilic archaea isolated from deep sea hydrothermal vents with emphasis on *Pyrococcus* Strain GB-D. *Appl. Environ. Microbiol.* **58**, 3472–3481.

Jones, W. J., Leigh, J. A., Meyer, F., Woese, C. R., & Wolfe, R. S. 1983 *Methanococcus jannaschii* sp. nov., an extremely thermophilic methanogen from a submarine hydrothermal vent. *Arch. Microbiol.* **136**, 254–261.

Jørgensen, B. B. 1983 Processes at the sediment-water interface. In *The major biogeochemical cycles and their interactions* (ed. B. Bolin & R. B. Cook), pp. 477–509. New York: Wiley.

Jørgensen, B. B., Isaksen, M. F. & Jannasch, H. W. 1992 Bacterial sulfate reduction above 100 °C in deep-sea hydrothermal vent sediments. *Science* **258**, 1756–1757.

Jørgensen, B. B., Zawacki, L. X. & Jannasch, H. W. 1990 Thermophilic bacterial sulfate reduction in deep sea sediments at the Guaymas Basin hydrothermal vent site. *Deep-Sea Res.* **37**, 695–710.

Juniper, S. K., Martineaux, P., Sarrazin, J. & Gelinas, Y. 1995 Microbial-mineral floc associated with nascent hydrothermal activity on co-axial segment, Juan de Fuca Ridge. *Geophys. Res. Letts.* **22**, 179–182.

Krouse, H. R., Christian, A. V., Eliuc, L. S., Ueda, A. & Halas, S. 1988 Chemical and isotopic evidence of thermochemical sulfate reduction by light hydrocarbon gases in deep carbonate reservoirs. *Nature* **133**, 172–177.

Kurr, M., Huber, R., König. H., Jannasch, H. W., Fricke, A., Trincone, H., Kristiansson, J. K. & Stetter, K. O. 1991 *Methanopyrus kandleri*, gen. and sp. nov. represents a novel group of hyperthermophilic methanogens growing at 110 °C. *Arch. Microbiol* **156**, 239–247.

Miller, J. F., Shah, N. N., Nelson, C. M., Ludlow, J. M. & Clark, D. S. 1988 Pressure and temperature effects on growth and methane production of the extreme thermophile *Methanococcus jannaschii*. *Appl. Environ. Microbiol.* **54**, 3039–3042.

Muyzer, G. E., de Waal, E. C. & Uitterlinden, A. G. 1993 Profiling of complex microbial populations by denaturing gradient gel electrophosesis analysis of polymerase chain reaction-amplified genes coding for 16S rRNA. *Appl. Environ. Microbiol.* **59**, 695–700.

Muyzer, G., Teske, A. P., Wirsen, C. O. & Jannasch, H. W. 1995 Phylogenetic relationships of *Thiomicrospira* species and their identification in deep-sea hydrothermal vent samples by denaturating gradient gel electrophoresis of 16S rDNA fragments. *Arch. Microbiol.* **164**, 65–172.

Nelson, D. C., Wirsen, C. O. & Jannasch, H. W. 1989 Characterization of large, autotrophic *Beggiatoa* at hydrothermal vents of the Guaymas Basin. *Appl. Environ. Microbiol.* **55**, 2909–2917.

Nelson, C. M., Schuppenhauer, M. R. & Clark, D. R. 1992 High-pressure- temperature bioreactor for comparing the effects of hyperbaric and hydrostatic pressure on bacterial growth. *Appl. Environ. Microbiol.* **58**, 1789–1793

Pace, N. R., Stahl, D. H., Lane, D. J. & Olsen, G. J. 1986 The use of rRNA sequences to characterize natural microbial populations. *Adv. Microb. Ecol.* **9**, 1–55.

Pagan, E. & Cohen, Y. 1982. Anoxic photosynthesis. In *The biology of cyanobacteria* (ed. N. G. Carr & B. A. Whitton), pp. 215–235. Berkeley: University of California Press.

Pfeffer, W. 1897 *Pflanzenphysiologie*, 2nd edn. Leipzig: Engelmann.

Pley, Y., Schipka J., Gambacorta, A., Jannasch, H. W., Fricke, H., Rachel R. & Stetter, K. O. 1991 *Pyrodictium abyssi* sp. nov. represents a novel heterotrophic marine archaeal hyperthermophile growing at 110 °C. *Syst. Appl. Microbiol.* **14**, 245–253.

Ravot, G., Ollivier, B., Magot, M., Patel, B. K., Crolet, J. L., Fardeau, M. L. & Garcia, J. L. 1995 Thiosulfate reduction, an important physiological feature shared by members of the order Thermotogales. *Appl. Environ. Microbiol.* **61**, 2053–2055.

Rona, P. A. & Trivett D. A. 1992 Discrete and diffuse heat transfer at ASHES vent field, axial volcano, Juan de Fuca Ridge. *Earth Planet Sci. Lett.* **109**, 57–71.

Ruby, E. G. & Jannasch, H. W. 1982. Physiological characteristics of *Thiomicrospira* sp. isolated from deep sea hydrothermal vents. *J. Bacteriol.* **149**, 161–165.

Ruby, E. G., Wirsen, C. O. & Jannasch, H. W. 1981. Chemolithotrophic sulfur-oxidizing bacteria from the Galapagos Rift hydrothermal vents. *Appl. Environ. Microbiol.* **42**, 317–342.

Schidlowski, M. 1984 Biological modulation of the terrestrial carbon cycle: isotope clues to early organic evolution. *Adv. Space Res.* **12**, 183–193.

Schönheit, P. & Schäfer, T. 1995 Metabolism of hyperthermophiles. *World J. Microbiol. Biotech.* **11**, 25–57.

Schultz, A., Delaney, J. A. & McDuff, R. E. 1992 On the partitioning of heat flux between diffuse and point source seafloor venting. *J. Geophys. Res.* **97**, 12 299–12 314.

Shah, N. N. & Clark, D. S. 1990 Partial purification and characterization of two hydrogenases from the extreme thermophile Methanococcus jannaschii. *Appl. Environ. Microbiol.* **56**, 858–863.

Stetter, K. O., Fiala, G. Huber, G., Huber, R. & Segerer, A. 1990 Hyperthermophilic microorganisms. *FEMS Microbiol. Rev.* **75**, 117–124.

Völkl, P., Huber, R., Drobner, E., Rachel, R., Burggraf, S., Trincone, A. & Stetter, K. O. 1993 *Pyrobaculum aerophilum* sp. nov., a novel nitrate-reducing hyperthermophilic archaeum. *Appl. Environ. Microbiol.* **59**, 2918–2926.

Wächtershäuser, G. 1988 Pyrite formation, the first energy source for life: a hypothesis. *Sys. Appl. Microbiol.* **10**, 207–210.

Wächtershäuser, G. 1990 Evolution of the first metabolic cycles. *Proc. Natn. Acad. Sci. USA* **87**, 200–204.

Winogradsky, S. 1887 Über Schwefelbakterien. *Bot. Ztg.* **45**, 489–507, 513–523, 529–539, 545–559, 569–576, 585–594, 606–610.

Wirsen, C. O., Jannasch, H. W. & Molyneaux, S. J. 1993 Chemosynthetic microbial activity at Mid-Atlantic Ridge hydrothermal vent sites. *J. Geophys. Res.* **98**, 9693–9703.

Woese, C. R. 1991 The use of ribosomal RNA in reconstructing evolutionary relationships among bacteria. In *Evolution on the molecular level* (ed. R. K. Selander, A. G. Clar & T. S. Whittam), pp. 1–24. Sunderland, MA: Sinauer Assoc. Inc.

Woese, C. R., Kandler, O. & Wheelis, M. L. 1990 Towards a natural system of organisms: proposal for the domains Archaea, Bacteria and Eucarya. *Proc. Natn. Acad. Sci. USA* **87**, 4576–4579.

Woodwell, G. M., Whittaker, R. H., Reiners, W. A., Likens, G. E, Delwiche, C. C. & Botkin, D. B. 1978 The biota and the world carbon budget. *Science* **119**, 141–146.

Index

Page numbers in *italic* indicate figures and those in **bold** indicate definitions

adiabatic upwelling 32, 39, 40, *41*, 42, 48, 50, 67, 68, 75, 79, 80, 83, 85, 88, 92, 95
Aegir Ridge 105
Alvin 155, 160, *166*, 229
amplitude
    modelling 20, 23, 24, 25, *27*, 29, 30, *31*
    splitting 29, 30, *31*
aspect ratio 6, 7, *8*, 94
asthenosphere 2, 13, **135**
    density 173
    fractures 83, 84, 85
    mantle 117, 120
    melt 81
        extraction 83, 91, 96
        region 32
        transport 86, 95
    potential temperature 103
    thermal models 142
    viscosity 90
asymptotic ray theory 7
Atlantis II fracture zone *74*
attenuation 6, 25, 105, *138*
Azores triple junction 215

backstripping 109, *111*, *112*
bathymetry
    data 20, 23, 103, *110*, 111, 142
        distance from spreading centre 176
        heat flow predictions 174
        SeaBeam swath *127*
    fields 105
basalt
    geochemistry 116, 117, 135
    glass 139
    halite coating 165, *166*
    liquid 92
    MgO number *138*, 139
        and eruption temperature 139
    permeability 205
    production 39
    resistivities 28
Bauer transform 133
Bauer microplate 133, 142
boehmite 165, *166*
Bouguer gravity anomalies *138*
boundary layer instabilities 116
BRAVEX 172
Broken Spur vent field 172

celadonites 179
chalcopyrite 165
Charlie–Gibbs fracture zone 18

Chile transform 129, *131*, 134
chlorite 165
chromite
    crystallization 83
    nodular 83
    nucleation 83
    orbicular 83
Clausius–Clapeyron equation 45
clinopyroxene 43, 55, 57, 58, 69, *70*, *71*, 72, 73, *74*, 75, *79*, 81, 82
CoAxial Ridge Segment 153, **154**, *155*, *167*
    event plumes 153, 154, *155*, 164, *167*
    floc site *155*, 160–2
        chlorinities 161, 164
        fluid chemistry *156*, 160, *161*, 162, 164
        fluid temperature 156
        heat source 166, *167*
        high-temperature reactions 161
        sulphate reduction 161
        sulphide oxidation 166
    flow site *155*, 160, 162, 164
        basalt alteration products 164
        dyke injection 164
        event plumes
            chemistry 164
        fluid chemistry 160, *161*, 162, 164, 166
        fluid phase 164
        fluid temperature 160, 164
        heat source 166
        iron oxidation *161*
    heat source *167*
    hydrothermal evolution *167*
    neovolcanic zone 162
    source site *155*, 157, 162, 163, 164
        fluid chemistry *156*, 162
        fluid pH 162
    volcanic event 167
coefficient of thermal expansion 44
conductive
    cooling 81, 90, 94, 118, 119, 142, 172, 173–8, *179*, 183
    geotherm 69, 81
conductively cooled plate model 173, 175, 176, 178, 180
    heat flow predictions 175, 176, *178*, 181
continental break-up 103, 105, 111, 112, *115*, 117, 118, 119, 120
continental crust *104*, *114*
controlled source electromagnetic sounding (CSEM) 17, 18, 25–31, *30*, 32
Costa Rica rift 205
cotectic 55

crenulations (seafloor) 141, 143
crust
    accretion 17, 18, 34, 83, 84, 93, 127, 128, 129,
        137, 141, 144
    formation 103, 104, 105, 108, *114*, 120, 129
        mantle temperature effect 117–19
    lower 119
    overthickened 103, 105, 108
    production rates 91
    residual height
        mantle temperature effect 108–16
    strength 118
    tectonic style 119
    thermal structure 142
    thickness 18, 20, 34, 40, 75, 104, 108, 109,
        111, *112*, *115*, 117, 119, 120, 136, 137,
        140, 143
    upper 75, 119, 128
crust–mantle transition zone 22, 73, 75, 83, 136,
    137, 143
crystal fractionation 81, *82*
crystal mush zone 144
    lateral flow 145
cumulates 77, 78, 81, 82, 180

Darcy's Law *84*, 87, 188, 190
DASI 27, *28*, 29
decompression (melting) 1, 40, *41*, 42, 68, 78, 86,
    87, 92, 93, 94, 103, 108, 112, 115, 117
deglaciation (Icelandic) *84*, 86, 90
diagenesis 83
diapirs 93
digital ocean bottom seismometers (DOBS) 20,
    21, *22*, *23*, *24*
dike *see* dyke
diopside 46, *47*, *49*, 50
    solidus *47*
    thermodynamic properties 51
    thermophysical properties *48*
diopside–hedenbergite system 51, 53
discontinuous reactions 55
Discoverer (ship) 154
Discovery transform fault *130*, 133
dissolution 84, 91, 92, 94, 95
    channels 79, *80*, 81, 83, 86, *93*, 94, 95
    reactions 73, 77, 80, 94
DSDP 116, 205
dunite **75**
    chemistry 82
    conduits 75–7, 84, 86, 87, 90, 91, 95
    dimensions 77
    discordant 75
    formation 77–8, 81–3, 84, 91
    fractures 83, 85
    grain size 87
    porosity channels 80
    reaction zones 83, 92
    reactive infiltration instability (RII) 79–81

signature 78
spinel composition *76*, 78, *79*
viscosity 89
zones 67, 92, *93*
dykes 18, 67, 75, 81, *82*, 93
    annealing model 180
    flow 144
    gabbronorite 81, *82*
    injection 162, 163, 164, *165*, 167, 180
    mantle 82
    sheeted 75, 128

earthquakes 118
    depths 181
    hypocentral determinations 119
Easter
    hot spot 125, 129, 134
    Island 129, *132*
    microplate *127*, 128, 129, *131*, *132*, 133, 134,
        135, 137, *138*, 139
East Pacific Rise **126**
    accretionary processes
        along-axis magma transport 142–4
        uniform 141–2
    axial depth 143, 145
    axial high 118
    axial region
        rheology 135–7
        structure 135–7
    cross-sectional area 140, 143, 144
    crustal thickness 143
    dunites 75, *76*
    extrusive layer 139, 142
    kinematic evolution 128, 129–33
    low-velocity zone 175
    magma chamber 34, 128, 142
    mantle Bouguer anomalies 140
    migration 135, 145
    mineralisation
        controls on 215, 219
        sulphides 2, 11
    overlapping spreading centres (OSC) 128,
        129, *130*, *131*, 132, 133, 134, 135, 137,
        139, 140, 143
    porosities 28, 74
    propagation 133
    resistivities 28
    ridge
        propagation 125, 135, 142, 143
        segments 137–41, 153 *see also* CoAxial and
            North Cleft Ridge Segments
    spreading rate 128, 135
    tectonic segmentation 133–5
    thermal structure 136, 142
    transform faults 133
    upwelling pattern 141, 142, 144
    vents 164, *184*
        ecosystems 269, 273, 274, 276

fluid salinities 233
Edoras Bank *111, 112, 114, 115*
elastic constants 6
electric field 25, *28*, 31
electrical resistivity 17, 18, 25, *30*, 33
electromagnetic data 25, 28, 29
ELF receivers 25, 27, *28*, 29
El Niño-southern oscillation (ENSO) 249, 250
Endeavour main field 164, 198
enthalpy (upwelling mantle) 39–62
entropy (upwelling mantle) 39–62
ERS-1 satellite *106*
eutectic 40, 51, 52, 55, 58, 59
event plumes see CoAxial Ridge Segment – event plumes
extension (tectonic) 19, 20, 103

Faeroe–Iceland Ridge *104*, 108, 109, *111, 114, 115*, 116
Faeroe islands 116
FAMOUS area *70*
faults 118
　normal 19, 20, 23
　scarps 32, 134
　throws 140
flowlines 109, 111, *112*, 140
fluid mechanics 96
forsterite–fayalite system 53
fractures 67, 77, 83, 84, 89, 91, 95
　brittle 91
　mechanics theory 134
　melt-filled 92, *93*
　propagation of 91
　crack-tip 89
fracture zones 103, *104*, 105, 106, 118, 120, 129, 132
　hydrothermal activity 215
　mineralisation 215
　offsets 105
　and spreading direction 105, 118
free-air
　anomaly 133, 174
　gravity field *106*, 116
fusion
　batch 39, 40, 46, 55, 56, 57, 58, 59, 60
　continuous 44
　fractional 39, 40, 43, 44, 46, 48, 52, 56, 57, 58, 59, 60

gabbro 67, 75, 82, 145, 180
　deformation 144
　dykes 81, 82, 86
　rheology 136
　rigidus 144
Galapagos 172
　microplate *127, 130*
　sulphide deposits 211
　triple junction *130*

vent ecosystems 273, 274
garnet 72, 86, 88
　lherzolite 68, 69, 72
　peridotite 69
　pyroxenite 86
garnet–spinel peridotite transition 86
Garrett transform fault 135, 136, 137, *138*, 139, 140, 143
geoid data *106*
geometric effect 57
Geosat *106*
glass 68, 69, *70*, 165
　inclusions 68, 81, 82
　MgO content 140
GLORIA data *130, 131*, 133
GLORI-B *131*
Gofar transform fault *130*, 133
grain boundaries 74
grain size 83, *84, 85*, 87, 88, 89, 90, 91, 95
granite 83
gravity 74, 103, 104, 128, 142, 143, 174
　anomalies 108, 135, 145
　constant 88
　fields 105
　lineations 116
Greenland *115*, 116, 120
　continental margin 116
Greenland–Iceland–Faeroe Ridge 103, *104*, 105, 108, *114*, 119, 120
Greenschist conditions 165
Griffith failure 89
Guaymas basin 197

harzburgite 68, 69, *70, 74*, 75, *76*, 77, 81, 82
Hatton Bank *111, 112, 114, 115*
Hawaii 117
　basalts 117
heat transport equation 174
Hess Deep 75, *76*, 82, 83
Hess–Langmuir melting path 42
hot spot 19, 34, 39 see also Iceland – hot spot
Hudson theory 4, 8
hydrofracture 67, 83, *85*, 88, 90, 91, 92, 93, 95, 96
hydrosweep survey *130, 131*, 133
hydrothermal chemistry
　chlorinities 154, 163, 166
　dissolution reactions 153
　exchange reactions 153
　high-temperature reactions 161, *166*
　low-temperature alteration 179
　microbial methanogenesis *165*
　pH 163
　reaction kinetics 163
　sulphate reduction 161
　sulphide oxidation 161, 166
　sulphur reduction *165*
　tectonic controls 153

hydrothermal chemistry (cont.)
  water–rock interaction 163, 167, 175
hydrothermal circulation
  age of lithosphere 176, *179*
  basement permeability 176
  chemical flux 191
  Darcy's law 189, 190
  dyke injection 162, 163, 164, 167
  extent of 13, 28, 118, 171, 175–9
  focusing 215, 217
  heat flux budget 172, 173, 175, *178*, *179*, 180, 181, *182*, 183, 184, 185, *186*, *187*, 191
  partitioning between on- and off-axis 182, 185, *186*, *187*
  interstitial melt 139
  magmatic
    budget 182–5
    sources 35, 180
  modelling 30, 180
  predictions from heat flow deficits 177
  Rayleigh number 189, 190
  reaction zone 165, 167, 175
    depths 185
    double diffusion convection model 183
    temperature 183
  sealing age 176, 179, 181, 182
  sinks 180
  stream function 188, 189, 190
  thermal models 172–87
  water flux 182, 185–8, 191
  water/rock ratio 185
  see also Trans-Atlantic Geotraverse
hydrothermal fluids
  biocatalytic transformations
    aerobic chemosynthesis 283–4
    anaerobic chemosynthesis 284–7
  brine-dominated 153, 154, 163, *164*, 165, 166, 183, 220, 221
  chemistry 153, *156*, *157*–9, 162, *163*, 154
    and heat output 164
  evolution 153
  heat capacity 183
  heat flux 153
  high-temperature 162
  iron concentration 163, *164*
  macrofaunal colonisation 153
  mass flux 153
  microbiology 153, 162
  path lengths 185
  phase separation 154, 163, 164, *165*, *166*, 262
    effect of pressure 186
    evidence for 183
  reservoir 167
  temperature 185
  vapour-dominated 153, 154, 163, *164*, 165, 166, 183, 220
hydrothermal mineralisation
  heat flux 194, *197*
    rate of 196, *197*
  stockwork 215, 220, 221
  sulphides
    caldera setting 219
    deposit size 212, *214*
      controls on 217–21
    fluid boiling 219–21
    fluid mixing 217, 221
    geological trap 220–1
    kuroko-type 217
    location 212–17
    regional controls 212, 215
      back-arc environment 216–17, 221
      fast-spreading ridges 215–17, 221
      slow-spreading ridges 212–15, 221
    seafloor permeability 217–19
    source rock 211, 212, *213*, *214*
    tectonic settings 211, 212, *213*, *214*, 215
    volcanic controls 215
hydrothermal plumes
  buoyancy 249, 252
    equilibrium point 253
    frequencies 250
    source 250
  circular slab model of spreading 255–8, *259*, *260*
  double diffusion 262
  Earth's rotation 249, 252
  entrainment of sea water 249, 258
  fluid
    supercritical expansion 262
  heat source 251, 261, 262
    duration 252
    magnitude 250, 251, 253, 254
    mechanism 251, 252
    and penetration height *254*, 262
  lens oscillation 262
  momentum source 250
  parameters *251*
  Red Sea brine plume 258–61
  regimes *251*
  ridge cross-sectional area 139
  rise of 249, 261
    through thermocline 249, 250, 252, *256*, 262, 263
  rise height 250, 252
    formula 253, *254*
  salinity anomalies 249, 262
  sea–surface temperature anomalies 249, 250, 252, 254
  standard model 253, *254*
  temperature anomalies 249, 252, 253, 262
    function of height *254*
    function of source strength *255*
  turbulent plume model 249
  velocity 250
  vertical structure 253
  and volcanic eruptions 249

volume flux 250
hydrothermal vents
  black smokers
    chemistry 171, *193, 194, 195*, 226
    diurnal periodicity 198
    production 217
    temperatures 175
    time-series measurements 201
  diffuse effluent
    chemistry 162, 171, 172, 191–7
    episodicity 197–200 see also tides
    heat flux *196*
    Mg concentration 156
    mineralisation 192, 197, *218*
    temperatures 172, 188–91, 193, 195, *196*, 198, *199*, 205
    velocity 172, 193, 195, *196*, 198, *199*, 200, 205
  white smokers
    chemistry *193, 194, 195*, 226
    temperatures 175
hydrothermal vent ecosystems
  biocatalytic reactions 281, 282, 283, 285
  Calvin–Bensen cycle enzymes *283*, 285
  chemosynthetic
    bacteria 283, 284
    foodchain *283*, 284
    microbes 267, 281, 282, 286, 287, 288
  convective heat flow *270*
  distribution 265
  ecosystem
    productivities 265, *266*, 267, 276
    properties *266*, 267
  enzymatic reactions 281
  faunal collection analyses *269*
  feeding relationships 267
  habitat *266*, 267
    abundance 265, 273, 274
    controls 265, 269, *270*, 273
    creation 272
    distribution 267–70
    heterogeneity 267
    stability 265, 267, 269, 270–2, 276
  historical factors 273–4
  island biogeographic theory 268, 275
  lithotrophic microbes 282
  origin of life 285
  oxygen 281, 282
  plume incidence *270*
  spatial heterogeneity 272–3, 274, 275
  species diversity 265, *266*, 267, 268, 269, 272, 273, 274, 275, 276
  symbiosis 267, 276
  trophic structure 267, 274, 275, 276

Iceland 67, *84, 85*, 86, 87, 89, 90, 105, 107, 109, *115*, 116, 117, 120
  basalts 116

basin *114*
hot spot 19, 34
neovolcanic zone 117, 119
plume 20, 33, 103–20
  tectonomagnetic regimes 104–8
  spreading direction 118
incompatible elements 57, 68, 69, 72, 75, 82
incongruent melting 46
intratransform spreading centres 125, *127*, 128, 129, 134, 137
  ridge segment propagation 129, 134
isentropic
  decompression *114*, 115
  melting
    multicomponent systems 49–56
      isentropic 53–5
      isobaric 51–3
    simple systems 43–9
      constant coefficients 45–7
      variable coefficients 47–9
isobaric productivity 50, 51, 53, 55, 57, 58, 59, 63
  batch melting 52
  fractional melting 52
isochrons 105, 109, 110, *111*, 112, *115*
isostasy 109
  Airy's *111*, 112
  axial highs 145
  compensation 142, *173*
  equilibrium 173, 174

Juan de Fuca Ridge 13, 153, 154, 175
  hydrothermal
    circulation 179, 201
    heat flux 181, 191
    fluid salinities 233
    phase separation 183
  vent
    ecosystems 268, 269, 273, 284
    fields *184*
Juan Fernandez microplate *127*, 128, 129, *131, 132*, 133, 134, 135

Kaolinite 165

Labrador sea
  opening of 120
latent heat of crystallisation 166, 173, 180
  hydrothermal circulation 182
Lau basin 34, 175
  mineralisation 217, 221
lava 18, 20, 75, 140
  series 75
  tholeiitic 75
Lever rule 39, *50*, 51
lherzolite 68, 69, *74*, 90
liquid–crystal equilibria 43
liquidus 39, 40, 43

lithosphere
  advective heat loss 175
  age 174, 176, 177, 178, *179*
  asymptotic heat flow 176
  brittle behaviour 118, 119, 120
  cooling *70, 111, 112,* 142, 171, 172, *173,* 174, 176
    models 180
  density 173, 176
  dunite 83
  dykes 67, 81, 82, 86
  extension 117
  fractures 91–2, 118
  heat flux 177, *178, 179, 182*
  hydrofracture 90
  isostasy 109
  melt flow 94, 95
  permeability 204, 205
  plates 117
  resistivity 32
  rheology 134
  spreading 2, 103, 135, 145
  thermal structure 13, 119, 125
low resistivity
  body 30, *31*
  layer 29, 32
low seismic velocity
  body *24,* 25, 29, *30*
  zone (LVZ) 23, *26, 30,* 33, 135, 143, 175

magma
  budget 140, 141, 173, 212
  chromite concentration 83
  crystallisation 48, 78, 81
  deglaciation in Iceland 86
  mantle source composition 136, 137, 139
  mass 79
  melt/rock reaction 69, 78
  migration 77
  porosity 79
  sediments 83, 84
  transport
    lateral 125, 142, 143, 144, 145
  see also melt
magma chamber 17, 30, 32, 140, 142, 145, 175, 180
  axial 17, *26,* 34, 128, 136, *138,* 139, 140, 143, 145, 181
  crustal 18, 35
  crystallisation 18
  depth 181
  latent heat 181
  models 180
magnetic
  anomaly 105, *107,* 108, *111, 115*
  data 103, 129, 132
  fields 31, 105
  lineations 133

magneto-telluric sounding 17, 19, 31–3
mantle
  anomalously hot 103, 105, 108, 109, 111, 112, 116, 119, 120
  convection 115, 117
  dehydration 93
  diapirs 143, 144
  dry *114, 115*
  dunites see dunite
  flow 2–4
    rate 119, 120
  mineralogy 90
  modelling 2, 44
  plume 103, 105, 109, *111, 115,* 116, 118, 119
    axis 119
    core 103, 104, 116, 117, 119
      diameter 117
      temperature 108
    instability 120
    migration 119, 120
    temperature 103, 104, 116
    see also Iceland plume
  porosity 73–5
  primitive 116
  structure *3*
  temperature *85,* 103, 104, 105, 108, 109, 110, 111, 112, *114, 115,* 116, 117, 120
  upper 116, 125, 128
  upwelling 2, *41,* 67
  viscosities 2, *3, 85,* 88, 89, 90, 91, *93,* 95
mantle Bouguer anomaly (MBA) **140**, 142, 143, 145
Medusa hydrothermal monitoring system 172, 192, 193, 195, *196, 199*
melt
  anhydrous 72
  batch 44, 45, 46, *47,* 48, *49,* 51, 53, *54, 56,* 57, 58, 59, 62, 63, 69, *71,* 72
  buoyancy 88
  column 70, 72, 74, 86, 89, 90, 117
  continuous 44, 69
  creep
    diffusion 89
    dislocation 89
  crystallisation 94
  extraction 39, 60, 67, 68, 69, 72, 75, 83, 86, 87, 91
  flow
    focused (conduits) 68, 75–86, 88–96
    porous 86–8
    reactive 72-3
    suction 88, 91, 94, 95
  flux *84, 85,* 87, 91
    equations *84, 85*
  focusing 81, 92–5, 141
  formation 72, 103
  fraction 68, 69, 81, *84,* 86, 91, 135, 142

fractional 4, 5, 46, *47, 49*, 51, *54*, 55, *56*, 57, 58, 59, 62, 67, 69, *71*, 72
  inclusions 68–9
  injection
    frequency 119
  isentropic *see* isentropic – melting
  near-fractional 67, 68, 69–72, 88, 95
  partial 4, 6, 32, 33, 39, 68, 72, *82*, 93, 139, 142, 143, 145
  pressure 90
  productivity 20, 34, 39, 40, 41, 42, 43, 45, 46, *47*, 48, *49, 50, 54*, 55, 56, 57, 58, 59, 60, 62, 91
  region 86, 87, 89, 90, 94, 117
  supply 125, 135, 141, 143, 144
    and upwelling 135,141
  thickness 115, 117
  velocities 67, *84*, 86, 87, 88, 89, 90, 92, *93*, 95
  viscosity 67, *80*, 81, 89, 90, 95
  volume 103, 117, 119
melt/rock
  ratio 78, *79*, 83
  reaction *70*, 78, 82, 95
MELTS 39, 43, 44, 50, *56*, 57, 58, 59
melt-squirt 6
metasomatism 73
microplates 125, 128, 129, *130, 131*, 133, 134
Mid-Atlantic Ridge
  accretion 141
  axis 141
  crustal thickness 141
  gravity anomalies 141
  magmatic flux 34
  mantle Bouguer anomalies 140
  melt 74
    inclusions 69, *70*
  peridotite *76*, 82
  ridge propagation 133
  spreading direction 133
  structure 34, 128, 141
  TAG hydrothermal field *see* Trans-Atlantic Geotraverse
  upwelling 125, 135
  vent ecosystems 269, 275
Mid-Ocean Ridge
  abandoned 129, *132*
  accretion 92–5
  axis
    cross-sectional area 137, *138*, 139, 140, 142, 143
    jog 129
  basalt production 39
  elevation 173
  flanks 128, 141, 142
  fracture zone 105
  Free Air Anomalies 174
  hydrofracture 90

  hydrothermal circulation *see* hydrothermal circulation
  jumps 104, 105, 108, 119
  median valley 103, 108, 118, 119
  melt *see* melt
  offsets 128, 129, 134
  peridotites 68
  plagioclase lherzolites 73
  propagation *131*, 133–5
  residual height *111, 112,*
  segments 128, 129, *130, 131*, 133, 137, 139, 141, 144, 153
    propagation 125, 128, 129, *130, 131*, 134, 139
    relief 134
  subsidence rate 135
  tips 134
  upwelling 39, 40, 75
  vent ecosystems *see* hydrothermal vent ecosystems
  v-shaped ridges 105, 116, 117, 120
Moho discontinuity *22, 23*, 68, 75, 77, 84, 108, *114*, 136, 143
MORB 67, 68, 69, *70*, 72, 73, 75, *76*, 77, 78, *79*, 81, 82, 84, 86, 93, 94, 95, 116, 117, 144, 180
  rigidus 136

nanoplate *130*, 133, 134
Nazca plate *126*, 129, *130, 132*, 133
  rotation 133
neovolcanic
  basalts 116
  ridge 212
  zone 108, 109, *115*, 120
North American plate *104*
North Atlantic 104, 109, 112, *115*, 116, 117, 120
  crustal
    generation 117, 119
    thickness *113*, 119
  mantle temperature 119
  opening 105, 116
  spreading axis 108, 119, 120
North Cleft ridge segment 153, 154
  fluids 164, 166
  volcanic event 164

obduction 144
Occam inversion 27, *28*, 32, *33*
oceanic
  basin 107, 118, 120
  crust *see* crust
  stratification 249, 250
  subsidence curve 109
ODP 172, 194, 198, *199*, 202, 220, 226, 227, 229, 244
olivine 57, 68, *70*, 73, *74*, 75, 77, 78, *79*, 80, 82, 83, 87
  alignment 2, 3, *5*

ophiolites 4, 17, 68, 73, *74*, 94, 144
   magma chamber models 180
   melt migration features 75–86
     Oman 73, 75, *76*, 77, 78, *79*, 81, *82*, 86, 93, 143, 144
orthopyroxene 68, *70*, *74*, 78, 81, 92
overlapping spreading centres 125, *126*, 128, 129, *130*, *131*, 133, 134, 135, 137, 140, 142, 143
   bisection 133
   left-stepping *126*, 132, 133, 134, 135
   migration *131*, 132, 133, 134
   propagation rate 143
   right-stepping *126*, 129, 135
   spacing 128

percolation threshold 90, 91
peridotite
   deformation 1
     brittle 89
     plastic 89
     viscous 95
   melting 1, 39–43, 50, 67–94
     isentropic model 56–9
peritectic 40, 51, 52, 55, 58, 59
permeability barrier 94, 95
plagioclase 57, 73, *74*, 81
   lherzolite 73
plate boundary
   geometry 128, 133
   linearity 128
     Pacific–Antarctic 128
     Pacific–Cocos 128
     Pacific–Nazca 125, 133, 134
potential temperature 41, 72, *114*
P-waves 2
   velocities 4, 6, *21*, 22, 23
pyroxene 69, 75, 77, 78, 79, 80, 81, 85, 87, 92
pyroxenite 67, 81, 83, 86
   dykes 81, 86

quartz solubility 175
Quebrada transform fault *130*, 133

rare earth elements (REE) 68, 69, *70*, *71*, *74*, 75, 78, *79*, 81, *82*, 88, 116, 117
reactive infiltration instability (RII) 79, 80, 81, 83, 84, 91, 92, 94
Réunion mantle plume 103
Reykjanes
   Peninsula 18, 19, 20, 119
   Ridge 17, 18, **19**, 20, 32, 33, 34, 35, 104, 105, 108, *114*, 116, 117, 118, 119, 120, 143
rift valley 19, 22, 33, 128
rifted margin 115
rigidus 136, 144, 145

satellite altimetry data 133, 141, 143
seafloor spreading

hydrothermal heat loss 181
   oblique 103, 105, 108, 118, 120
   orthogonal 103, *104*, 105, 106, 108, 118, 119, 120
   with fracture zones 103, 105, 106, 107, 108, 109, 118, 119
   without fracture zones 103, 104, 105, 107, 108, 118
seamounts 135
sea water
   advection 171, 173, 189
   density 173
   entrainment 156, *165*, 191, 194
   evolution into hydrothermal fluid 153
   specific heat *173*, *184*, 185, 186
   temperature *184*
seismic
   modelling effect of melt
     on velocities 6–7
     on travel times 7–12
   reflection profiles 17, 18, 24, 25, *26*, 112
   reflector 17, 18, 136, *138*
   refraction 103, *111*, *112*, 136
   tomography 143
   velocity 8
   wide-angle 17, 18, 20–4, 25, 26, *27*, 108, 111
shadow zone *21*, *22*, 23
shear zones 81
side-scan sonar imagery 19, 20
silica geobarometry 175
sills 93
   gabbroic 82
SOEST
   fracture zone 129
   transform fault 129, *131*, 134
SOSUS array 154
spinel *70*, *74*, 75, *76*, 78, *79*, 83
spinel-lherzolite *71*, 73
spinel-peridotite 69, *70*, 86
spinel–plagioclase transition 55, 90
spreading centre 12, 103, 104, *112*, 117, 118, 135, 142
   extinct 104
Stevenson instability 81
stresses
   compressive 92, 93
   concentrations 91
   differential 88, 89
   field 93
   steady-state 88
swath sonar survey 129, *130*, *131*, 133
S-wave 2
   splitting 7, *9*, 12, 13
   triplications 5
   velocities 4, 6, 74
synthetic seismogram modelling 20, *21*, 22, 24, 108

tensile strength 88, 89, 91, 95
thermal
 anomalies 103, 108, 111, 115, 117, 135, 137
 diffusivity 174
 expansion coefficient 173, 175, 176
tides
 compressional stress 204, 205
 effects on continental aquifers 200, 201
 effects on hydrothermal systems 197–205
 geoid anomalies 203
 lithosphere permeability 206
 sea floor deformation 202, 203
towed-camera surveys 126, 139
trace elements 68, 73, *74*, 75, 82, 135
Trans-Atlantic geotraverse 171, 172, *184*, 191
 anhydrite
  dissolution 229, 233
  distribution 227
  evidence of sea water entrainment 225, 226, 229
  fluid inclusions 225, 229, 230–3, 237, *238*, 243
   homogenisation temperatures 230, *231*
   ice-melting temperatures *231*, 232
   trapping temperatures 229, 230, *231*, *232*, 236, 237, *238*
  Mg partitioning 225, 229, *234*, 235, 236, 237, 244
  precipitation 229
   rate 243
  retrograde solubility 227
  Sr isotopes 225, 228, 230, 233–6
   evidence of fluid mixing 229, 230, 235
  Sr partitioning 225, 229, *234*, 235, 236, 237, *238*, 239, 240, 244
   closed system 239–41, 244
   replenished system 237–9, 244
  veining 227
 black smoker complex (BSC) 193, 226, 227
  conductive heat flow 227
  sea water entrainment 227
 black smokers
  fluid composition 226
   cooling 225
   temperature 230
  conductive
   heat flow 196, 227
   heating 241–4
  diffuse effluent 192, *193*, 195, *196*, 197, 198, *199*, 226
  fluid
   chemistry 195, 226
   evolution 236
   flow 227
   mixing 225, 227, 235, 236, *241*, 244
  heat flux 192, 195, 243
  hydrology 172, *191*

mineralisation 194, 197, 212, 219, 220, 226, 227, 244
 rates *197*
permeability *191*, 225
phase separation 225, 243
pore fluids 226
sea water entrainment 225, 227, 238, *241*, 242, 243, 244
tectonic setting 215, 226
tidal variability 198, *199*
time series measurements 197–206
white smokers 193
 fluid composition 226, 237
 fluid formation 227
 isotopic signature 235
 Kremlin area 226, 237, 244
zone refining 194
transform faults 125, 128, 129, 134, 137
 discontinuities 133
 domain 134
 multiple 128, *130*, 134
triple junction *131*
Troodos ophiolite 179

ultra-depleted melts (UDM) 68, 81, 82, 95
upwelling 2, 12, 32, 34, 39–46, 63, 67, 75, 77, 86, 87, 89, 90, 91, 93, 94, 95, 103, 116, 117, 120, 125, 128, 135
 buoyant 93, 141, 142
 column 86
 diapiric 143
 focusing 125, 128, 143
 intensity 125, 145
 passive 108, 119, 120, 141
 pattern 120, 141, 143, 145
 rates 3
 region 108
 three-dimensional 141–5
 two-dimensional 141–5
 velocities *84*
uranium series disequilibrium 72, *84*, 86, 88, 95

volcanism
 constructive 19, 20
 phase separation 153
 post-glacial 86, 87, 89, 90, 95

Wadi Tayin massif 77, 81, 82
wall rock 92
Walvis Ridge 103
wavefront 7
wave-guide phenomenon 29
Wilkes transform fault *130*, *131*, 133, 134, 140

Yaquina transform 133, 134

zero-age depth variations *127*, 137, *138*